Computer Science Fundamentals

an
algorithmic
approach
via
structured
programming

Computer Science Fundamentals

E. A. UNGER

NASIR AHMED

KANSAS STATE UNIVERSITY

to
Sam, Mark, Michelle, and Kirsten Unger
and
Esther and Mike Ahmed

Published by
Charles E. Merrill Publishing Company
A Bell & Howell Company
Columbus, Ohio 43216

This book was set in Optima
The production editor was Ann Mirels
Computer graphic courtesy of William J. Kolomyjec
Cover design by Dodrill Design

International Standard Book Number: 0-675-08301-X
Library of Congress Catalog Card Number: 78-61859

1-2-3-4-5-6-7-8-9-10-88-87-86-85-84-83-82-81-80-79

Printed in the United States of America

Preface

In recent years a growing interest in the introductory computer science course has been expressed by representatives of a variety of rather diverse disciplines—the social and life sciences, agriculture, business, clothing and textiles, computer science, and engineering, to name just a few. The major difficulty in developing such a course arises from the fact that it must be useful to all students, and yet meet the varied needs of their respective disciplines. To this end, we have attempted to provide a balance between such usefulness and desired learning objectives of the disciplines involved.

This book is directed to anyone who wishes to (a) acquire a working knowledge of general computing concepts, (b) learn a problem-solving methodology, (c) develop a working computer science vocabulary, and (d) begin to appreciate the social impact of computer technology.

A problem-solving approach, using structured programming, is used throughout the text. The resulting solutions are expressed in a graphical form called flowcharts, which are developed using a flowchart language. The motivation for using the flowchart language emanates from three sources. First, from a pedagogical point of view, flowcharts are very helpful to the novice in gaining insight into the subject matter. Second, the flowchart language is universally understood. Third, the use of flowcharts is currently mandatory in some large computing communities; for example, those affiliated with computer installations at the various United States government agencies.

In our classes we instruct students in a particular programming language, using a variety of language manuals. These include

☐ FORTRAN: P. Cress, P. Dirkson, and J. W. Graham, *Fortran IV with Watfor and Watfiv* (Englewood Cliffs, N.J.: Prentice-Hall, 1970).

☐ COBOL: F. Stuart, *Introduction to Standard Cobol Programming* (New York: Harcourt Brace Jovanovich, 1974).

v

☐ PL/I:
M. Davidson, *PL/I Programming with PL/C* (Boston: Houghton Mifflin, 1973).

☐ APL:
L. Gilman and A. Rose, *APL 360—An Interactive Approach,* 2nd ed. (New York: Wiley, 1970).

☐ BASIC:
J. Sacks and J. Meadows, *Entering BASIC* (Chicago: Science Research Associates, 1973).

☐ PASCAL:
K. Jenson and N. Wirth, *Pascal User Manual and Report* (New York: Springer-Verlag, 1976).

Problems for solution in a computer language are presented at the end of Chapters 3, 4, 6, 8, 9, and 10 of this book.

Computer Science Fundamentals consists of eleven chapters, eight of which are devoted to general concepts, problem-solving methodology, and vocabulary. One chapter is concerned with common applications of the computer; it includes an assortment of examples. The remaining two chapters are concerned with computer terminology, business data processing concepts, and the history of computers and their impact on society.

An overview of computers is presented in Chapter 1, followed by a consideration of problem solving via computers in Chapter 2. Problem solving is discussed in terms of three phases: problem analysis, algorithm design for solving the given problem, and computer implementation of the algorithm.

In Chapter 3 the basic components of computer languages are introduced and related terminology is defined. To familiarize the reader with the use of this terminology, several illustrative examples are included. A discussion of computer calculations involving numeric and nonnumeric data is presented in Chapter 4.

In the fifth chapter the input/output aspects of a computer are considered in detail, including the important technique of formatting. Commonly used formatting schemes are presented and illustrated by examples.

Chapter 6 introduces the concept of data structures. A general discussion is followed by a relatively specific treatment of one- and two-dimensional arrays. In Chapter 7 the notion of loops is introduced, and their use in the manipulation of arrays is considered.

The eighth chapter addresses the problem of representing two classes of algorithms, known as function subalgorithms and general subalgorithms. In Chapter 9 an assortment of application examples are discussed, while Chapter 10 introduces terminology related to the areas of computer systems and business data processing.

The final chapter in the book is devoted to a survey of the computer revolution and its impact on society, institutions, and individuals. A historical overview of the important events that have led to the present "computer age" is included in this chapter.

We are grateful to the faculty members of the Department of Computer Science of Kansas State University for their constructive criticism, and to Dr. Paul S. Fisher, department head, for his support and encouragement. We also wish to thank the students who used the text in draft form for their suggestions for improvements. Three individuals, Mr. M. J. Buschlen, Ms. Rachel Moreland, and Ms. Barbara North deserve special thanks for their efforts. And to Ms. Jan Gaines we pay special tribute for typing the manuscript rapidly and accurately. We would also like to express our appreciation to Ms. Ann Mirels, Production Editor at Charles E. Merrill Publishing Company, and Professors Ted Lewis, Ron Danielson, Bob Ducharme, and James Powell who reviewed the manuscript.

Finally, personal notes of gratitude go to our spouses and children, without whose encouragement, patience, and understanding this book could not have been written.

Contents

8

SUBALGORITHMS

CONTENTS

SOME COMMONLY USED ALGORITHMS

COMPUTER AND DATA PROCESSING SYSTEMS

THE DEVELOPMENT OF COMPUTER SYSTEMS
AND THEIR IMPACT ON SOCIETY

CONTENTS

xii

1

Computers:
An Overview

The computer is a tool . . . whose continued improvement, and subsequent extended application, can be of immeasurable benefit to mankind. (Bertram Raphael, 1976)

KEYWORDS

analog
algorithm
coder
computer
concurrent
CPU
data
digital
firmware
hardware
hybrid
input
memory
programs
programmer
output
sequential
software
stored program
system analyst

The computer is perhaps our most useful modern-day tool yet developed. It was only in 1951 that the first commercially available computer was marketed. Today we are using computers in ever-increasing numbers, in ways never imagined just a few years ago. Computers are now an integral part of the information needs of many societies. They are used world-wide to store a variety of information—financial, health, tax, credit, insurance, and so forth. They have made complex telephone and communication systems a reality and have contributed to more accurate planning and forecasting systems. In addition, they have helped us to explore new frontiers, such as outer space.

We are only just beginning to feel the impact of computers on our lives. The future will increase the interaction between people and computers all over the world. In the United States, for example, it is estimated that by the year 2000 most jobs will involve the use of computers either directly or indirectly. It is important, then, that we understand (a) how computers are made to perform specified tasks, (b) what computers are capable of accomplishing, and (c) their impact on society and the individual. These topics will be addressed in forthcoming chapters.

This chapter serves as an introduction to the basic parts of a computer; it also provides the necessary terminology to become conversant in this area. Further, it addresses the familiar question: For what purpose should a computer be used?

1.1 BASIC STRUCTURE OF A COMPUTER

A **computer** is an automatic device that performs calculations and makes decisions. It has four basic parts (input, storage, central processor unit, output), as illustrated in Figure 1.1-1.

1. **Input** is the part of the computer that enables us to communicate with it. That is, it provides a means of reading **programs** and related **data** into the computer from some external source.[1]

   Punched card input is still widely used in computing. The most common type of card is shown in Figure 1.1-2. Several other types are also used; for example, the one in Figure 1.1-3. The holes shown on these cards are punched using devices called **keypunches,** an example of which is shown in Figure 1.1-4.

---

[1] *Programs* are sets of instructions to the computer, directing it to perform a specified task. *Data* consist of names, numbers, or other such information to which meaning is assigned.

2

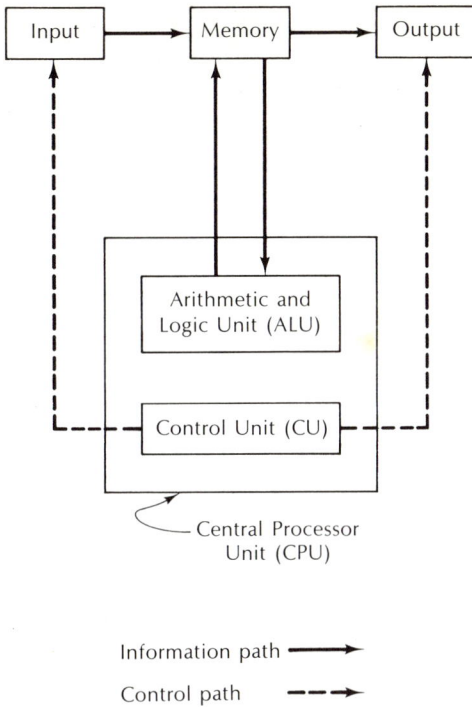

FIG. 1.1-1. Structure of a computer.

Other forms of input include pencil marks on cards (see Figure 1.1-5), magnetic tape, or disk-stored information prepared by a key entry device similar to the keypunch (see Figure 1.1-6). Nowadays, it is quite common to find penlike devices capable of reading a set of "marks" on a variety of items in department and grocery stores. Such devices serve as input media for sets of "marks" that identify the product with a code. Typewriterlike and televisionlike (or video) devices called **terminals** are also commonly used to communicate information to or from a computer (see Figure 1.1-7).

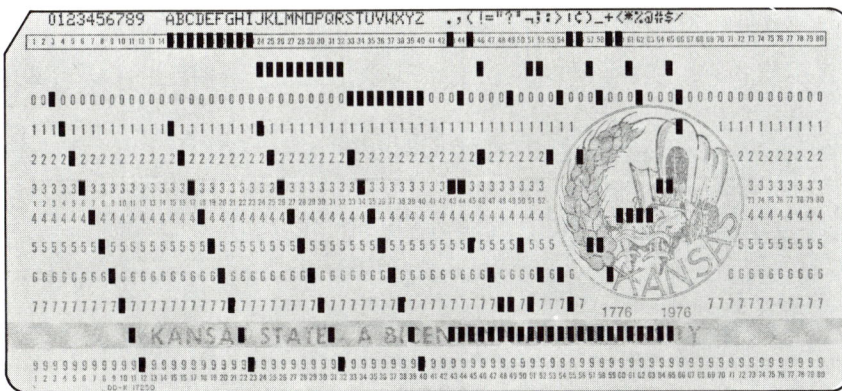

FIG. 1.1-2. An 80-column punched card with rectangular holes (Hollerith card).

3

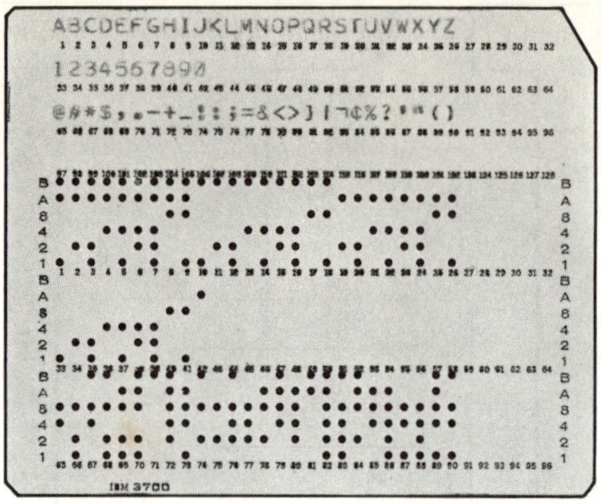

FIG. 1.1-3. A 96-column punched card with round holes.

FIG. 1.1-4. A keypunch—a device for punching holes into 80-column Hollerith cards. (Courtesy of International Business Machines Corporation.)

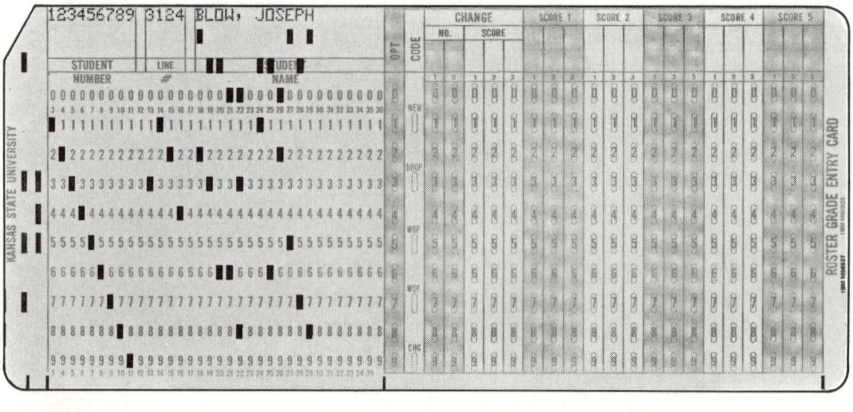

FIG. 1.1-5. A punched and pencil-marked input card.

4

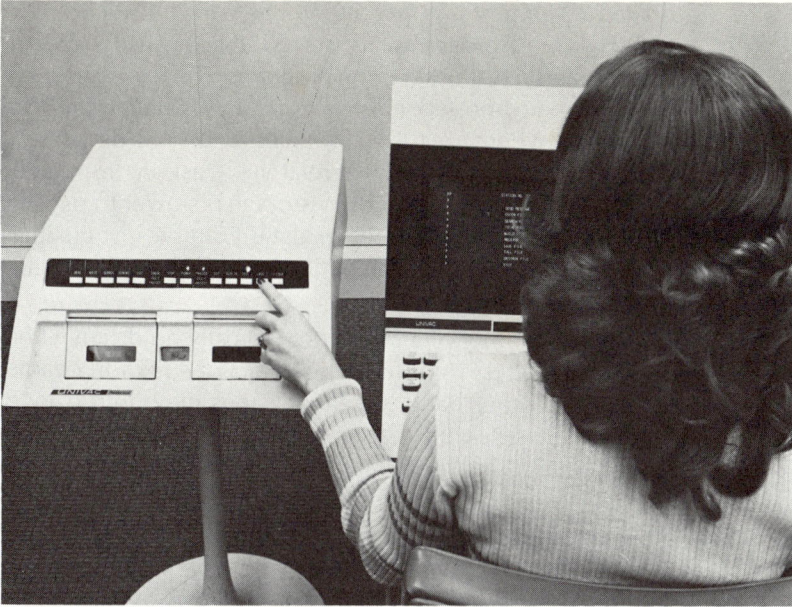

FIG. 1.1-6. A key entry device that can create and change data on cassette-type magnetic tape. (Courtesy of Sperry Univac Corporation.)

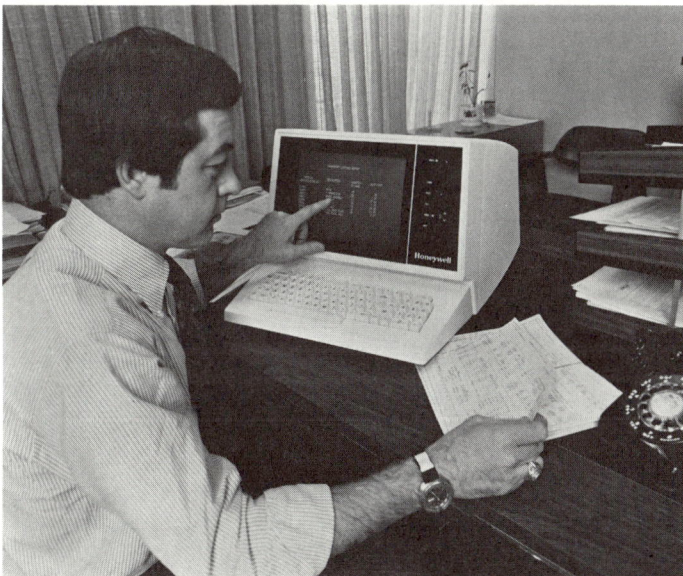

FIG. 1.1-7. A visual display terminal being used by a manager to check current inventory levels of products distributed by a company. (Courtesy of Honeywell, Inc.)

2. **Storage or Memory** is conceptually the part of the computer where data and programs are stored. It may also store the results obtained from executing programs. There are two types of storage: main and secondary (or auxiliary). Main memory is that memory where executing programs and some data reside. Data can be stored/retrieved relatively fast in/from main memory. In contrast, a secondary storage is slower to use, but is able to store large volumes of data. Examples of auxiliary storage media include magnetic tape, disks, [see Figure 1.1-8(a)–(c)], and drums.

The main memory of a computer typically holds anywhere from 256 characters (of information) to 3 or 4 million characters.[2] On the other hand, magnetic tapes may hold 6 to 20 million characters, while magnetic disks are capable of storing up to about 250 million characters per disk pack.

FIG. 1.1-8(a). An auxiliary storage medium—a magnetic tape drive with a 2400-foot reel of tape mounted on it. (Courtesy of Sperry Univac Corporation.)

[2] *Characters* refer to letters and/or numbers.

FIG. 1.1-8(b). An auxiliary storage system—a four-unit disk drive with removable disk packs mounted in them. A disk pack is shown in Figure 1.1-8(c). (Courtesy of International Business Machines Corporation.)

FIG. 1.1-8(c). A removable disk pack which can be mounted in one of the four units shown in Figure 1.1-8(b). (Courtesy of International Business Machines Corporation.)

3. **Central Processor Unit (CPU)** is the heart of the computer. Figures 1.1-9 and 1.1-10 show examples of CPUs. A CPU consists of two components; namely, the **arithmetic and logic unit (ALU),** and the **control unit (CU).** ALU is the portion of the CPU responsible for arithmetic (e.g., addition/subtraction) and logical (e.g., comparison of two numbers) operations that are to be carried out. The CU causes the various computer actions to occur in the correct sequence.

4. **Output** is that part of the computer that provides a means of presenting information; i.e., data, programs, and related results from memory. Devices for computer output include line and page printers for printing pages of computer-created infor-

7

FIG. 1.1-9. A general purpose computer system. The CPU and memory are located in the box on the left; a card reader is shown in the center. (Courtesy of Sperry Univac Corporation.)

mation. Printers vary in speed, but speeds of 300 lines per minute (lpm), 1200 lpm, and 8000 lpm are common (see Figure 1.1-11). Other output devices include video tubes (like television screens) for graphic displays (see Figure 1.1-12), punched cards, and punched paper tape.

FIG. 1.1-10. A large-scale computer system. Two CPUs and main memory appear at the center in the back, in light-colored boxes. (Courtesy of International Business Machines Corporation.)

FIG. 1.1-11. A page-printing system capable of printing 18 000 lpm.[3] It is also able to separate and collate the printed pages. (Courtesy of Honeywell, Inc.)

FIG. 1.1-12. A video tube used for graphic displays of line drawings. (Courtesy of Tektronix, Inc.)

---

[3] Following the recommendations of the American National Metric Council, commas will not be used to group digits in this text, since the comma has been traditionally used as a decimal marker in other countries.

There are seven characteristics by which computers are usually classified. We will now enumerate and discuss these in some detail.

1. *Data Representation*. Computers are classified as being either **analog** or **digital,** depending upon the manner in which they represent data. Analog computers represent data in a continuous form. This is because such computers establish an analogy between an electrical signal and some measure of a physical process. A thermostat, for example, is a very simple analog computer, where an input value corresponds to the temperature of the air molecules. A continuous range of temperature values is represented by a temperature-sensitive bimetallic strip.

   Digital computers handle data in a discrete (noncontinuous) form. A simple type of digital computer is a desk or pocket calculator. The fact that a calculator is able to handle only discrete data is evident when we consider that an eight-digit calculator is not able to represent any number between a pair of eight-digit numbers, such as 19659643 and 19659644.

   Analog computers can be used in conjunction with digital computers. Such a combination leads to a third category called **hybrid** computers.

2. *Hardware Features*. By **hardware** is meant the physical components and pieces of equipment involved in a computer. Thus, **mechanical computers** consist mainly of such devices as gears and levers, while **electromechanical computers** are made up of an assortment of electrical and mechanical devices. Computers that were built during the early 1940s were of the electromechanical variety. **Electronic computers** are those that employ electronic devices, such as vacuum tubes and transistors. Most of the computers in current use are electronic. The more recent ones consist of electronic devices that are available in the form of **integrated circuit "chips"** (see Figure 1.2-1). The manufacture of such chips involves an extremely sophisticated technology that is rapidly advancing.

3. *Algorithm Implementation*. An **algorithm** consists of a set of unambiguous rules specifying a sequence of operations that provides the solution to a given problem in a **finite** number of steps.[4] The manner in which algorithms are implemented in computers enables us to categorize them. In a **manual computer,** instructions representing algorithms are fed to it one at a time by an operator—for example a cash register in a grocery store. If the **program** representing an algorithm is available in

---

[4] An algorithm always stops after a finite number of steps. If it cannot be guaranteed that an algorithm always stops, it is a *procedure*.

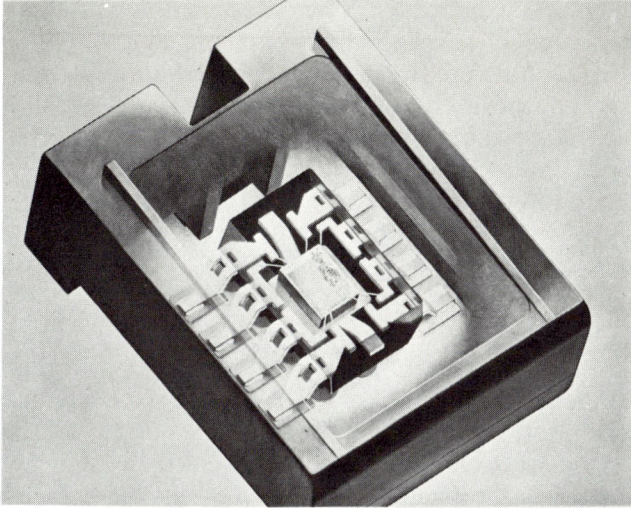

FIG. 1.2-1. A rectangular chip consisting of miniturized computer logic is shown in the center of the photograph. Such chips may contain programs or data. (Courtesy of Texas Instruments, Inc.)

some permanent physical form, such as a board of circuitry, then the corresponding machine is called a **hard-wired computer,** an example of which is shown in Figure 1.2-2. In contrast, **stored-program computers** operate by first storing a program **(software)** in their memory and then executing those instructions.[5] Such computers provide great flexibility, since the programs to be executed can readily be changed. An integral combination of hardware and software used to instruct a computer system is known as **firmware.**

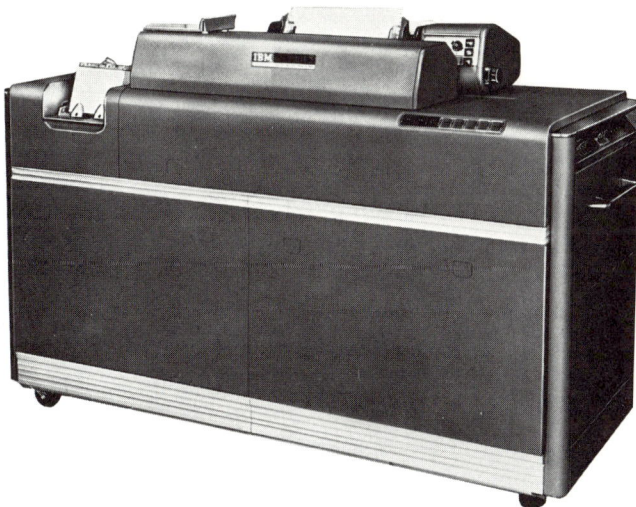

FIG. 1.2-2. A hard-wired tabulating machine (IBM 407). (Courtesy of International Business Machines Corporation.)

---

[5] *Software* refers to programs for a computer system.

4. *Scope of Application.* Computers can be classified into two categories according to the scope or range of applications for which they are designed: (a) **special-purpose** and (b) **general-purpose.** For example, a computer used for navigational purposes aboard an aircraft is a special-purpose computer; it can be used to do little else. On the other hand, computers that can be programmed to accomplish a variety of tasks are called general-purpose computers.

General-purpose computers are sometimes classified into two subclasses: (a) business computers and (b) scientific computers. Business computers are capable of handling large volumes of data of interest to the business world. As such, they are said to be used in **data processing applications.** In contrast, scientific computers are generally not capable of handling large volumes of data efficiently, but are usually much faster than business machines in executing arithmetic operations.

5. *Internal Mode of Operation.* Most computer systems available today operate in a **sequential mode,** in the sense that the first instruction is executed first, followed by the second instruction, and so on. The notion of computers operating in a sequential mode is introduced in this text in Chapter 3 using a programmable mechanical robot, which is essentially a very simple computer. A few of the newer computers, however, are able to accomplish more than one such task at a time using various schemes which discover computations that can be executed simultaneously. Such computers are said to operate in a **concurrent mode,** as opposed to those that operate in the sequential mode.

6. *Performance.* The speed with which certain operations are performed in computers enables us to classify them into three broad categories:

    (a) **millisecond computers,** which execute about one thousand instructions per second (1 millisecond = 10^{-3} second).

    (b) **microsecond computers,** which execute about one million instructions per second (1 microsecond = 10^{-6} second).

    (c) **nanosecond computers,** which execute about a billion instructions per second (1 nanosecond = 10^{-9} second).

7. *Size.* Computers may be classified with respect to their size in that their related systems can loosely be referred to as being small-scale, medium-scale, or large-scale. Small-sized computers include minicomputers, microcomputers, and pocket calculators.

A **minicomputer** is usually a small-sized general-purpose stored-program computer (see Figure 1.2-3), while **microcomputers** are most often small special-purpose computers. The recently developed **pocket calculators,** as well as some

FIG. 1.2-3. A system consisting of a minicomputer, operator console, punched-card reader, and line printer. (Courtesy of Interdata, Inc.)

microcomputers, are individual-user devices which may be manual or programmable (examples are shown in Figure 1.2-4). Pocket calculators are becoming increasingly powerful; some present models easily surpass the capability of early computers, although the physical sizes of these computers were thousands of times greater.

A summary of the preceding discussion related to the classification of computers is presented in Table 1.2-1.

TABLE 1.2-1.   Classification of computers

Criterion	Categories
Data representation	Analog, digital, hybrid
Hardware features	Mechanical, electromechanical, electronic
Algorithm implementation	Manual, wired, stored-program
Scope of application	Special-purpose, general-purpose
Internal mode of operation	Sequential, concurrent
Performance	Millisecond, microsecond, nanosecond
Size	Small-scale, medium-scale, large-scale

13

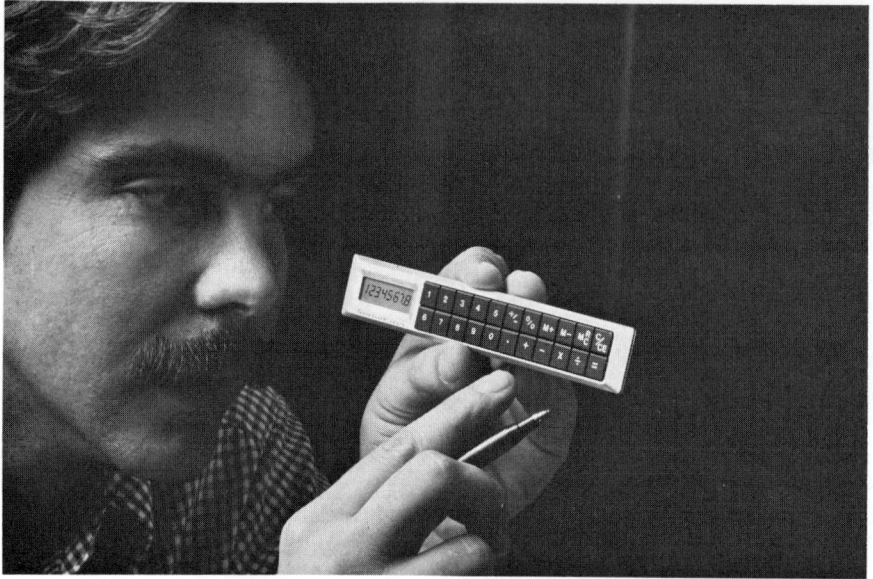

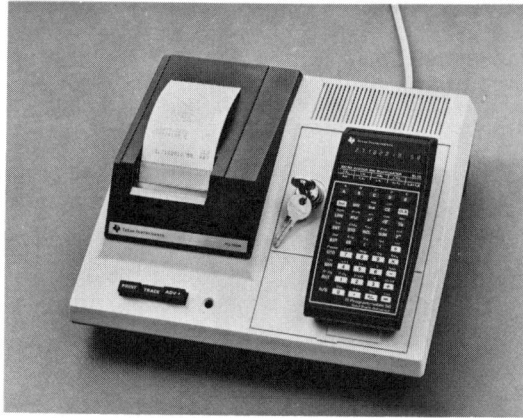

FIG. 1.2-4. Manual and programmable pocket calculators. (Courtesy of Texas Instruments, Inc.)

1.3 INSTRUCTION OF COMPUTERS

The process of instructing computers to perform a specified task involves three levels of instruction that are carried out by three types of personnel. First, a **systems analyst** defines the problem to be solved and develops a clear, well-defined solution to the problem. Next, a **programmer** works with part or all of this solution by adding the necessary detail so that a computer can be used to solve the problem. Finally, a **coder** takes the programmer's solution and writes it in a specific language that is understood by the computer.

Use of a computer to solve a problem is justified if one or more of the following instances occur:

☐ If the solution must be repeated a number of times, as is the case, for example, with preparing payroll checks. This is because computers are capable of performing such repetitive calculations rapidly and accurately.

☐ If the solution requires some calculations to be performed repeatedly, such as approximating the solution to a set of equations.

☐ If the solution must be obtained rapidly to be of any value, as, for example, in the control of chemical processing equipment after automatically sensing temperatures and pressures related to the process.

☐ If the solution requires a large amount of storage; for example, weather predictions from data gathered via a satellite.

☐ If the problem-solving process can facilitate a clearer understanding of the given problem. The effects of pollution on the environment, for instance, may be better understood.

It is important to note that although computers are very fast and reliable, they need to be presented with precise solutions to problems they are to solve. The development of such precise solutions generally takes more time than the related process of instructing the computers, as outlined in Section 1.3.

SUMMARY 1.5

In the overview presented in this chapter, we have looked upon the computer as an automatic device that performs calculations and makes decisions. Seven characteristics were used to classify computers into several categories.

The process of instructing computers to perform a specified task was associated with three levels of instructions. Five criteria were cited for determining whether a computer should be used to solve a given problem.

In the context of the preceding overview, a useful vocabulary of related computer terminology was also presented.

1-1. Secure a pocket calculator. Classify it according to the characteristics given in Table 1.2-1.

problems

1-2. Analyze one computer application you are aware of in terms of justifying the use of a computer for this application.

1-3. For each of the reasons given for using a computer, suggest a computer application from the current literature that is primarily justified for that reason.

1-4. Visit a computing center (possibly at your institution), and classify the available equipment according to Table 1.2-1.

2

Problem Solving

Programming is more difficult than commonly assumed . . . an urgent advice is to restrict ourselves to simple structures. (E. W. Dijkstra, 1972)

KEYWORDS

compiler
computer programming language
debugging
echo
flowchart
initial solution
procedure-oriented language
program
programming language
semantics
structured algorithm
structured programming
syntax
trace table
translator

Problem solving is a daily human experience, most of which is done on a rather informal basis. Informal problem solving may work very well, especially when a solution is communicated to another person who can add missing steps, interpret ambiguous statements, and correct errors. Although computer systems can perform calculations very rapidly and with great precision, they are not equal to human beings in the realm of creative activity. Thus, our informal methods of solving problems must be replaced with those that are more formal, guiding the problem-solver to avoid solutions that are incomplete, incorrect, and ambiguous.

This chapter introduces a formal method that has proven effective in solving problems with computers. It consists of three phases as follows:

1. problem analysis;
2. algorithm design for solving the given problem; and
3. computer implementation of the algorithm.

2.1 THE STRUCTURED METHOD OF PROBLEM SOLVING

Problem solving can be achieved by systematically following a set of steps, as summarized in Table 2.1-1. The arrows in this table indicate the sequential manner in which these steps are executed. The process of going back from Step 7 to Step 6 and then returning to Step 7 represents a **loop** in the sense that the steps could be executed repeatedly if necessary.

The steps summarized in Table 2.1-1 provide a correct solution to a given problem. The reader should note, however, that in practice it is necessary to backtrack when an error occurs. This aspect of problem solving is presented in a more detailed, tabular form at the end of this chapter (see Figure 2.6-1).

The notion of structured problem solving was developed by Dijkstra, Wirth, and others during the late 1960s and early 1970s. **Structured programming** is a well-organized method for designing an algorithm. It has been used in production programming groups by business and industry. Related preliminary studies have shown spectacular increases in productivity.

We normally associate two phases with the life of a program: initial creation and maintenance. By *maintenance,* we refer to the process of changing programs to keep them functioning correctly as external changes occur. It has been reported that structured programming could double a programmer's productivity in the initial creation phase, and also substantially reduce the maintenance programming time—to as little as 10 percent in many cases. This is particularly significant, since the maintenance cost of a program typically averages to 70

TABLE 2.1-1. Summary of steps involved in problem solving

Problem-solving phase	Steps
Problem analysis	1. Define the problem 2. Eliminate all ambiguous words or decisions in the problem statement 3. List the available inputs 4. List the desired outputs
Algorithm design	5. Write down a 3- to 7-step *initial solution* in general terms using a natural language; e.g., English. 6. Refine each of the steps in the previous solution into a sequence of more detailed steps. 7. Repeat Step 6 until sufficient detail is provided.
Computer implementation	8. Develop a flowchart representation for the algorithm obtained via the above seven steps 9. Convert the flowchart representation of the algorithm into steps in a computer language; the resulting product is called a *program*.

percent of its total cost. In addition, structured algorithms are much easier to understand and change than are conventional algorithms. This aspect also contributes to an increase in maintenance efficiency.

A **top-down approach** is employed. It requires the entire solution to be defined initially at the highest level of abstraction. This is called the **initial solution,** and is stated in terms of a set of steps (from three to seven in number).

For the purpose of illustration, let us consider the following initial solution:

1. Read necessary input values.
2. Process input.
3. Output the answers.

It is worth commenting that the process of "finding" a correct initial solution is, in general, a difficult task. This is because there is no scientific way of doing so at the present time. The initial solution just presented, however, is a good place to start for developing the desired algorithm. For convenience, we will not include incorrect initial solutions and the process of backtracking to correct them.

19

Next, the initial solution is subjected to a **stepwise refinement process,** which means that each of the steps is refined in such a way that no "overlap" occurs between any of them. The notion of overlapping between steps in an initial solution is best illustrated by a simple example, such as the calculation of payroll checks. Then Step 1 might be refined as follows:

    1.0   Read necessary input values
        1.1   Read a value for hours worked.
        1.2   Read a value for rate of pay.

An examination of this refinement shows that there is no overlap between Steps 1, 2, and 3 of the initial solution, in that *all* the information in Steps 1.1 and 1.2 relates to exactly *one* step (i.e., Step 1) in the initial solution. However, if the next step in the refinement process is

        1.3   Multiply rate of pay by hours worked,

then we observe that an overlap does occur between Steps 1 and 2, since the information in Step 1.3 relates to Step 2 in the initial solution, while that in Steps 1.1 and 1.2 relates to Step 1. In other words, an overlap occurs due to the fact that the overall information in the refinement refers to *more than one step* in the initial solution.

From the preceding discussion it is apparent that the essence of the refinement process is that each set of refinements is small (one to seven steps) and completely replaces the step being refined. The refinement process is repeated until the desired level of detail is attained. This approach to problem solving is known as a top-down structured solution.

In order to illustrate the manner in which the algorithm design phase follows the problem analysis phase (Steps 1–7 in Table 2.1-1), let us now consider a simple example.

Example 2.1-1 Perform the steps associated with problem analysis and algorithm development related to the process of replacing a broken pane of glass in a door. That is, remove the old glass pane, and put in a new pane of the best glass available.

Solution: First we must clarify what we imply by the term *best*. Several possibilities exist. For example, do we mean the strongest, the cheapest—or should it be a thermopane? Let us assume that the *best* implies the cheapest in this case.

The inputs to this problem are the resources to buy the window, (say, in dollars), the tools required, and the old frame. The output is the door with an unbroken glass pane.

Next, we attempt an initial solution:

   1.   Secure a new glass pane.

   2.   Clean the frame.

   3.   Put in the new glass pane.

A refinement of Step 1 would be

1. Secure a new glass pane.
    1.1 Measure the opening for the pane.
    1.2 Call a glass store and secure a price.
    1.3 Called all stores? If not, go back to 1.2; otherwise proceed
         to 1.4.
    1.4 Determine the lowest price.
    1.5 Buy the pane at the store that offers the lowest cost.

A further refinement can be attempted if desired. For example, a refinement of Step 1.1 could be

   1.1.1 Obtain a device for measuring length.
   1.1.2 Measure the height and width of the window opening.
   1.1.3 Add 1/4 in. (63.5 mm) to each measurement to account for
          the molding. (Note: this step is necessary, as we chose not
          to clean the window frame first.)

Clearly, one can proceed through several stages of refinement. At some point in the refinement process, we decide to stop and say that a **structured algorithm** has been realized.

We will now discuss some aspects of a technique called **flowcharting.** This technique enables us to realize Step 8 of the problem-solving process given in Table 2.1-1. An additional set of examples, the solutions of which include Step 8 in Table 2.1-1, will be presented in Section 2.3.

FLOWCHART SYMBOLS 2.2

Flowcharting is a useful technique, since it provides a **graphical representation** of structured algorithms in a nearly universal language. Flowcharts are required for most computer programming projects of the federal agencies in the United States. Other organizations may or may not require flowchart representations of algorithms. Since graphical representations of algorithms have been found to be very effective teaching devices, we will use them throughout this text. To develop a flowchart, it is necessary to use a set of symbols. These are summarized as follows:

☐   **Flow direction symbol.** This symbol denotes the direction of
     information flow, which is generally from top to bottom, left
     to right. In order to avoid confusion, flowlines are usually
     drawn with an arrowhead at the point of entry to a symbol.
     When possible, the crossing of flowlines should be avoided.

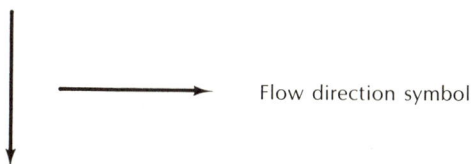

Flow direction symbol

☐ **Processing symbol.** Indicates general processing functions that are not represented by other symbols. These functions generally contain the information processing operations of the program, such as addition and subtraction.

Processing symbol

☐ **Input/Output symbol.** Represents any function of an input/output device that is employed.

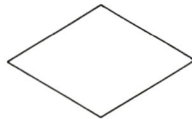

Input/Output symbol

☐ **Decision symbol.** Indicates a point at which a branch to one of two or more alternative symbols is possible.

or

Decision symbol

standard   alternate

☐ **Terminal symbol.** Represents any point at which information flow either originates or terminates.

Terminal symbol

☐ **Connector symbol.** Clarifies flowline patterns in a flowchart. There are two types of connector symbols.

In-connector symbol

Out-connector symbol

☐ **Annotation symbol.** Provides a means of including comments or explanations in flowcharts.

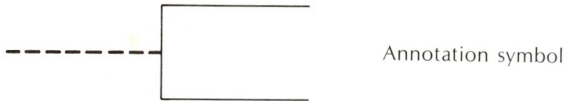

Annotation symbol

ADDITIONAL EXAMPLES 2.3

Example 2.3-1 Saturday night is an excellent night for students to relax. Assume that Tom Jones is a student who has access to a little black book of telephone numbers of prospective dates. Develop a structured algorithm for Tom Jones to secure a date for some Saturday. Show the corresponding flowchart.

Solution: The input will be telephone numbers from Tom's little black book. There really is no output, except perhaps an elated, satisfied, or dejected Tom. Consider the following initial solution:

1. Select and read input from Tom's little black book.
2. Place the telephone call.
3. Decide action upon receiving the answer.

Refining the above initial solution, we obtain the following structured algorithm:

1. Select and read input from Tom's little black book.
    1.1 Select the characteristics desired in a date for Saturday evening.
    1.2 Use little black book to read phone number of person with desired characteristics.
2. Place the telephone call.
    2.1 Find a telephone.
    2.2 Dial the telephone number.
    2.3 If the prospective date answers, proceed to Step 3.
    2.4 If she does not answer, determine a course of action.
        2.4.1 If she does not answer, then let the phone ring awhile.
        2.4.2 If waited long enough, return to Step 1.2; else return to 2.4.1.
3. Ask the person for a date.
4. Decide action upon receiving the answer.
    4.1 If she accepts, then stop.
    4.2 Else see if there are any more numbers and make a decision.
        4.2.1 Else see if there are any more numbers.

23

4.2.2   If there are more numbers, select one and return to Step 1.2.

4.2.3   Else think of some other activity for Saturday night.

Using the flowchart symbols in Section 2.2 to graphically represent the preceding structured algorithm, we obtain Figure 2.3-1.

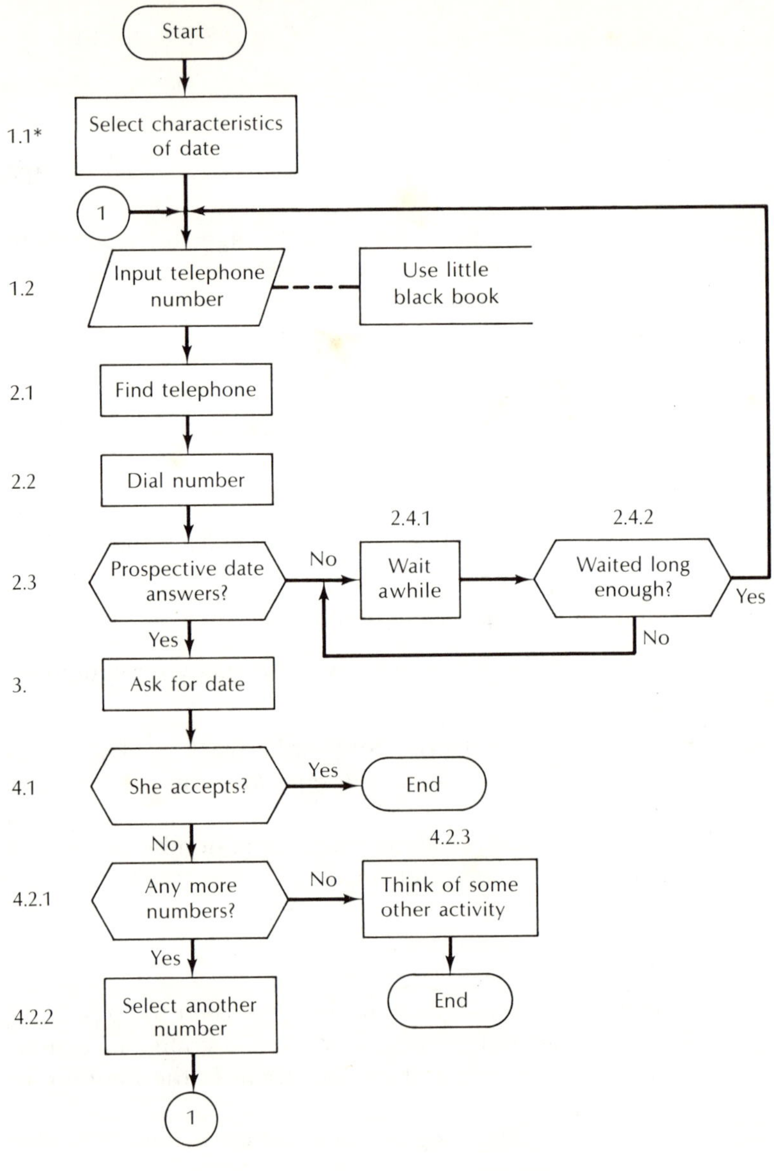

*These numbers correspond to the structured algorithm statements.

FIG. 2.3-1.   Flowchart pertaining to Example 2.3-1.

Example 2.3-2   An instructor determines a student's grade for a course from the sum of four test scores and a project score. Each student was required to do a class project, but this score is not used to improve the class score unless a reasonable effort was given to the project. If the project score exceeds 70 points, the points in excess of 70 are added to class score. If the project score falls short of 70 points, no points are added to the class score. Create a structured algorithm and flowchart for this problem.

Solution: The problem has no ambiguous words, but it is not stated what should be done if the project score is exactly 70 points. Assume that the instructor was contacted, and it was learned that in this case no points would be added to the test scores.

The input for a student would be Test Scores 1 through 4 and the project score. Let us call these TEST1, TEST2, TEST3, TEST4, and PROJ, respectively. The output is a sum that we will call RESULT.

A structured algorithm to solve this problem is

1. Input the data values and print them.
   1.1 Input the values for TEST1, TEST2, TEST3, TEST4, and PROJ.
   1.2 Output the values of TEST1, TEST2, TEST3, TEST4, and PROJ.
2. Calculate the sum of the test scores, RESULT.
3. Determine the final class score.
   3.1 If PROJ score is less than or equal to 70, do not change RESULT.
   3.2 If PROJ value exceeds 70, then add (PROJ − 70) to RESULT.
4. Output the value of RESULT.

The flowchart corresponding to the preceding algorithm is shown in Figure 2.3-2. We can make the following observations with respect to this flowchart:

☐ Two verbs, INPUT and OUTPUT, are used in conjunction with the input/output symbols. The verb INPUT implies that one or more **inputs** to the algorithm are to be obtained. On the other hand, the verb OUTPUT implies that one or more values are to appear as **outputs** of the algorithm, and subsequently recorded on some medium. (A more detailed discussion of the corresponding INPUT and OUTPUT statements is presented in Chapter 3.)

☐ The annotation **"Echo check"** implies that the values of TEST1, TEST2, TEST3, TEST4, and PROJ, which are inputs to the algorithm (see Symbol 1.2 in Figure 2.3-2), are now to appear as output. Clearly, this process serves as a check. If any of these input values is in error, then the user should be made aware of it. Otherwise the computations that are carried out would be futile.

25

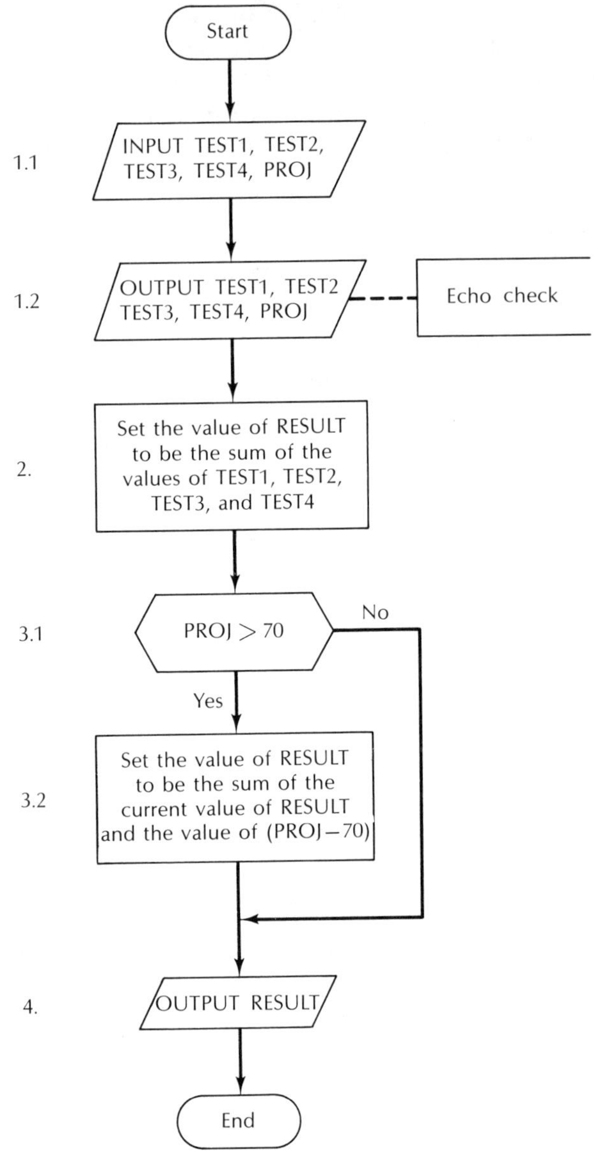

FIG. 2.3-2. Flowchart pertaining to Example 2.3-2.

☐   We also observe that each of the flowchart symbols is numbered so that we can conveniently "trace" through the flowchart to check its correctness in solving the problem. To trace
the flowchart, we execute the steps involved and list the
corresponding values of TEST1, TEST2, TEST3, TEST4, PROJ,
and RESULT.

Let us now consider the following test data:

TEST1 = 70   TEST2 = 60   TEST3 = 90   TEST4 = 80   *PROJ* = 91

Tracing through the flowchart in Figure 2.3-2, we obtain the following
trace table. The flowchart symbol numbers are listed in the leftmost

column as each flowchart symbol is executed. The quantities we wish to trace (TEST1, TEST2, etc.) are used as headings for the table columns. The corresponding values taken by TEST1, TEST2, etc., when the respective symbols are executed, are then entered sequentially in these columns. The term *arbitrary* under RESULT implies that as far as the user is concerned, the value of RESULT is *unknown* when Symbols 1.1 and 1.2 are executed. However, when Symbol 2 of the flowchart is executed, RESULT takes the known value 300.

TABLE 2.3-1. A trace table

Flowchart symbol #	TEST1	TEST2	TEST3	TEST4	PROJ	RESULT
1.1	70	60	90	80	91	Arbitrary
1.2	70	60	90	80	91	Arbitrary
2.	70	60	90	80	91	300
3.1	70	60	90	80	91	300
3.2	70	60	90	80	91	321
4.	70	60	90	80	91	321

It is now necessary to adopt some convention for developing trace tables, in the interest of avoiding confusion. We will use the following:

In the columns associated with TEST1, TEST2, TEST3, TEST4, PROJ, *and* RESULT, *their current values will be indicated only when they change. Thus, a blank space will imply that the value of the related quantity has not changed.*

To illustrate, let us apply this convention to the line of Table 2.3-1 which concerns Symbol 1.2 of the flowchart. We observe that the values of TEST1, TEST2, TEST3, TEST4, PROJ, and RESULT need not be listed, since they have not changed in the move from Symbol 1.1 to Symbol 2.2. The same is true of TEST1, TEST2, TEST3, TEST4, and PROJ in the next line of the table. The value of RESULT, however, changes from "Arbitrary" to 300. Hence the values of TEST1, TEST2, TEST3, TEST4, and PROJ are omitted while that of RESULT is left as is. Proceeding in a similar manner with the remaining three lines of the table, we obtain Table 2.3-2. It is apparent here that the final value of RESULT is 321.

TABLE 2.3-2. Revised version of Trace Table 2.3-1

Flowchart symbol #	TEST1	TEST2	TEST3	TEST4	PROJ	RESULT
1.1	70	60	90	80	91	Arbitrary
1.2						
2.						300
3.1						
3.2						321
4.						

This is what one would expect, since PROJ is greater than 70; hence,

$$RESULT = RESULT + (PROJ - 70)$$
$$= 300 + (91 - 70) = 321.$$

It should be noted that the process of tracing a flowchart provides one means of checking for errors in an algorithm while using test input data. It is not, however, a means of proving that a given algorithm is correct in general.

2.4 COMPUTER PROGRAMMING OF ALGORITHMS

We now turn our attention to the last step in problem solving—computer programming—Step 9 in Table 2.1-1. Implementation involves a process called **coding,** which converts the steps in an algorithm into corresponding steps in a computer programming language. The result of this conversion is called a **program.** There are basically three levels of programming languages which are used to implement algorithms on computers:

1. machine-oriented programming languages;
2. procedure-oriented programming languages; and
3. problem-oriented programming languages.

Machine-oriented languages consist of languages called **machine** and **assembly languages.** These are generally associated with a specific computer. In contrast, **procedure-oriented languages** relate to the programming of procedures and can be used on a variety of computers. Procedure-oriented languages are translated into the machine language of a given computer, as illustrated in Figure 2.4-1. The desired translation is achieved by a **translator** or **compiler.**

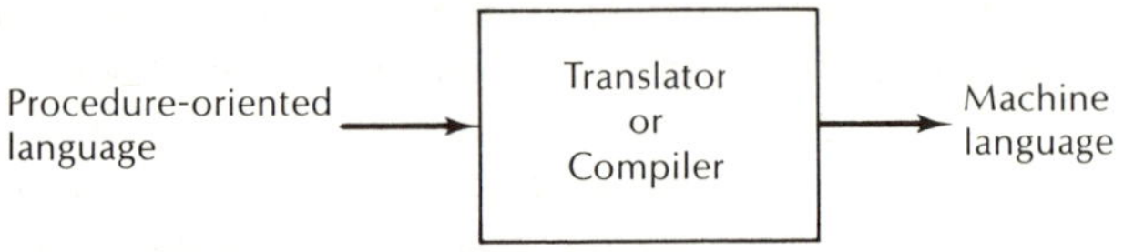

FIG. 2.4-1.   Pertaining to a translator or compiler.

There are a number of procedure-oriented programming languages, the following of which are used quite extensively:

☐ APL (A Programming Language) directly incorporates a large number of commonly used mathematical operations. Consequently, it is quite useful in the area of scientific computing.

☐ ALGOL (ALGOrithmic Language) is a powerful language oriented towards the scientific community; it is very popular in Europe. ALGOL is an accepted language for communication of algorithms among people in various nations.

☐ BASIC (Beginners All-Purpose Symbolic Instruction Code) is used for problem solving in small- to medium-scale systems in a variety of disciplines. It is easily read and understood when used in problem solving.

☐ COBOL (COmmon Business-Oriented Language) is well adapted to data processing problems, and is hence used widely in business applications. It is supported by the U.S. Federal Government for primary use in federal agencies.

☐ FORTRAN (FORmula TRANslation) is the most widely used language for numerical computing related to a large number of mathematical applications.

☐ PASCAL is a general purpose language developed recently. It is currently available on many minicomputer systems, and also on some medium- to large-scale systems.

☐ PL/I (Programming Language One) is a powerful general purpose language that combines various features of COBOL, ALGOL, and FORTRAN.

Several versions of some of these languages that are often available are summarized in the following table:

Language	Version
COBOL	American National Standard COBOL, WATBOL
FORTRAN	WATFIV, FORTRAN IV, FORTRAN II, WATFOR
PL/I	PL/C, PL/I
ALGOL	ALGOL-60, ALGOL-68

Problem-oriented languages provide a means of programming narrower classes of problems than are solved by procedure-oriented languages. In this case, only the exact problem to be solved must be described, rather than the entire procedure associated with the solution. These languages are used in a variety of rather diverse areas, including business, statistics, computer-aided instruction, and engineering. An example of this type of language is RPG (Report Program Generator), which is used in business applications since it yields desired information in the layout of a business report.

Once a program has been developed using an appropriate computer programming language, it is necessary to carry out two final steps in the overall computer implementation stage. These are referred to as program testing and debugging.

Program testing involves running the computer program using carefully selected sets of input data. The resulting outputs are then examined to determine whether or not they satisfy one or both of the following conditions:

1. They coincide with answers that have previously been obtained via hand calculations (perhaps using a trace table), or some other method.

2. They agree with answers that are considered to be reasonable on the basis that they have been obtained via theoretical or other considerations.

The process of removing errors, or bugs, from an algorithm is known as **debugging.** There are basically three sources of bugs:

1. Poor definition of the problem, in which case the entire problem-solving process must be repeated.
2. Logical errors present in the algorithm.
3. Errors incurred while coding. Such errors are due to (a) improper translation of algorithm steps into programming language steps, or (b) violation of **syntactic** or **semantic rules** pertaining to the programming language.[1]

2.5 ADDITIONAL CONSIDERATIONS

By the term *flowchart language,* we imply a language made up of the set of symbols summarized in Section 2.2. An important property of the flowchart language is that it yields algorithm representations that are *independent* of any particular computer or computer programming language. In addition, there are other languages that also can be used to represent algorithms. These include natural languages and decision tables.

Natural languages, such as English, suffer from several basic disadvantages:

1. They are ambiguous.
2. Statements used in their representation of algorithms tend to be too long.
3. The logical flow associated with the algorithm is difficult to follow.

Decision tables represent decision processes by systematically listing the actions to be taken for all possible conditions that are valid in complex decision processes. An example of a decision table for a portion of an algorithm is given in Table 2.5-1. It may be used to determine the insurance rate for individuals. The symbol "—" implies

TABLE 2.5-1. Example of a decision table

Characteristics of person applying for insurance		Rule number				
		1	2	3	4	5
Condition	Age < 19?	Yes	No	No	No	—
	Number of accidents in the last three years	—	0	<3	≥ 3	≥ 3
	Previous policies	—	No	Yes	Yes	No
Action	Rate	$+10\%$	$+0\%$	$+10\%$	$+25\%$	—
	Refuse to issue a policy	—	—	—	—	X

---

[1] *Syntactic rules* state the legalities for constructing statements in a language, while *semantic rules* define the legal meaning of language features.

that the indicated condition or section is ignored, while "X" denotes action to be taken. For example, if an individual is under 19 years of age, then Rule 1 applies, since all of its conditions are met. The action indicated for Rule 1 is to add 10 percent to the standard rate and issue a policy to the individual.

Computer languages are available for translating decision tables directly into computer programs.

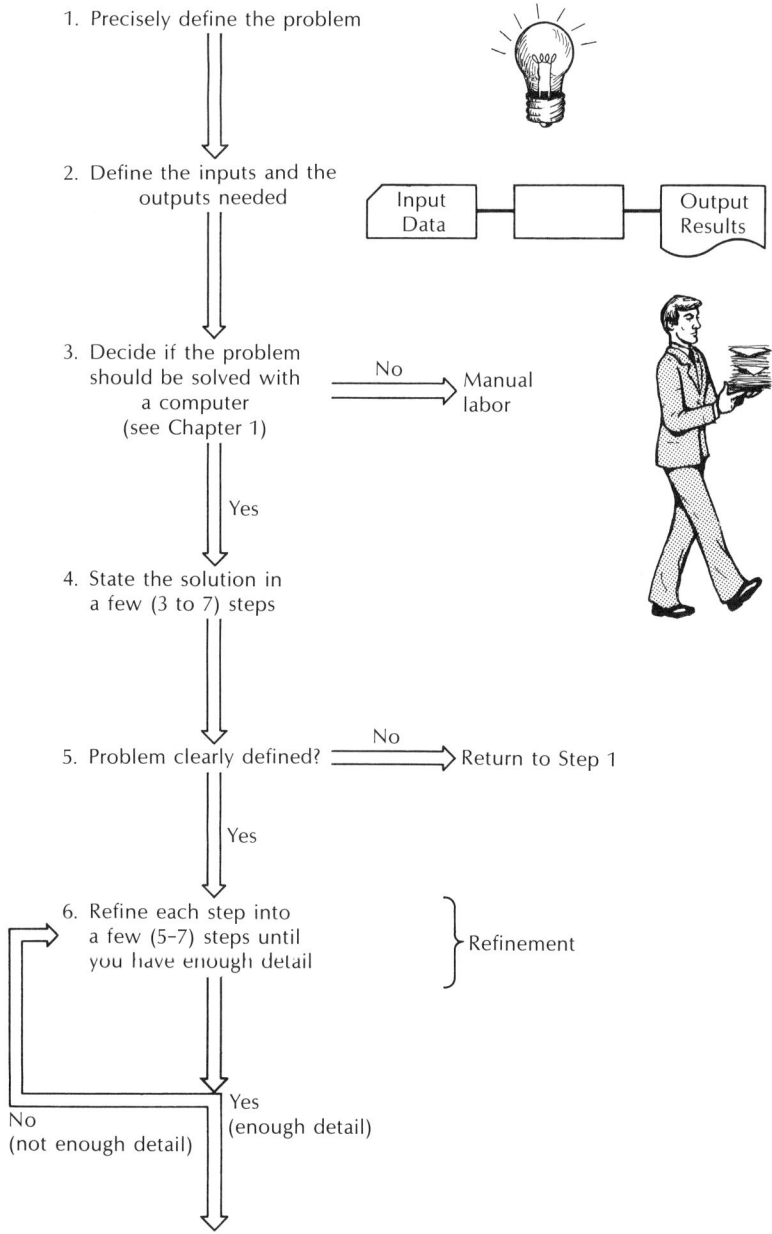

1. Precisely define the problem

2. Define the inputs and the outputs needed

Input Data — Output Results

3. Decide if the problem should be solved with a computer (see Chapter 1)

No → Manual labor

Yes

4. State the solution in a few (3 to 7) steps

5. Problem clearly defined?

No → Return to Step 1

Yes

6. Refine each step into a few (5–7) steps until you have enough detail

} Refinement

No (not enough detail)

Yes (enough detail)

FIG. 2.6-1. Problem solving using computers—a summary.

7. Graphically present the solution as a flowchart with sufficient detail to create a program from it

8. Trace a flowchart (debugging)

9. Solution correct? ———— No ————▷ ————▷ Return to Step 4

Yes

10. Code the solution using a computer programming language

11. Punch it on cards or type on a computer terminal

12. Run it on the computer

Input Data

Program

Machine

Output Results and Errors

Language Translator, "Compiler"

13. Language errors? ———— Yes ————▷ Correct syntax

No

Return to Step 11

14. Result correct? ———— No ————▷ Correct logic

Yes

Return to Step 4

⎱ Debugging

15. Relax

FIG. 2.6-1. Continued.

Details pertaining to the various phases of problem solving were systematically introduced in this chapter in accordance with the nine steps listed in Table 2.1-1. Additional information about the individual phases was also considered. Consequently, the steps listed in Table 2.1-1 can be expanded to obtain the summary given in Figure 2.6-1.

Some aspects of the various languages that are used to represent algorithms were considered. In the remainder of this text, we will restrict our attention to the flowchart language and use it to implement structured algorithms.

problems

2-1. Suppose while driving your car, you have flat tire. Perform Steps 1 through 7 in Table 2.1-1 to develop a structured algorithm for changing the flat tire.

2-2. Develop a flowchart representation of the algorithm in Prolem 2-1.

2-3. Perform Steps 1 through 8 in Table 2.1-1 to develop an algorithm for inputting a number, X, squaring it, finding its absolute value, and then outputting X, X^2, and $|X|$.

2-4. Perform Steps 1 through 8 in Table 2.1-1 for a structured algorithm to place a person-to-person call to Mr. Jones at his home in Baltimore, Md.

2-5. Consider the following sequence of positive integers, which is known as the **Fibonacci sequence:**

$$0,1,1,2,3,5,8,13,21,34,55, \ldots$$

Observe that if the first two terms 0 and 1 are given, then the rest can be constructed according to the rule that each number in the sequence is the sum of the two preceding ones.

Develop a structured algorithm and its flowchart representation (Steps 1 through 8 in Table 2.1-1), so that the Fibonacci sequence can be generated.

2-6. Suppose you are given a sentence and asked to tabulate the number of vowels in it. Perform Steps 1 through 7 in Table 2.1-1 to develop a structured algorithm for this process.

2-7. Develop a flowchart representation for the algorithm in Problem 2-6.

2-8. Suppose you are given a sentence and asked to tabulate the number of *T*'s in it. Perform Steps 1 through 7 in Table 2.1-1 to develop a structured algorithm for this process.

2-9. Develop a flowchart representation for your solution in Problem 2-8.

2-10.   Assume that you have selected an item to purchase at a store. Further, assume that you have a credit card and some cash. Perform Steps 1 through 7 in Table 2.1-1 to develop an algorithm for the process of paying for the item.

   Comment: Note that you will have to decide whether you want to pay cash or use the credit card, and then make decisions related to the card being acceptable or your having sufficient cash.

2-11.   Flowchart the solution to Problem 2-10.

2-12.   Create a structured algorithm that will add two 4-digit decimal numbers, digit by digit, and flowchart the algorithm.

2-13.   Create a structured algorithm that will subtract two 4-digit decimal numbers. Flowchart this algorithm.

2-14.   Develop a structured algorithm that will multiply two decimal numbers, each of which is not more than three digits long.

2-15.   Explain to a foreign student (new to your campus) how to use the city phone directory. Do this by performing Steps 1 through 7 in Table 2.1-1. Flowchart the solution.

   Comment: Do not forget the Yellow Pages. Also consider that an individual or business may not be listed in the directory.

2-16.   Assume you are the manager of a variety story. Develop a structured algorithm that describes to your employee how to make change for customers. Assume that no one will purchase more than 10 dollars worth of merchandise, and that no customer will pay with a bill exceeding 50 dollars. Flowchart your solution.

2-17.   Repeat Problem 2-16 so that the customer is given the minimum number of coins.

2-18.   Explain in detail how to use an abridged dictionary to determine the meaning of a word, assuming the correct spelling for the word is known. Express this process in the form of a structured algorithm.

2-19.   Develop a structured algorithm that explains to a player, X, in tic-tac-toe, how to play the game. Flowchart this solution.

2-20.   Develop a structured algorithm that explains how to find the largest value in a set of three numbers. Include a flowchart for your solution.

2-21.   Develop a structured algorithm that a new student at your university can use to clarify the procedure for evaluating a semester grade-point average. Flowchart your solution.

2-22.   Develop an English language representation of an algorithm for making a long distance telephone call. Allow for alternatives normally involved in placing a long distance call. Assume you are using a touch-tone telephone and already know the number you wish to call. Flowchart your solution.

3

Computer Language Fundamentals

Language shapes the thought and culture of those who use it. (Benjamin L. Whorf, 1939)

KEYWORDS

alphabet
assignment
branch
constant
control
declare
input
label
numeric
output
string
variable

On a world-wide basis, there are somewhere between 120 and 130 computer programming languages that are currently being used. It is obvious that it would take an enormous effort to become familiar with all these languages. However, a large subset of them have similar characteristics; hence, they can be discussed in terms of a universally understood language, commonly referred to as the **flowchart language.** In this chapter we will introduce some of the basic components of computer programming languages via the flowchart language.

The flowchart language, like all of the common programming languages, is sequential in nature; i.e., the first statement is executed first, followed by the second, and so on. To further illustrate this concept, let us consider a simple robot that can be programmed to move in a maze, as shown in Figure 3.0-1. Thus, we look upon the robot as a simple computer.

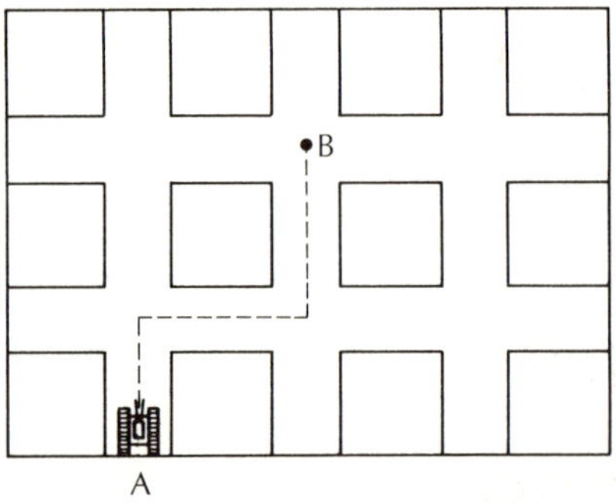

Program

MOVE
TURN
MOVE
TURN
TURN
TURN
MOVE

FIG. 3.0-1. Robot movement related to a specific program.

We assume that the language the robot understands consists of two commands; namely, MOVE and TURN. A MOVE command causes the robot to move one maze space, while a TURN command causes it to turn 90 degrees to the right. According to the program of Figure 3.0-1, it is straightforward to verify that the robot moves from A to B as shown. The specific path traversed is indicated by a dashed line. It is important to note that the statements of the program that cause the robot to traverse the path are executed *sequentially;* i.e., in the order MOVE, TURN, MOVE, TURN, TURN, TURN, MOVE.

3.1 FLOWCHART LANGUAGE

There are three basic ingredients that are necessary in developing any language. They are as follows:

1. *Alphabet*—a set of symbols or characters which can be combined to form words.

2. *Syntax*—a set of rules by which words can by combined in conjunction with punctuation symbols to form sentences **(statements)** and compositions **(programs).**

3. *Semantics*—a set of rules to determine the meaning of words, statements, units of programs, and programs.

In the flowchart language we wish to develop, the alphabet will consist of the following symbols:

1. The 26 capital and lowercase letters of the English alphabet.

2. A set of punctuation and grouping symbols which consists of

    $$. \;) \; , \; (\; ' \; \_ \; \% \; : \;] \; ? \; [\; - \; \$ \; \&$$

    The use of these symbols will be explained as they are introduced in this chapter.

3. The decimal digits $0, 1, 2, \ldots, 9$ and the following mathematical operators:

    ☐ **Arithmetic operators:** $+, -, /, *$ represent addition, subtraction, division, and multiplication, respectively.

    ☐ **Comparison operators** are the following:
    (a) $=$ and $\neq$ mean "equal" and "not equal," respectively—e.g., $x \neq y$ is read, "the value of x does not equal the value of y."
    (b) $>$ and $<$ mean "greater than" and "less than," respectively—e.g., $x > y$ is read, "the value of x is greater than the value of y."
    (c) $\geq$ and $\leq$ mean "greater than or equal to " and "less than or equal to," respectively—e.g., $x \leq y$ is read, "the value of x is less than or equal to the value of y."

    ☐ **Assignment operator** is denoted by the symbol $\leftarrow$. Thus, $x \leftarrow 3.2$ means that x is given the value 3.2; $x \leftarrow y$ means that the value of y is given to x also.

4. The seven flowchart symbols summarized in Section 2.2.

As is the case in any language, characters belonging to the flowchart language alphabet can be combined to form words. Some examples of words that are legal in the flowchart language are (a) Declare, (b) DO, (c) dog, (d) 12, and (e) 6.3.

Using mathematical symbols and appropriate punctuation, words belonging to the flowchart language can be combined to obtain statements. There are four basic types of statements:

1. **Processing statements** cause information to be altered in some way. In flowcharts, such statements appear in processing symbols. The statement $B \leftarrow 5$ in the flowchart shown in Figure 3.1-1 is a processing statement, since its execution results in the value 5 being assigned to B.

37

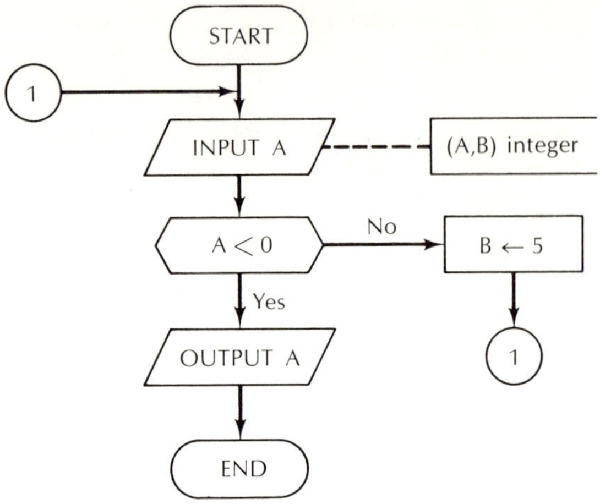

FIG. 3.1-1. Pertaining to the various types of statements.

2. **Control statements** are used to sequence the steps in a solution to be executed. The statement $A < O$ in Figure 3.1-1 is a control statement, since it causes the statements OUTPUT A and END to execute if the current value of A is negative. Otherwise, the action $B \leftarrow 5$ is performed. Such comparison statements as $A < O$ often appear in the decision symbols of a flowchart.

3. **Declarative statements** provide information pertaining to the names that appear in flowcharts, and also other comments that are included to assist a flowchart user. Such statements are written in the annotation boxes of a flowchart. For example, the statement (A,B) integer in Figure 3.1-1 is a declarative statement, since it specifies that A and B take integer values only.

4. **Input/Output statements** are used to communicate information between a flowchart or a computer and its user. Such statements appear in the input/output symbols of a flowchart, as is the case with the statements INPUT A and OUTPUT A in Figure 3.1-1.

3.2 CONSTANTS AND VARIABLES

Constants and **variables** are two types of words in the flowchart language that play an important role in the design of algorithms. Constants have values that are *fixed*, as is the case with numbers such as 3 and -4.5. On the other hand, values assigned to variables can be *changed*. For example, consider the problem of finding the average high temperature for the days of a given week. A reasonable way to approach this problem is to form the sum of the high temperatures for Sunday and Monday. Then add the high temperature for Tuesday to the current sum to obtain a new sum, and so on:

High Temperatures (°F)

Sunday:	60
Monday:	62
Tuesday:	58
Wednesday:	64
Thursday:	63
Friday:	61
Saturday:	52

$$
\begin{array}{r}
60 \\
+62 \\
\hline
122 = \text{Sum} \\
+58 \\
\hline
180 = \text{Sum} \\
+64 \\
\hline
244 = \text{Sum} \\
\vdots \\
+52 \\
\hline
420 = \text{Sum}
\end{array}
$$

Finally, divide by 7 for a specific set of temperatures:

$$
\begin{aligned}
\text{Average} &= \text{Sum}/7 \\
&= 420/7 \\
&= 60
\end{aligned}
$$

It is apparent that "Sum" can be viewed as a variable, since it is capable of taking different values as the calculations are carried out. A flow-chart for the solution and the related trace table are shown in Figure 3.2-1 and Table 3.2-1, respectively.

In order to use the computer memory to store a value that may change, we identify a place or location in memory. For example, we could use the name SUM to identify a memory location and adopt a convention that would reflect such an action. To this end, the following convention will be used in the remainder of this text:

In the preceding notation, the integer 122 is considered to be the current *value* of a *variable* whose *name* is SUM. The value taken by this variable can be changed in any of several ways. For example, if 58 is added to the current value of SUM, then the new value it takes is 180, and hence denoted as

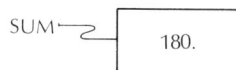

Thus, we come to the important conclusion that *a variable has both a name and a value, and while its name is fixed, its value can change.*

Having introduced the notion of constants and variables, we can now discuss them in more detail.

39

```
                          ( START )
                             |
                             v
         +--------------------------------+
         | SUNDAY  <- 60                  |        (SUNDAY, MONDAY, . . . ,
         | MONDAY  <- 62                  |         SATURDAY, SUM) integer
         | TUESDAY <- 58                  |- - - -
    1    | WEDNESDAY <- 64                |         AVERAGE real
         | THURSDAY <- 63                 |
         | FRIDAY <- 61                   |
         | SATURDAY <- 52                 |
         | SUM <- 0                       |
         +--------------------------------+
                             |
                             v
    2    +--------------------------------+
         | SUM <- SUNDAY + MONDAY         |
         +--------------------------------+
                             |
                             v
    3    +--------------------------------+
         | SUM <- SUM + TUESDAY           |
         +--------------------------------+
                             |
                             v
    4    +--------------------------------+
         | SUM <- SUM + WEDNESDAY         |
         +--------------------------------+
                             |
                             v
    5    +--------------------------------+
         | SUM <- SUM + THURSDAY          |
         +--------------------------------+
                             |
                             v
    6    +--------------------------------+
         | SUM <- SUM + FRIDAY            |
         +--------------------------------+
                             |
                             v
    7    +--------------------------------+
         | SUM <- SUM + SATURDAY          |
         +--------------------------------+
                             |
                             v
    8    +--------------------------------+
         | AVERAGE <- SUM/7.              |
         +--------------------------------+
                             |
                             v
    9    /      OUTPUT  AVERAGE          /
                             |
                             v
                          ( END )
```

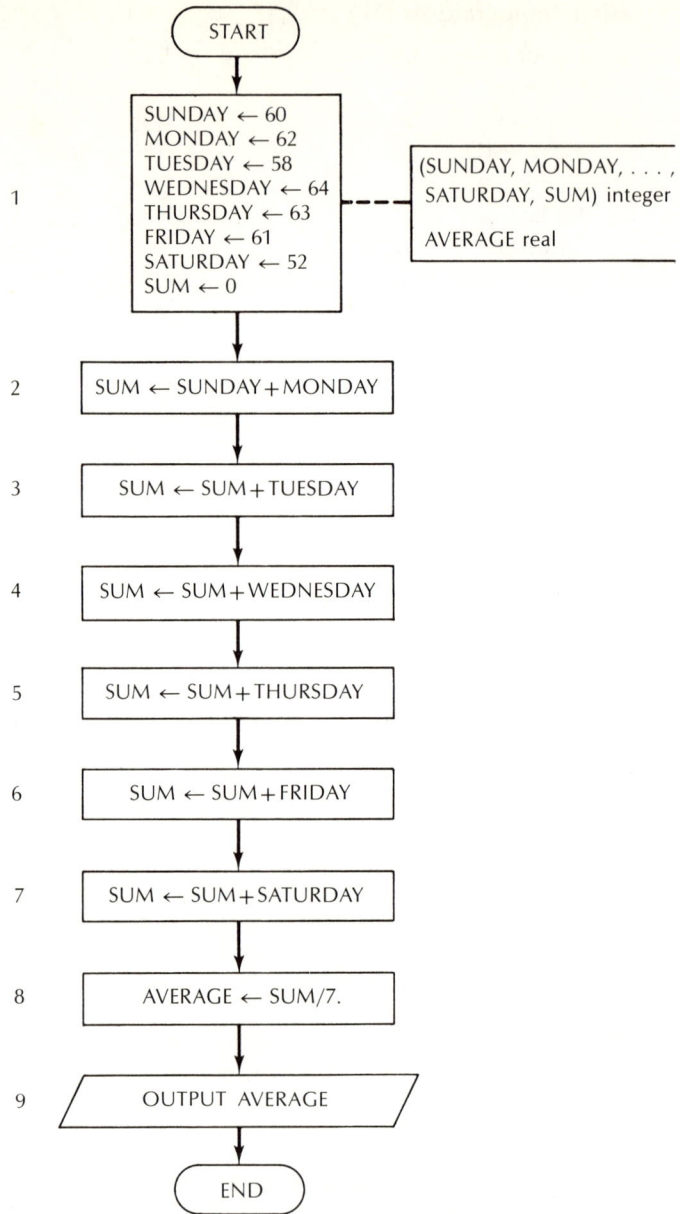

FIG. 3.2-1. Flowchart for finding the average high temperature.

CONSTANTS

There are three types of constants:

    1. Numeric constants
        (a) integer constants
        (b) real constants
        (c) fixed decimal constants
    2. String constants
        (a) character string constants

TABLE 3.2-1. Trace table of flowchart shown in Figure 3.2-1

Symbol #	SUN.	MON.	TUES.	WED.	THURS.	FRI.	SAT.	SUM	AVERAGE
1	60	62	58	64	63	61	62	0	Arbitrary
2								120	
3								180	
4								244	
5								307	
6								368	
7								420	
8									60.
9									

      (b)   bit string constants[1]

  3.   Label constants

Numeric constants consist of strings of the digits 0 through 9; these may be preceded by a plus or minus sign. In addition, such strings may contain a decimal point. Thus, numeric constants appear either as integers such as 1, -100, 14, or as real numbers such as 1.62, -100.009, $+0.921$. Correspondingly they are referred to as **integer** and **real constants,** respectively. Fixed decimal constants will be considered at the end of this section.

We will assume that there is no limit on the magnitude or precision of numeric constants when used in conjunction with flowchart language, and that a blank indicates the end of a numeric constant.

Character string constants consist of a sequence of symbols enclosed in apostrophes. Some examples of string constants are

    (a) 'SMITH, JOHN. A.'   (b) '10113'
    (c) '$1001.24'   (d) 'ISN''T'
    (e) '$_\wedge$'(Note that the symbol $_\wedge$ will be used in the text to represent a blank character.)

Since the apostrophe is used to indicate the point at which a character string constant starts or terminates, it cannot occur as a symbol in a string constant. We therefore adopt the convention that *two consecutive* apostrophes represent *one* apostrophe within a character string constant, as illustrated in (d) of the preceding examples.

To further elucidate the concepts we have just discussed, see Table 3.3-1. Here, the type of constant is identified in each example, and those that are illegal in our flowchart language are identified as being invalid.

The third type of constant is known as a **label constant.** Such constants are unsigned positive integers and may appear within connector boxes in flowcharts. For example, the 1 that appears as ① in Figure 3.1-1 is a label constant.

---

[1]Bit string constants are beyond the scope of this book.

TABLE 3.3-1. Examples related to numeric constants and character string constants

Constant	Type
1. $+100.01$	real
2. 'COMPUTER∧SCIENCE'	character string
3. 'IT'S∧OUR∧CLASS'	invalid; only one internal apostrophe
4. '$246.72'	character string
5. '∧'	character string (of one blank)
6. ''	character string (of zero length)
7. 'BILL''S∧BOOK'	character string
8. 100.∧01	invalid; blanks cannot occur in a number
9. $-.76321712910041692$	real
10. -421.62	real
11. $-19.102.16$	invalid; 2 decimal points in a number
12. 16^2	integer
13. $-100.01+102$	invalid; 2 signs in a number
14. 100.	real
15. 179∧182	invalid as a numeric constant in a flow chart or program, since the blank would indicate two constants; namely, 179 and 182, respectively.
16. 'ANSWER=∧∧'	character string
17. 142001	integer
18. 1	integer

VARIABLES

There are two basic concepts that are associated with variables. These are (a) variable names and (b) values of variables. The term *variable name* merely refers to a sequence of characters, the first of which must be alphabetic, while the rest may be alphabetic, numeric, or both. Thus,

A, SUM, MATTER, JOHN, BX12

are all valid variable names. The purpose of using variable names is to be able to refer to specific locations in the storage or memory of a computer. The term *value of a variable* refers to the specific value taken by a variable at a particular time. For example, the variable SUM may take the value 251, while the variable ⌐NAME takes the value 'MARY':

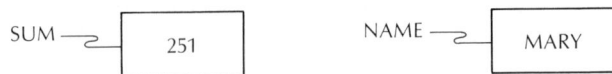

SUM ⟶ [251]   NAME ⟶ [MARY]

It should be noted here that when a string constant is stored in memory, the enclosing apostrophes are not stored. However, for each pair of apostrophes that is part of a string constant, one apostrophe is stored in memory.

There are essentially three types of variables: (a) numeric, (b) character string, and (c) label. As the names suggest, numeric variables take values which are numeric constants, while the values of character string variables are restricted to character string constants. Variables that take values that are label constants are called label variables. For example, SW in Figure 3.2-2 represents a label variable. The reader should observe that the inconnectors 5 and 14, which are label constants, can be looked upon as labels of the statements SW ← 16 and SW ← 42, respectively.

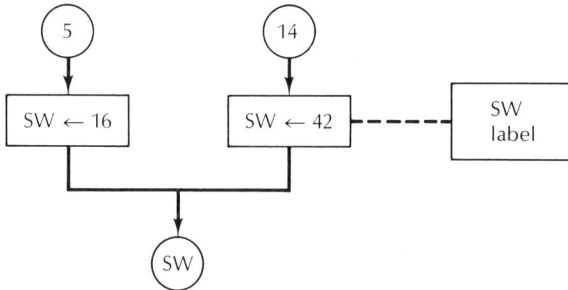

FIG. 3.2-2.   Illustration of a label variable.

Specific information related to the type of variable can be provided via declarative statements, which are not executed. For example, to indicate that the variables A and B can take only **real** and **integer values,** respectively, we can use the following declarative statement:

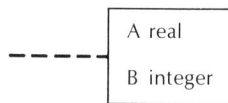

Note that an **integer variable,** such as B, is a special case of a numeric variable. Thus, if a statement such as

$$B \leftarrow 7/2$$

is executed, then the value assigned to B would be the integer 3. The fractional portion of the quotient is dropped, since B has been declared an integer variable. In general, fractional parts are discarded **(truncated)** in integer arithmetic operations. The declaration

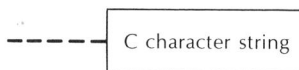

conveys the information that C can take only character string constants for its values. Both of the preceding declarations cause distinct locations in the computer memory to be reserved for the said variables and tagged with their names. In addition, the type of variable is also indicated, as follows:

43

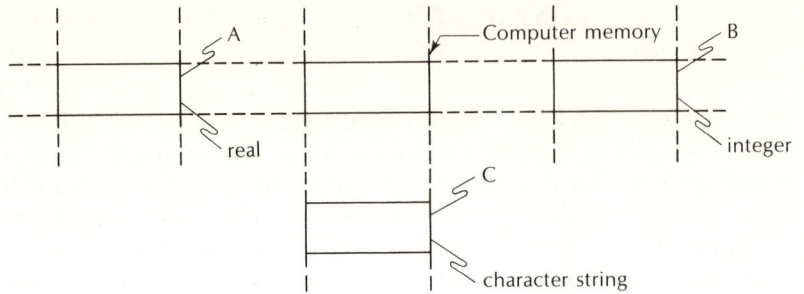

In flowchart language, we will assume that if a variable is *not* declared, then it will automatically (i.e., by default) be treated as a numeric variable. An additional default condition is that variable names that start with any one of the letters I, J, K, L, M, or N are treated as integer variables. A variable name starting with any other letter will be treated as a real variable by default. This default condition is common to several languages. However when using a language, the reader should determine the default. Some languages have no defaults, while others default to a specific type; for example, real.

It is important to note that the defaults we have mentioned can be overridden by resorting to annotation boxes. To illustrate, consider the following case:

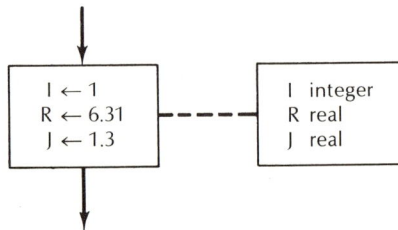

Then the action taken can be summarized as follows:

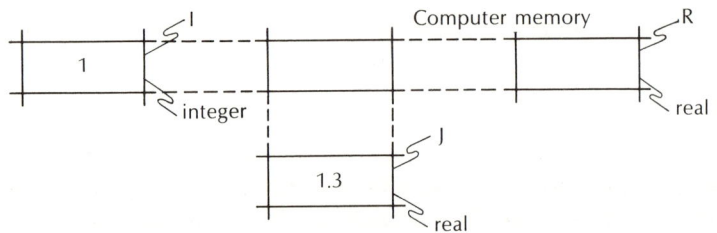

The default condition that J is an integer variable is overridden by declaring it a real variable in the annotation box.

We emphasize that a variable possesses three components: (a) name, (b) type, and (c) value. Conceptually we may identify the name and type with a unique location in memory, and the value as a numeric, label, or character string constant which is stored in that location. When using a variable, the name specified is tagged to some location. Until that location is initialized by a current user, the value available in that location is attributed to a previous user of the com-

44

puter memory; therefore it is considered to be undefined or **arbitrary.** For example, the value of the variable A in the following illustration is arbitrary until Symbol 3 is executed.

```
        ┌─────────┐
        │  START  │
        └─────────┘
             │ ┄┄┄┄┄┄┄┄┄┄ ┌──────────────────────────────┐
             │            │ The value of A at this point  │
             ▼            │ is undefined or arbitrary     │
 1     ┌──────────┐       └──────────────────────────────┘
       │ C ← 10.1 │
       └──────────┘
             │                    ╭───╮
             ▼ ◄──────────────────│ 1 │
 2     ┌──────────┐               ╰───╯
       │ C ← C+1. │
       └──────────┘
             │
             ▼
 3     ┌──────────┐ ┄┄┄┄┄┄ ┌──────────────────────────────┐
       │  A ← 1.  │        │ It is not until this symbol that│
       └──────────┘        │ A takes the value 1            │
             │             └──────────────────────────────┘
             ▼
```

In some cases it is possible to control the size of the memory location associated with a variable. For instance, the size of a location assigned to a character string variable can be specified as follows:

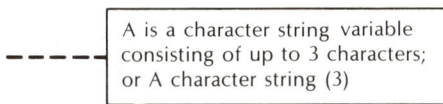

```
              ┌──────────────────────────────────┐
  ┄ ┄ ┄ ┄ ┄ ┄ │ A is a character string  variable │
              │ consisting of up to 3 characters; │
              │ or A character string (3)         │
              └──────────────────────────────────┘
```

which is equivalent to

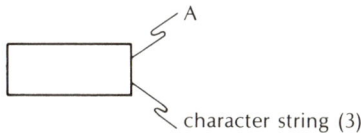

```
                         ╱ A
           ┌─────────┐  ⌐
           │         │
           └─────────┘  ⌐
                         ╲ character string (3)
```

This procedure provides a variable named A that can hold up to three characters, as illustrated below:

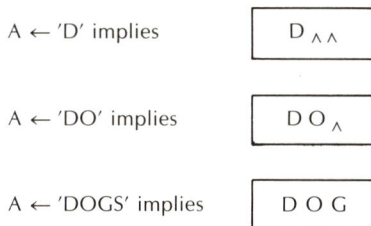

A ← 'D' implies   ┌─────────┐
                   │ D ∧ ∧   │
                   └─────────┘

A ← 'DO' implies   ┌─────────┐
                   │ D O ∧   │
                   └─────────┘

A ← 'DOGS' implies ┌─────────┐
                   │ D O G   │
                   └─────────┘

We can now make the following two observations with respect to the preceding examples:

1. If fewer than three characters are assigned to A, blanks are added to the right.

2. If the number of characters assigned to A exceeds three, then the rightmost extra characters are lost.

45

Again, it may be possible to define a special type of numeric variable by creating a fixed decimal number, which is stored differently than either integers or numeric reals.[2] We specify the size by annotating the total number of decimal digits to be allocated in the memory location and the number of these digits that are to be to the right of the decimal point. The sign and decimal point are not counted as part of the storage required and are indicated below the location, thus:

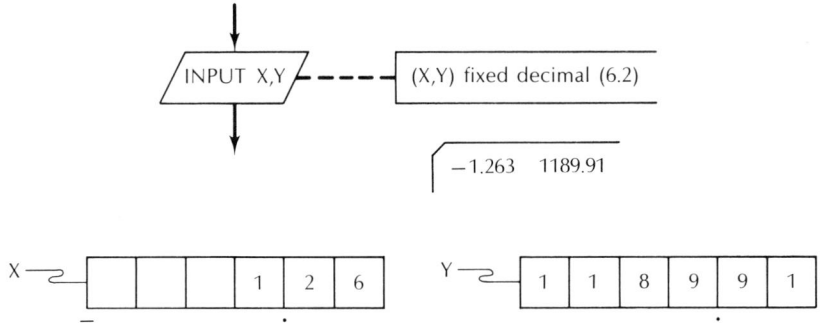

We observe that the (6.2) specification implies that only *two* digits following the decimal point in a six-digit number are stored. Hence the digit 3 in the number -1.263 is ignored and the value stored in the memory location assigned to X is -1.26. This way of storing numbers is commonly used for monetary accounting purposes.

3.3 ASSIGNMENT STATEMENTS

The process of assigning new values to variables involves the use of **assignment statements.** The syntax associated with this class of statements is

variable ← expression

where the expression must be a value, or must result in a value, the type of which is the same as that of the variable. Thus,

(a) A ← 13   (b NAME ← 'JOHN'
(c) A ← B   (d) B ← 13
                            A ← B + 1

are all assignment statements, where ← is the assignment operator that was introduced in Section 3.1. In (a), the numeric constant 13 is assigned as the value of the numeric variable A, while the statement in (b) causes the string variable NAME to take the string constant value 'JOHN'. The statement in (c) causes the current value of the variable B to be assigned to the variable A, whereas the two assignment statements in (d) cause A to take the value 14. These actions can be summarized as follows:

---

[2] A discussion of the manner in which numeric integers and reals are stored is given in Appendix A.1.

(a) | 13 | ⌐ A / ⌐ integer

(b) | JOHN | ⌐ NAME / ⌐ character string

(c) | (○) | ⌐ A | (○) | ⌐ B

(d) | 13. | ⌐ B / ⌐ real

| 14. | ⌐ A / ⌐ real

In flowcharts, assignment statements appear in processing symbols. It is important to note that our convention does not allow a constant to appear to the left of the assignment operator. The corresponding statements would be undefined, as is the case with the statement $32 \leftarrow B$, since one cannot assign the value of the variable B to the numeric constant 32.

More than one variable can appear to the left of the assignment operator. For example, the statements

$$A \leftarrow 15$$
$$B \leftarrow 15$$

can be replaced by the single statement

$$A,B \leftarrow 15$$

This multiple assignment, however, is not available in all programming languages. In conclusion, we add that in some programming languages, the assignment operator $\leftarrow$ is represented by other symbols (i.e., $=$ or $:=$). In some cases, the symbol $=$ may represent two operators—assignment, and comparison for equality.

INPUT/OUTPUT OPERATIONS 3.4

These are operations which enable a user to communicate with a computer system. Many types of media are employed, such as punched cards, magnetic tape and diskettes,[3] typewriter terminals, and voice. Output information can be displayed on a cathode ray tube for temporary viewing or preserved in a variety of ways—printed pages or photographs, for example. Figure 3.4-1 shows the computer output being displayed on a cathode ray tube to a team of researchers in a

---

[3] Also called "floppy disks"; they are similar to phonographic records.

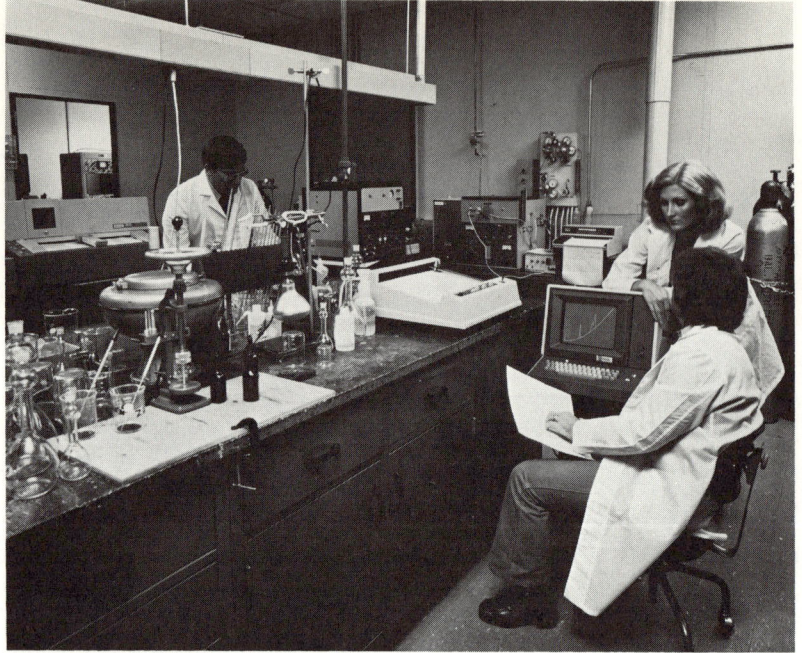

FIG. 3.4-1. Graphical display of computer information to aid researchers. (Courtesy of Textronix, Inc.)

laboratory environment. Input/output operations are carried out by means of statements that are available in computer programming languages.

In flowchart language, the statement associated with the input operation is called the **INPUT statement;** it enables input values to be stored in the computer memory. For example, consider Figure 3.4-2(a). Since X is declared to be a real variable, the computer expects a real constant for its value on the input medium—a data card, for example—being used. Execution of the INPUT X statement causes the value 13.2 to be stored in a memory location that is reserved for the variable X. On the other hand, if the value 'DOG' was found on the card instead of 13.2, then this error would probably terminate execution of the program. Figure 3.4-2(b) relates to reading in a value for a character string variable Y. In this case, the character string constant 'FEVER' is stored in the location reserved for the character string variable Y when the INPUT Y statement is executed.

More than one variable can be read in via a single INPUT statement, as is the case in Figure 3.4-3. Here, X and NAME are real and character string variables, respectively, due to the declarative statement. Thus, execution of the statement INPUT X, NAME causes the values 2.57 and 'JOHN' to be stored in the memory locations reserved for X and NAME, respectively.

We observe that when more than one name is in the INPUT list, the first value read is stored in the location assigned to the first variable listed, while the second value is stored in the location assigned to the second variable listed, and so on. The specific input medium from

(a)

(b)

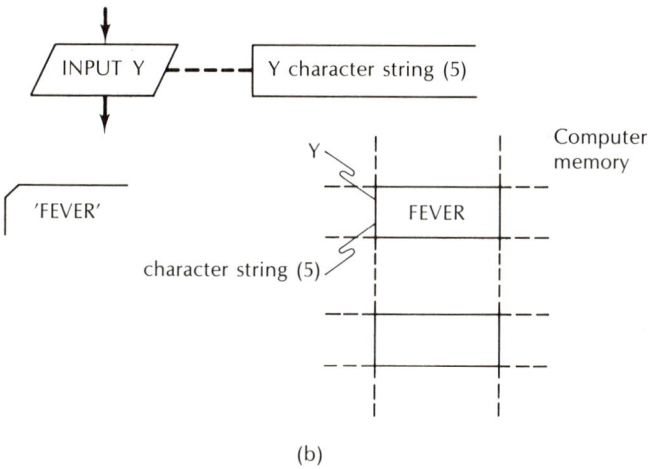

FIG. 3.4-2. Input of a single value.

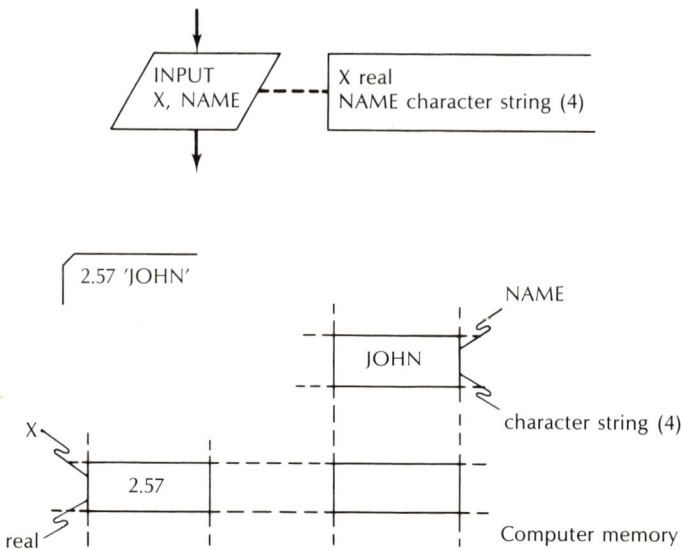

FIG. 3.4-3. Input of more than one value.

49

which these values are read is determined by the computer installation being used. Most commonly, the INPUT statement causes data to be read from cards that have been placed in a card reader.

The statement associated with the output operation is called the **OUTPUT statement;** it enables us to retrieve results from the computer memory and write them on some output device, such as a line printer. As illustrated in Figure 3.4-4(a), the statement OUTPUT X causes the

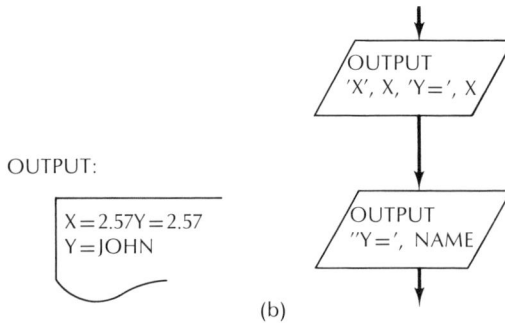

(a)

(b)

FIG. 3.4-4. The output operation.

current value of X (in this case, 2.57) to be printed. Similarly, in Figure 3.4-4(b), the statements

OUTPUT 'X = ',X, 'Y = ',X
OUTPUT 'Y = ',NAME

cause the *current values* of the variables X and Y to be printed, along with the string constants 'X = ' and 'Y = '. In other words, any constant appearing in the OUTPUT statement will appear as part of the output. Note that the apostrophes associated with a string constant are not printed.

Figure 3.4-5 is a further example in which the two values 12 and 15 are read in from a card. Execution of Symbol 1 in the flowchart leads to

X and Y being assigned the values 12 and 15, respectively. As a result, when Symbol 3 is executed, the variable SUM takes the value 27, and the value of X (which is 12.) is now replaced by X + 2. (14.).[4] We observe that prior to execution of Symbol 3, the value of SUM is arbitrary.

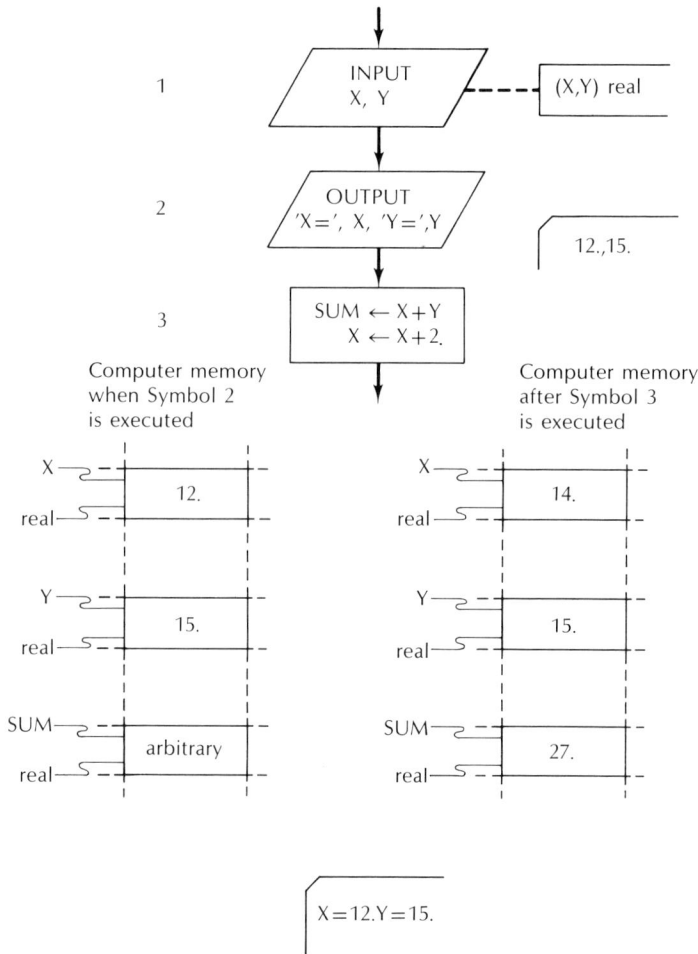

FIG. 3.4-5. An input/output example.

Due to the statement in Symbol 2, the output is as indicated at the bottom of Figure 3.4-5.

To summarize, the following important features associated with input/output operations are as follows:

1. When data are written in a memory location via an INPUT operation, the previous data in that location are *destroyed.*
2. When data are read from a memory location via an OUTPUT operation, the data in that location are *preserved.*

___
[4]Note that since SUM is not declared, it is treated as a real variable.

Branching is a process by means of which the next step to be executed in an algorithm is not necessarily the one that follows the current step being executed. Branching enables us to execute statements in a nonsequential manner, as opposed to sequential processing. The latter is a basic characteristic of most programming languages. There are two types of branching: (a) unconditional and (b) conditional.

An **unconditional branch** is one that is taken every time a certain step in an algorithm is reached, as illustrated in the flowchart examples of Figure 3.5-1. The label variable is useful to vary the place to

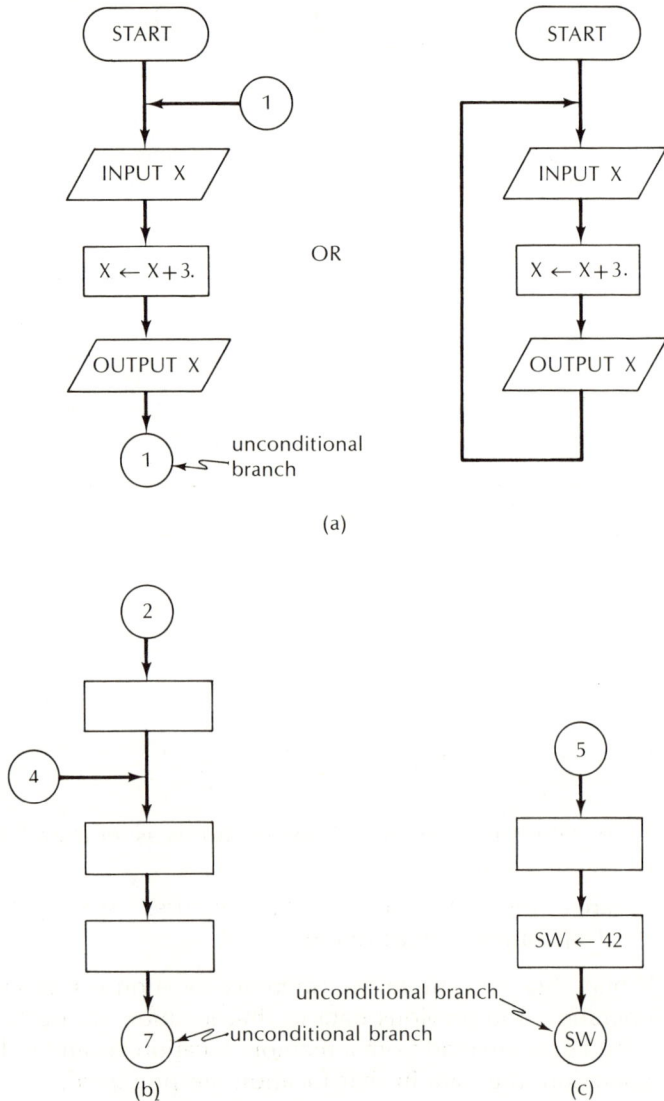

FIG. 3.5-1. Examples of unconditional branches.

which an unconditional transfer of control is made. To illustrate, consider the following case:

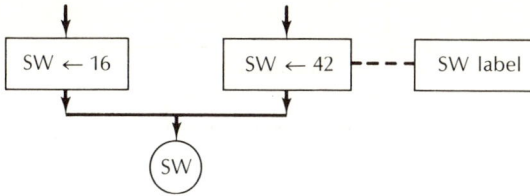

```
    ↓                      ↓
 ┌────────┐            ┌────────┐          ┌──────────┐
 │ SW ← 16│            │ SW ← 42│----------│ SW label │
 └────────┘            └────────┘          └──────────┘
    ↓                      ↓
    └──────────┬───────────┘
               ↓
             ( SW )
```

Here, the flowchart indicates an unconditional transfer of control to the inconnector, the value of which is the value of the label variable SW. If we reach the outconnector via the righthand path, processing will continue at the inconnector 42. Conversely, if the left path is traversed, then processing will continue at the inconnector 16.

A branch that is taken only when certain conditions are satisfied is called a **conditional branch** (see Figure 3.5-2). With respect to this

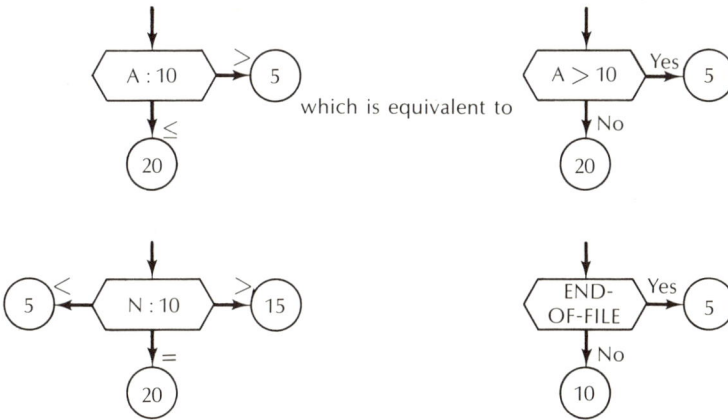

```
        ↓                                            ↓
   ╱─────────╲  >                             ╱─────────╲  Yes
  ⟨  A : 10   ⟩───( 5 )    which is          ⟨   A > 10  ⟩───( 5 )
   ╲─────────╱       equivalent to            ╲─────────╱
        ↓ ≤                                        ↓ No
      ( 20 )                                     ( 20 )


        ↓                                            ↓
   <  ╱─────────╲  >                          ╱──────────╲  Yes
 ( 5 )─⟨ N : 10  ⟩───( 15 )                  ⟨   END-     ⟩───( 5 )
   ╲─────────╱                                ╲  OF-FILE  ╱
        ↓ =                                        ↓ No
      ( 20 )                                     ( 10 )
```

FIG. 3.5-2. Examples of conditional branching.

figure, we can make the following two remarks:

1. The A : 10 notation is equivalent to the statement, "A is compared with 10."

2. The term *end-of-file* is a condition which may be used to determine whether or not the end of the input data has been reached. This condition can only be detected following the execution of an INPUT statement for which sufficient input data are not available. The action associated with the end-of-file condition is executed immediately upon detection of insufficient data.

Let us now return to the programmable robot that we introduced in Section 3.0. To its existing commands (MOVE and TURN), we now add a third command, implementing a conditional branch statement of the form

IF condition THEN command

53

which has the following flowchart representation:

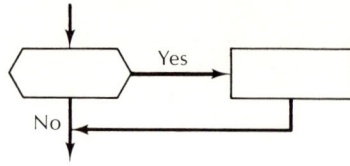

Note that the robot has no unconditional transfer of control commands, and assume that "command" is not a conditional branch statement.

Let us suppose that the maze has a stoplight at a corner (see the illustration below). We seek a program to get the robot to Point A or B such that

(a) if the light is green, then the robot must take the dashed path to Point A;

(b) if the light is red, then the robot must make a right turn and take the dotted path to Point B.

It is straightforward to verify that the robot traverses the desired path if it is programmed as follows:

MOVE
IF light red THEN TURN
IF light green THEN MOVE
MOVE

FURTHER CONSIDERATIONS

Using the "IF condition THEN command" statement, we add two features to the robot's language:

1. An additional statement of the form

    IF condition THEN command
        ELSE command

(assume "command" is not a conditional branch statement), which has the following flowchart representation:

Yes

No

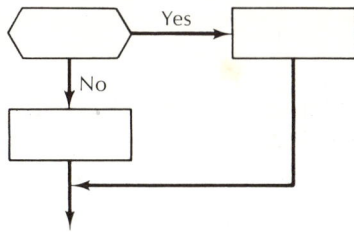

2. The command portion of the statement "IF condition THEN command" can be made more powerful by replacing it with:

IF condition THEN BEGIN
        command 1
        command 2
        ⋮
        command n
END

Here, BEGIN and END have been added to specify a series of commands to be accomplished where only one command is normally legal. With $n = 2$, for example, we obtain the following sequence and related flowchart:

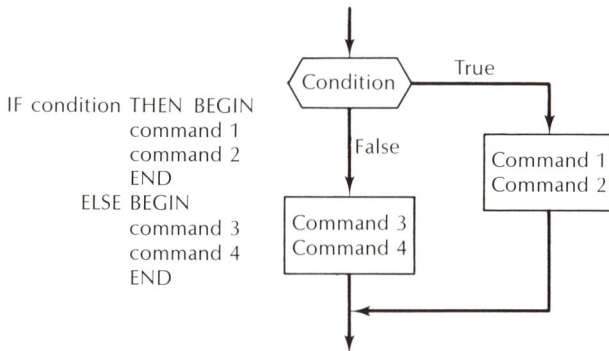

IF condition THEN BEGIN
    command 1
    command 2
    END
ELSE BEGIN
    command 3
    command 4
    END

Condition — True — Command 1 / Command 2

False — Command 3 / Command 4

To illustrate the manner in which the preceding features are used, the robot is now programmed to start from the location shown, and move to Point C in such a way that the two conditions cited earlier are satisfied. The resulting program is as follows:

MOVE
IF light red THEN BEGIN
        TURN
        MOVE
        TURN
        TURN
        TURN
        MOVE
        TURN
        END
    ELSE BEGIN
        MOVE
        TURN
        MOVE
        END
MOVE

55

Example 3.6-1 X and Z are real variables, A and B are integer variables, and Y is a character string variable (5). The contents of the corresponding memory locations are as shown in Table 3.6-1.

TABLE 3.6-1. Memory contents before execution

real — X	16.
character string (5) — Y	BIG∧∧
real — Z	arbitrary
integer — A	6
integer — B	2

(a) What are the contents of the above memory locations after the statement

$$\text{INPUT } X,Y,Z,A,B,X$$

is executed, if the input data are as follows:

$$1, \text{ 'HI', } 3.2, 6, 9, 3$$

(b) What would the output be if the statement

$$\text{OUTPUT 'X=', X, X, 'Y=', Y}$$

is executed immediately after the INPUT statement in (a)?

Solution:

(a) Execution of the INPUT statement causes each element of the input data to be matched with the corresponding variable in the input list X, Y, Z, A, B, X. Thus, the memory contents after the INPUT statement is executed are as shown in Table 3.6-2.

TABLE 3.6-2. Memory contents after execution

X	~~16~~.~~1~~.,3.
Y	~~BIG∧∧~~, HI∧∧∧
Z	~~arbitrary~~,3.2
A	~~6~~,6
B	~~2~~,9

(b) We observe that the value of X first undergoes a change from 16 to 1. However, since X appears twice in the input list, the value 1 is erased (indicated by ~~1~~); X then takes the value 3, which is the element in the data that matches with the second appearance of X in the input list. From Table 3.6-2 it follows that

$$\text{OUTPUT 'X=', X, X, 'Y=', 16, Y}$$

yields the output

$$X = 33Y=16HI_{\land \land \land}$$

Example 3.6-2 Identify the label constants, label variables, inconnectors, and outconnectors in the following flowchart. Assume that the flowchart is correct and appropriate type declarations have been made.

Solution: Label constants are 1,4,5,6,91.

Label variables are L, CAT

Inconnectors are (4), (91), (1), (5), (6)

Outconnectors are (5), (L), (CAT), (91)

Example 3.6-3

(a) In the flowchart shown below, identify the label, string, and numeric variables and the corresponding types of constants. Assume that the flowchart is correct and appropriate type declarations have been made.

(b) By means of a trace table, determine the number of times Symbols 1 through 4 are executed.

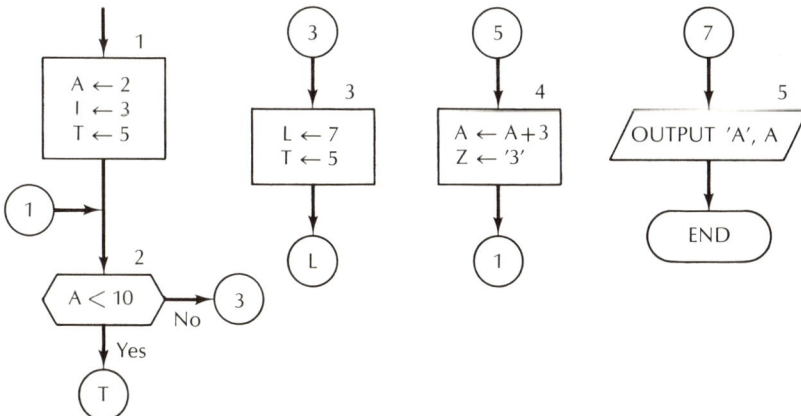

57

Solution:

(a)   Examining the flowchart, we find that the variables and constants of interest are as follows:

Label variables: T,L
Label constants: 1,3,5,7
Character string variables: Z
Character string constants: 'A=', '3'; ('A=' is from Symbol
                                                               5, '3' is from Symbol 4)

Numeric variables: A
Numeric constants: 2,3,10; (2 is from Symbol 1; 3 from
                                                               Symbol 4, 10 from Symbol 2)

(b)   It follows that the desired trace table is as follows:

Symbol #	A	L	T	Z
1	2	3	5	Arbitrary
2				
4	5			3
2				
4.	8			3
2				
4	11			3
2				
3		7	5	
5				

Output is A = 11

From this trace table it is apparent that Symbols 1 through 4 are executed 1, 4, 1, and 3 times, respectively. It is also evident that A has the value 11 when Symbol 5 is executed.

There are several observations that are worthwhile noting with respect to Example 3.6-3:

1.   Each variable has but one value at any time.

2.   There are three constants with value 3: (a) the numeric constant 3 in the statement $A \leftarrow A + 3$, (b) the character string constant '3' from $Z \leftarrow$ '3', and (c) a label constant 3 from $L \leftarrow 3$ or from ③.

3.   When one gets to Ⓣ, the value of the variable T must be determined, and then transfer to the inconnector with that constant value must be made. The first time Ⓣ is used, it has the value 5; hence, control transfers to Symbol 4.

Example 3.6-4 Develop an algorithm (along with a corresponding flowchart) for determining the maximum value in a list of N numbers.

Solution: The input consists of N, the number of elements in the list, followed by N numbers that form a list. The output is the maximum value found in the list.

We define three numeric variables: MAX, COUNTER, and VALUE.

58

Our objective is to place the largest number in a given list in the memory location assigned to MAX. This will be accomplished by scanning the list, starting with the first member and moving toward the last member. The largest value through any stage in the scanning process is assigned to MAX. The role of the COUNTER will be to keep track of which member of the list is being read. The list is exhausted when the value of COUNTER equals the value of N. Each time a value is read from the list, we will store it in the variable VALUE.

We consider the following initial solution:

1. Input a value for N, the number of values in the list.
2. Initialize a COUNTER of elements processed, and let MAX denote a storage location for the current largest value encountered, as the list is processed.
3. Process an element in the list.
4. Test whether Step 3 has been completed.
5. Output MAX.

The preceding solution is now subjected to the following refinement:

1. Input a value for N, the number of values in the list.
    1.1 Input a value for N.
    1.2 Echo N.
2. Initialize a COUNTER of elements processed, and let MAX denote a storage location for the current largest value, as the list is processed.
    2.1 Assign the value 1 to COUNTER; i.e., COUNTER ← 1.
    2.2 Input the first value and store it in MAX.
3. Process an element in the list.
    3.1 Input a value from the list; call it CURRENT.
    3.2 If the value of CURRENT is larger than the value of MAX, then assign the value of MAX to CURRENT.
4. Test whether Step 3 has been completed.
    4.1 Add 1 to COUNTER.
    4.2 If the value of COUNTER equals the value N, then go to Step 5.
    4.3 Else return to Step 3.1.
5. Output MAX.

From this refinement, it follows that the desired flowchart is as shown in Figure 3.6-1.

Example 3.6-5 Suppose you are given a specific integer to be found in a given list. Input this value into a variable called FIND. Develop an algorithm that enables you to read a list of unknown length, consisting of values as illustrated below, until you find a value in the list that matches the given value of FIND. Output the sequence number of the element in the list when this match occurs. The algorithm should be applicable to any type of list elements, as long as the value to be located is of the same type as the list members.

59

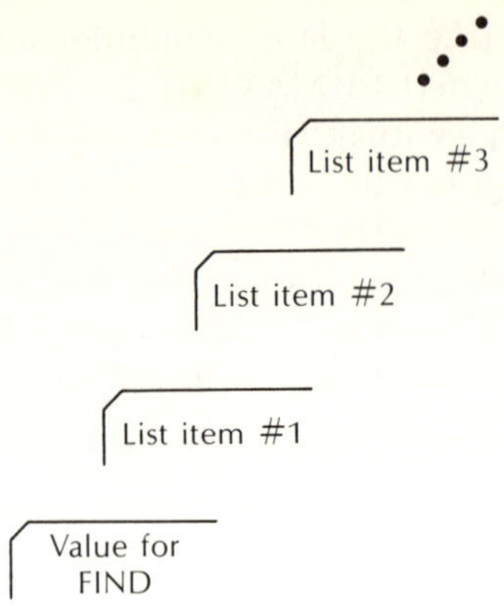

List item #3

List item #2

List item #1

Value for
FIND

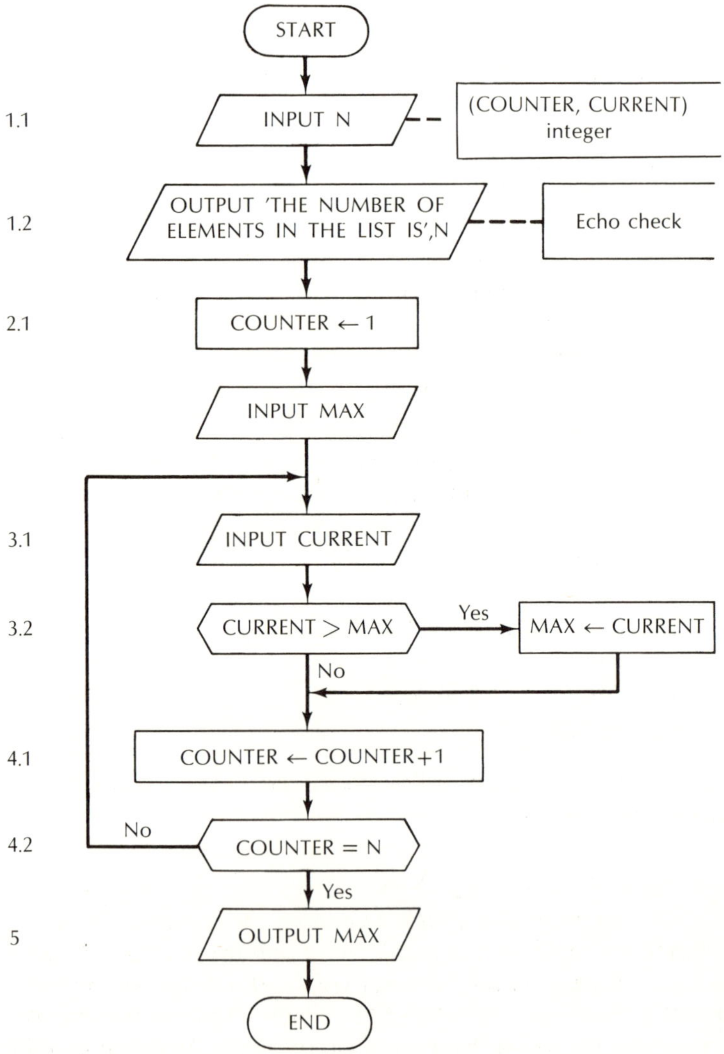

FIG. 3.6-1. Flowchart pertaining to Example 3.6-4: Search for maximum value in a list.

FIND: the value to be found—this value is generally referred to as the **search argument.**

LISTVAL: stores a value from the list as it is being processed.

The output variable is

COUNTER: the number of the element in the list when a match is found with the search argument.

The initial solution proposed is the following:

1. Input a value for FIND (i.e., the search argument).
2. Provide a way to count the list items to be input.
3. Input values until the value of FIND is matched.
4. Output the value of COUNTER.

A refinement of the initial solution results in

1. Input a value for FIND (i.e., the search argument).
   1.1 Input a value for FIND.
   1.2 Echo FIND.
2. Initialize a counter to count the list items to be input.
   2.1 COUNTER ← 0.
3. Input values until the value of FIND is matched.
   3.1 Increment COUNTER by 1.
   3.2 Input a value from the list.
   3.3 If value of the list element is not equal to the value of FIND, then return to Step 3.1.
   3.4 If the value of the list element equals the value of FIND, then go to Step 4.0.
4. Output the value of COUNTER.

The flowchart that results from this refinement is shown in Figure 3.6-2, where the variable LISTVAL is used to store a value that is being read from the given list.

Example 3.6-6 A florist has a standing order for roses each week. Based upon the number sold last week, he changes his order using the following criteria: (a) if the number sold last week was less than the standing order, then reduce next week's order to one-half the standing order; (b) if the number sold last week was greater than or equal to the standing order, then add 10 percent to the standing order.

Develop an algorithm that enables the florist to compute the cost of an order in dollars, assuming roses cost 50 cents each. Flowchart your solution.

Solution: We define the following input and output variables:

Input:

ORDER: represents a standing order
NOSOLD: represents the number of roses sold

61

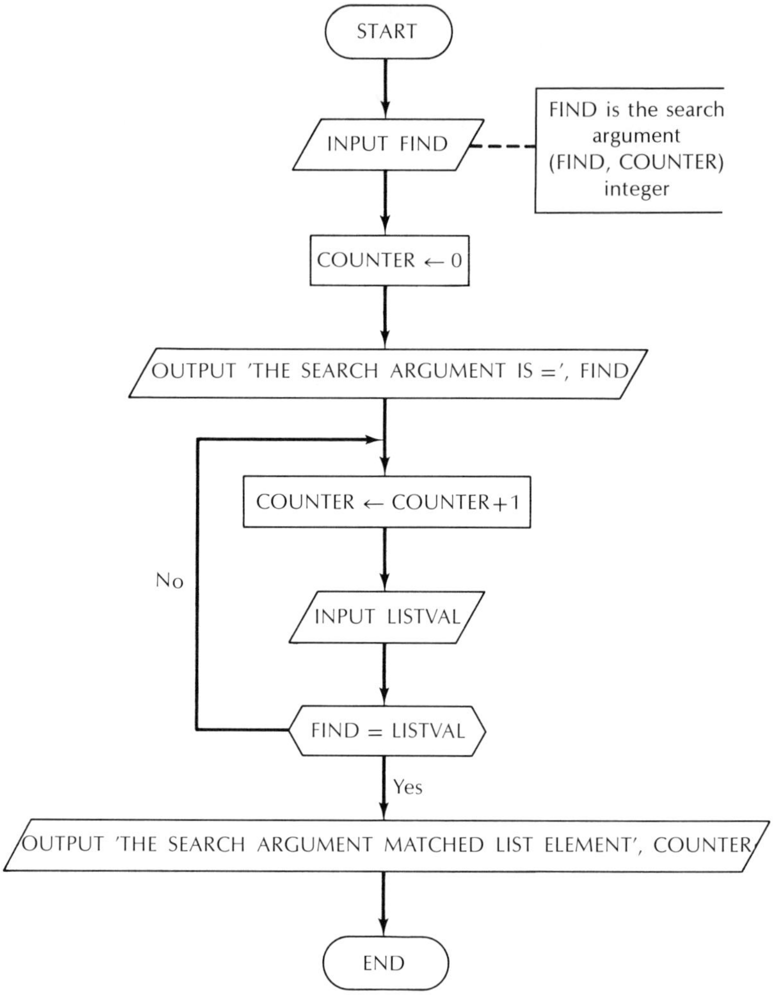

FIG. 3.6-2. Flowchart pertaining to Example 3.6-5: Locating a value in a given list of elements.

Output:

COST: represents the total cost of an order

Then, a structured algorithm may be considered to be as follows:

1. Input values for standing order and the number of roses sold last week.
2. Determine the number of roses to order.
   2.1 If number sold is less than the standing order, then reduce standing order to one-half its value.
   2.2 If the number sold is greater than or equal to the standing order, then add 10 percent to the standing order.
3. Calculate the total cost of an order.
4. Output values of total cost and the number ordered.

A flowchart representation of the preceding algorithm is given in Figure 3.6-3.

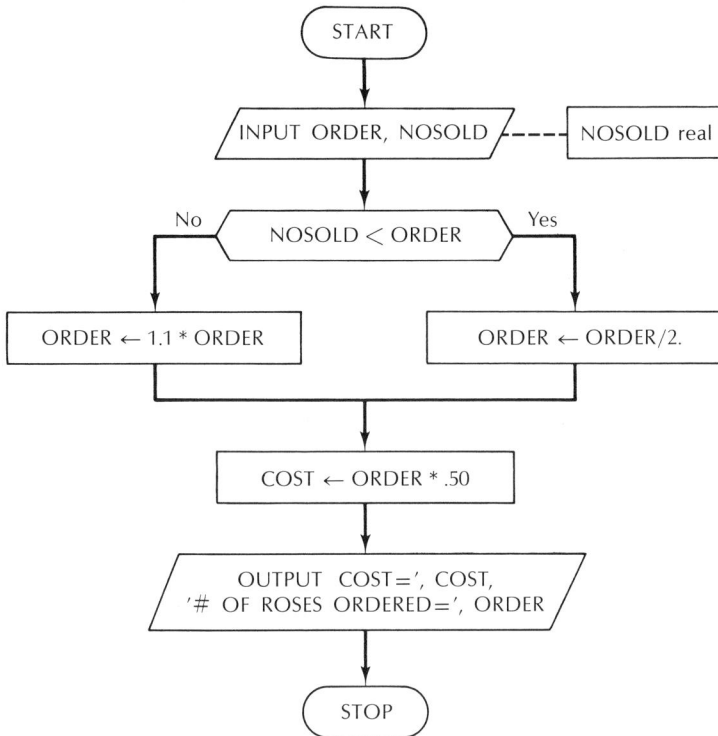

FIG. 3.6-3. Flowchart pertaining to Example 3.6-6: Florist ordering scheme.

Example 3.6-7 You are to calculate the mean score and find the high and low scores for an examination with a maximum of 100 points. Develop a structured algorithm that is capable of processing a given set of examination scores, prefaced by the name of the instructor and followed by the scores of students who took the examination. (Note that the number of students involved can vary from examination to examination.) Flowchart your solution.

Solution: The input consists of

NAME: name of instructor
SCORE: list of examination scores.

The quantities associated with the output are:

COUNTER: number of students who took the examination
MEAN: mean score
MAX: high score
MIN: low score

A solution in refined form is as follows:

1.   Initialize internal variables used for processing.
   1.1   COUNTER ← 0, COUNTER is a counter for number of scores.
   1.2   MAX ← 0, MAX holds current high score.
   1.3   SUM ← 100, MIN holds current low score.

63

1.4 SUM ← 0, SUM is a running sum of scores as they are processed.

2. Label output.
2.1 Read name of instructor (i.e., NAME).
2.2 Echo name of instructor.

3. Process the scores.
3.1 Input a SCORE.
3.2 If all scores have been input, then proceed to Step 4.
3.3 Process SCORE just input.
3.3.1 Add SCORE to SUM.
3.3.2 If SCORE is greater than MAX, then replace value of MAX with SCORE value.
3.3.3 If SCORE is less than MIN, then replace value of MIN with SCORE value.
3.4 Increment number of scores input by 1 (i.e., COUNTER).
3.5 Return to Step 3.1 to repeat for another score.

4. Calculate mean score.
4.1 Divide value of SUM by value of COUNTER to obtain mean score.

5. Output results.
5.1 Output number of students who took examination.
5.2 Output mean score.
5.3 Output high score (i.e., MAX).
5.4 Output low score (i.e., MIN).

The flowchart corresponding to the preceding algorithm is given in Figure 3.6-4.

3.7 SUMMARY

Commencing with the notions of alphabet, syntax, and semantics, we developed a flowchart language. A variety of related terminology—words, variables, constants, statements—was defined and illustrated with examples. In addition, some aspects of input/output operations and branching were also introduced. In the interest of consolidating the various concepts, Section 3.6 was exclusively devoted to several illustrative examples.

problems

3-1. Given pairs of two-digit numbers called NUM1 and NUM2, develop a structured algorithm which will read the numbers as two-digit numbers, echo them, and then calculate and output the sum. Assume that only single digit addition is available, but variables may hold more than one digit. Double digit subtraction is possible. Refer to the two digits associated with a variable NUM1 as NUM1LD and NUM1RD, respectively. For example, if NUM1 is 46, then NUM1LD is 4 and NUM1RD is 6. Perform Steps 1–8 in Table 2.1-1.

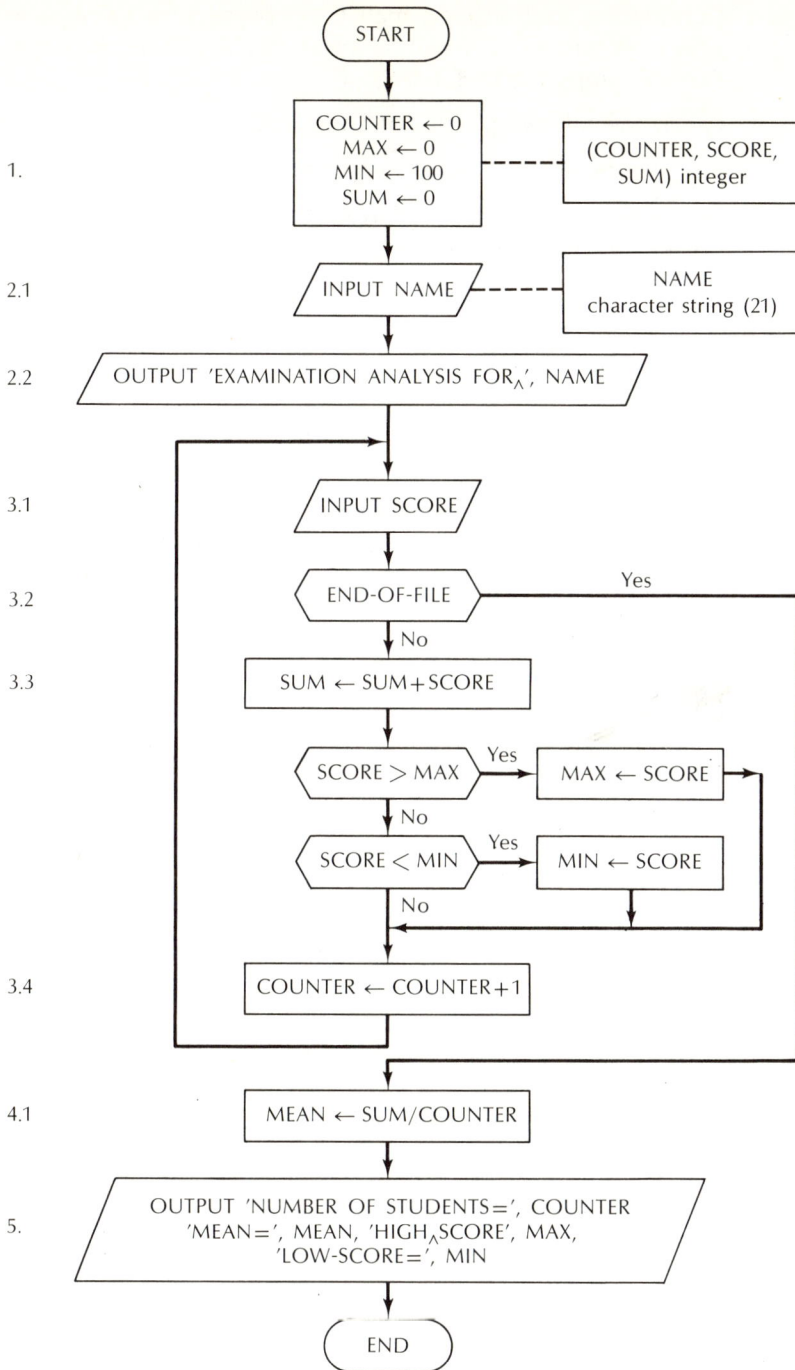

FIG. 3.6-4. Flowchart for Example 3.6-7: An examination processing program.

3-2. A car dealer determines the price quoted to a prospective customer by determining his cost (COST) for the car under consideration. If the COST is greater than $3000, then he adds 11 percent of COST to COST to obtain PRICE. If the COST is less than or equal to $3000, then he adds 15 percent of COST to

65

COST to obtain PRICE. Formalize this process for a new salesperson. Assume COST will be input, PRICE will be an output.
Perform Steps 1–8 in Table 2.1-1.

3-3. Given the following types of constants:

numeric (real or integer) and character string

classify the following list of constants with respect to their type,
if valid (if invalid, state why):
(a) IBM∧CORPORATION   (b) −0.543211
(c) 2²   (d) 4∧4∧82.2
(e) 'CS200∧-∧ASSIGNMENT∧4'   (f) 654
(g) 'JIM'S'   (h) +16.237965211962
(i) 0.0   (j) −16.2+1.106
(k) 'TINKER'∧'BELL'   (l) 'TINKER''BELL'

3-4. Identify the valid flowchart language variable names in the
following list (if invalid, state why):
(a) TODAY   (b) 4THOFJULY
(c) Z4221   (d) K
(e) IBM'S∧360   (f) PL/1
(g) I   (h) 'I'M'
(i) X*Y   (j) SALARY
(k) 200TH   (l) COST∧OF∧ITEM

3-5. Consider the following flowchart, where annotations have
been omitted intentionally:

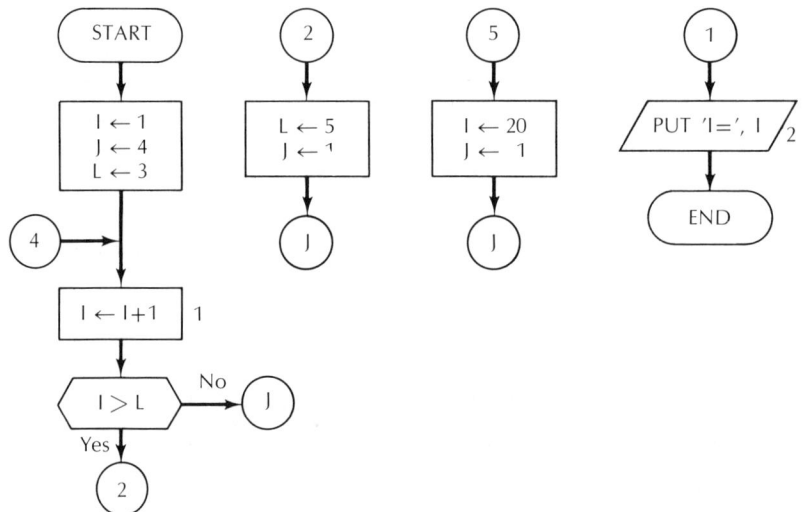

(a) Identify label con-   (b) Identify label vari-
stants.   ables.
(c) Identify string con-   (d) Identify string vari-
stants.   ables.
(e) Identify numeric vari-   (f) How many times is
ables.   Symbol 1 executed?
(g) What is the value of   (h) Identify the numeric
I when Symbol 2 is   constants.
executed?

3-6. Given the following input data:

8,24, 'YESTERDAY', 11.21, .05, 'TOMORROW', 'TODAY', 44
---

what values are assigned to variables in the following three sets of input statements? Assume that you start at the beginning of the input data for each set and that there are no variable-constant type conflicts.
(a)   INPUT A,B,C, STRING, TAX, DATE
(b)   INPUT X,Y,R,S
(c)   INPUT E,F, RATE, STRING, TAX, F

3-7. Given: A = 'TEMPERATURE=$_\wedge$'   B = 77.0
              C = 25.0   D = 'NICE'

Assume that the values associated with the variables A,B,C, and D are available in computer memory. Show the output resulting from the statement

OUTPUT A, B, 'F', '$_\wedge$=$_\wedge$', C, 'C', '$_\wedge$IT"S$_\wedge$A$_\wedge$', D, '$_\wedge$DAY'

Note: Use a form similar to the following:

⌊⊥⊥⊥⊥⊥⊥⊥⊥⊥⊥⊥⊥⊥⊥⊥⊥⊥⊥⊥⊥⊥⊥⊥⊥⊥⊥⊥⊥⊥⊥⊥⊥⊥⊥⊥⊥⊥⊥⊥⌋

3-8. Identify the type of each of the following flowchart language constants. When a constant may be of more than one type, then identify both types. If a constant is not legal, explain why.
(a)   +12.6   (b)   'HELLO'
(c)   'SHE'S$_\wedge$HERE'   (d)   1272.30312
(e)   166.21$_\wedge$5   (f)   4
(g)   '4'   (h)   HELLO
(i) 2^2 (j)   −102.459

3-9. Given the input data:

5, 6, 'LITTLE$_\wedge$TIME', 22.73, 'NO$_\wedge$DEAL'

what values would be associated with each of the variable names after executing the following INPUT statement:

INPUT X, Y, CAT, X, STRING

X = _____

Y = _____

CAT = _____

STRING = _____

3-10. Given the following values of the variables stored in memory:

A = 26, D = '$_\wedge$', F = 'THAT"S$_\wedge$ALL', C = 'FOLKS'

what is the output that results from

OUTPUT 'A=', A
OUTPUT F, D, C

67

3-11. Consider the following flowchart sequence, where pertinent annotations have been omitted intentionally:

(a) Identify the label constants, label variables, character string constants and variables, and numeric variables and constants.
(b) What is the value of A when Symbol 4 is executed?
(c) How many times is each of the symbols 1, 2, 3, and 4 executed?

3-12. Consider the following input:

$$-10.4, 12, \text{'EXAM'}, 16, 33.1$$

(a) What values will be associated with the variable names after execution of the following statement:

INPUT A1, A2, A3, A4, A2

(b) Given the following values in the computer memory:

Note: In a flowchart JOE and TOM'S would be written as 'JOE' and 'TOM''S', respectively.

what is the output from the following statement?

OUTPUT 'Y=', A, Y, 'SAIL', Z, W

3-13. Given: the input data

3,4

7,10

and the following flowchart:

```
              ┌─────────────┐
              │    START    │
              └──────┬──────┘
                     │
    ┌────────────────┤
    │                ▼
1   │      ╱ INPUT  X, Y ╱ ─ ─ ─   (X,Y) integer
    │                │
    │                ▼            5
    │                       Yes
2   │    〈 END-OF-FILE? 〉────────▶ (  END  )
    │                │ No           6
    │                ▼     Yes   ┌───────────┐
3   │      〈  X < Y  〉────────▶│  X ← X+Y  │
    │                │          └───────────┘
    │                │ No
    │                ▼
4   │      ┌───────────────────┐
    │      │  SUM ← X+Y+Z      │
    │      └───────────────────┘
    └────────────────┘
```

Develop the corresponding trace table.

3-14. Consider the flowchart shown in Figure P 3-14 on the following page. Suppose the input data is as follows:

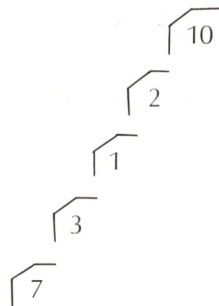

```
        ⌐10
      ⌐2
    ⌐1
   ⌐3
  ⌐7
```

Develop a trace table to find the value of the variable TEMP when Symbol 9 is executed.

3-15. A simple robot is placed in a maze, as shown in Figure P 3-15 on the following page. This robot responds to the three commands, MOVE, TURN, and STOP, as follows:

MOVE: moves the robot forward by one cell
TURN: turns the robot 90 degrees to the right.
STOP: robot stops and does not respond to further commands.

Write a set of instructions (a program) that will enable the robot to leave the maze via the path indicated in Figure P 3-15, assuming that it starts at Position A.

3-16. Using the description in Problem 3-15, get the robot out of the maze via the path shown, if it is given that the robot starts at Position B. Give the corresponding structured algorithm, flowchart, and a set of instructions.

69

FIG. P 3-14

A   B

FIG. P 3-15

3-17. Write a set of instructions that will cause the robot to go around the square to the upper right, using the infomation in Problem 3-15.

3-18. Assume that the robot in Problem 3-15 now has access to two additional commands of the form:

INPUT variable

and

IF condition THEN command

where the variable may take a value 'TURN$_\wedge$RIGHT' or 'TURN$_\wedge$LEFT', and may be named by any single letter. "Condition" refers to some condition that results in a yes-or-no decision. That is, we can have statements of the form

INPUT A
IF A = 'TURN$_\wedge$LEFT' THEN . . .

Write a set of instructions that will cause the robot to move through the maze if the robot is to read a "turn command" before each move. Move the robot one move at a time until it bumps a wall or until it accomplishes three moves. Create a flowchart corresponding to your solution.

3-19. Solve Problem 3-18 using an additional instruction of the form

IF condition THEN command
ELSE command

The learning objectives for this set of poblems are as follows:

☐ to develop skill in using computer input devices;

☐ to develop an awareness of system job control language or protocol required for class assignments;

☐ to appreciate the distinction between program and data and understand sequential program execution;

☐ to gain familiarity with the basic details of a programming language;

☐ to better understand the correspondence between structured algorithms, flowcharts, and program code; and

☐ to learn the basic documentation format required.

CS 3-1. A consumer wishes to find the total purchase price for any three items he/she may purchase. The tax rate is 5 percent in the state of purchase. Create a well-documented computer solution to this problem.

Solution: A structured algorithm to solve this problem is
1. Label output.

2. Read and Echo input.

    2.1 Read three numbers into variables called COST1, COST2, COST3.

    2.2 Print the message THE THREE COSTS ARE, followed by the values which were read for COST1, COST2, COST3, respectively. (This is an echo check.)

3. Calculate the cost of purchase as:

$$\text{TOTAL} \leftarrow (\text{COST1} + \text{COST2} + \text{COST3}) * 1.05$$

4. Print the results.

    4.1 Print PURCHASE PRICE IS, followed by the value of the variable TOTAL.

    4.2 Print NORMAL TERMINATION.

A flowchart corresponding to the preceding algorithm is shown in Figure CS 3-1.

FIG. CS 3-1

CS 3-2. Create a flowchart and computer program corresponding to the following structured algorithm:

1. Read a number into a variable called XYZ.

2. Print the message THE VALUE OF XYZ IS, followed by the value of XYZ which was read. (This is an echo check.)

3. Calculate

$$TAX \leftarrow XYZ * 0.045$$
$$SALE \leftarrow XYZ + TAX$$

4. Print SALE =, followed by the value of the variable SALE
    and TAX=, followed by the value of the variable TAX.
5. Print NORMAL TERMINATION.

CS 3-3. Create a flowchart and computer program for the following
         structured algorithm:
         1. Read and echo four test scores.
         2. Compute the average of these test scores.
         3. Print the average and pass or fail.
             3.1 If the average is greater than or equal to 61, then
                  print the average and the grade is pass.
             3.2 If the average is below 61, then print the average and
                  the grade is fail.
         4. Print NORMAL TERMINATION.

CS 3-4. A student wishes to have a computer solution to the problem
         of averaging four numbers. A structured algorithm for the
         problem is
         1. Read four numbers called A, B, C, D.
         2. Print THE FOUR NUMBERS ARE, followed by an echo
             check of the original four numbers.
         3. Calculate

$$Y \leftarrow A - B + C * D$$
$$X \leftarrow (A + B + C + D)/4$$

         4. Print Y=, followed by the value of Y.
             Print THE AVERAGE IS, followed by the value of X.
         5. Print NORMAL TERMINATION.
         Develop a flowchart and computer program for the preceding
         algorithm.

73

KEYWORDS

concatenation
functional operator
precedence
string comparison
subexpression
substring

4

Arithmetic and String Expressions

*He that will not apply New Remedies,
must expect New Evils: For Time is the
greatest innovator. (Francis Bacon, 1625)*

Once data are available in computer memory, they can be manipulated in many ways using arithmetic expressions (involving numbers) and string expressions (involving nonnumeric data).[1] Arithmetic expressions have a wide variety of applications ranging from preparing payroll checks to enhancing pictures obtained from a satellite. Similarly, string expressions also have diverse applications—e.g., creating blocked columns of newspaper pages and translating one natural language to another. Thus, arithmetic/string expressions play a fundamental role in computer use.

Arithmetic/string expressions are made up of numeric/string variables and constants, and are formed using a variety of numeric/string operators. This chapter is devoted to the study of the properties and evaluation of such expressions.

4.1 ARITHMETIC EXPRESSIONS

An **arithmetic expression** is a sequence of numeric variables and constants that is formed using arithmetic operators (numeric functional operators) in conjunction with a set of rules to govern their evaluation. In addition, such expressions may include parentheses to explicitly indicate the order in which operators are to be performed.[2]

We now define some arithmetic operators associated with flowchart language, and present some of the related rules for forming arithmetic expressions. All operators must be explicitly shown in flowcharts and most are binary infix; i.e., they require two values to operate upon, and the operator appears between them.

1. *Addition and subtraction* are indicated by the operators $+$ and $-$, respectively. As examples of such operations, we have

$$C \leftarrow A + B$$
$$A \leftarrow A + 2 - C + D$$
$$B \leftarrow B + 3$$

---

[1] One form of nonnumeric data is character string data. Another form of it includes *alphanumeric* or *alphameric* data; i.e., data consisting of letters of the English alphabet, blanks, decimal digits, and special characters.

[2] Numeric functional operators differ little from numeric operators. Instead of a single character representation, such as $+$ for addition, they have names like SQRT. To provide them information on which to act, the values required are enclosed by parentheses. Thus, SQRT (9) would take the square root of 9. They return only one value; for example, SQRT (9) returns the value 3.

2. *Multiplication* is indicated by the operator *, as in the statement

$$Z \leftarrow X * Y$$

which implies that the value taken by the variable Z is equal to the product of the current values of X and Y. We caution the mathematically oriented reader with respect to the possible confusion arising from the implied *product* of A and B, which is usually denoted by AB, and a *variable* AB. Notice that in flowchart we cannot write AB to mean the product of A and B.

3. *Division* is indicated by the operators / or —. For example, the division of the value of NUM by the value of DEN can be indicated as

$$Z \leftarrow NUM/DEN$$

or

$$Z \leftarrow \frac{NUM}{DEN}$$

4. *Exponentiation* is the operation of raising a value to a power. For example, the statement

$$X \leftarrow Y^Z$$

causes X to be assigned the value of Y raised to the power of Z. That is, Z copies of Y are to be multiplied together. Thus, if Y = 2 and Z = 3, we obtain

$$X = 2^3 = 2 \cdot 2 \cdot 2 = 8$$

The operations we have just defined are illustrated in Figure 4.1-1. If the input data are 4,2, then the resulting trace table is as shown in Table 4.1-1.

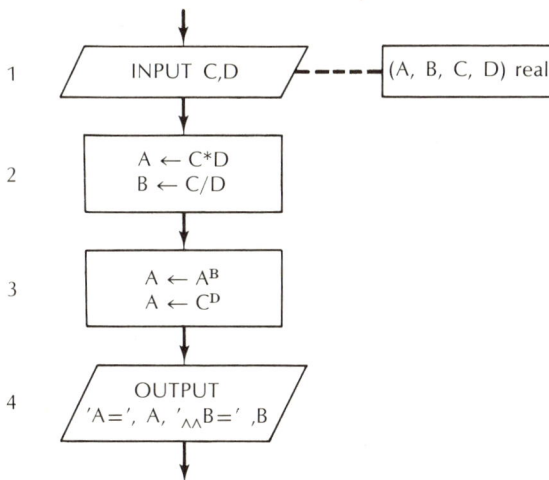

FIG. 4.1-1. Flowchart involving multiplication, division, and exponentiation.

TABLE 4.1-1. Trace table (see Figure 4.1-1)

Symbol #	A	B	C	D
1	Arbitrary	Arbitrary	4.	2.
2	8.	2.		
3	64.	16.		
4				

Thus, the output that results from the OUTPUT statement in Figure 4.1-1 is as follows:

$$A = 64_{\wedge\wedge}B = 16$$

It is important to note that two operators cannot occur in consecutive positions in an expression. For example, the expression

$$C \leftarrow A + -B$$

is incorrect. The correct version is obtained by using parentheses as follows:

$$C \leftarrow A + (-B)$$

4.2 NUMERIC FUNCTIONAL OPERATORS

There are several functional operators that are available for use, as summarized in Table 4.2-1. In each case, X is a variable and x is its value.

TABLE 4.2-1. Numeric functional operators

Operator	Mathematical functional operator	Mnemonic functional operator
Absolute value	$|x|$	ABS(X)
Exponential	e^x	EXP(X)
Natural logarithm	$\log_e x$	LOG(X)
Base-ten logarithm	$\log_{10} x$	LOG10(X)
Square root	$\sqrt{x}$	SQRT(X)

To illustrate, consider the flowchart shown in Figure 4.2-1 and assume that the input data are −19, 81. Execution of Symbol 2 leads to

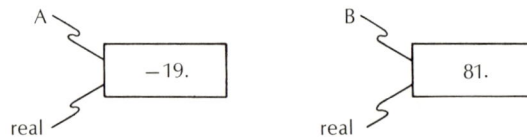

A

    −19.

real

B

    81.

real

Consequently, when Symbols 3 and 4 are executed, it follows from the definitions given in Table 4.2-1 that the following output is obtained:

$$X = 19_{\wedge\wedge}Y = 9_{\wedge\wedge}Z = 2$$

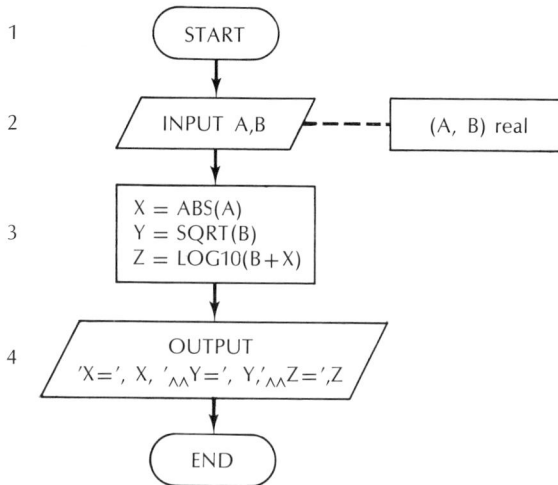

FIG. 4.2-1. Flowchart related to numeric functional operators.

EVALUATION OF ARITHMETIC EXPRESSIONS 4.3

A programmer must know how a computer language will interpret the meaning of an arithmetic expression. This section defines the most common (but not universal) interpretation of expressions.

Arithmetic expressions are evaluated by successively reducing sub-expressions to constant values until the given expression reduces to a numeric constant. While evaluating expressions, it is necessary to follow a set of **precedence rules** which determine the order in which operations are performed. The precedence rules associated with flowchart language are summarized in Table 4.3-1.

TABLE 4.3-1. A set of precedence rules

Precedence level	Operation	Order of evaluation
1.	Reference functions (ABS, SQRT, etc.)	Any order
2.	Exponentiation	Right to left* ←
3.	Multiplication Division	Left to right →
4.	Addition Subtraction	Left to right →

*If, for example, $Z = 2^{3^2}$, the quantity 3^2 is evaluated first to obtain 9. This is followed by the evaluation of 2^9, which yields $Z = 512$ as the final result.

A given expression can be thought of as being scanned *four* times, and all operations on one level are carried out on a scan in the order indicated. Table 4.3-1 indicates that reference functions (i.e., numerical functional operators [see Table 4.2-1]) are evaluated in any order.

79

Subexpressions are enclosed within parentheses, and must be evaluated before their result can be used. A subexpression is evaluated by using the entire set of precedence rules. For example, if A = 10 and B = 20, then in the subexpression

$$((A + B)/A)$$

(A + B) is evaluated first to obtain 30/A, which yields 3 as the final result.

It is important to note that if the argument of an arithmetic function is a subexpression, then that subexpression must be evaluated first. For example, in the expression

$$ABS(X*Y - Z)$$

the subexpression X*Y − Z must first be evaluated to a numeric constant, and then the absolute value of the resulting constant is taken.

Let us now consider some examples illustrating the systematic manner in which arithmetic expressions are evaluated via the precedence rules in Table 4.3-1. The values taken by the various variables used in these examples are given in Table 4.3-2.

TABLE 4.3-2. Values for variables (see Examples 4.3-1 through 4.3-5)

Variable	Value	Variable	Value	Variable	Value
A	−2	C	4	F	4
B	0.5	D	5	X	3
BAD	15	E	3	Y	2

Example 4.3-1 Evaluate the expression

$$X + A - 2*BAD$$

Solution:

Scan 1. No operations, since no reference functions are present.

Scan 2. No operations, since no exponentiation is involved.

Scan 3. The product 2*BAD is evaluated to obtain 2*15 = 30. The resulting reduced expression is

$$X + A - 30$$

Scan 4. The additions and subtractions indicated in X + A − 30 are carried out from left to right to obtain

$$3 - 2 - 30$$
$$= 1 - 30 = -29$$

Thus, the desired result is

$$X + A - 2*BAD = -29$$

Example 4.3-2 Evaluate the expression

$$(X - A/C)^Y * F$$

    Scan 1. No operations, since no reference functions are involved.

    Scan 2. The function $(X - A/C)^Y$ is evaluated, since exponentiation is involved. Since there is a subexpression $X - A/C$, it is first evaluated to a constant using the precedence rules in Table 4.3-1; then Scan 2 is initiated. Thus, we have

        Scan 2.1 n.o.
        Scan 2.2 n.o.
        Scan 2.3 $A/C = -2/4 = -1/2 = -0.5$
        Scan 2.4 $X - (-0.5) = 3 - (-0.5) = 3.5$

$$(3.5)^2 = 12.25$$

where "n.o." implies no operations.

    Scan 3. 12.25*F is evaluated to obtain

$$12.25*F = 49.0$$

(The computer will represent such numbers as real numbers in a specific manner. The representation, however, is not always exact. Some aspects of how numbers are represented in computers are discussed in Appendix 2.)

    Scan 4. n.o.

Hence, we have the result

$$(X - A/Y)^Y * F = 49.0$$

Example 4.3-3 Evaluate the expression

$$A - B^2/C*D*ABS(A)$$

Solution:

    Scan 1. Evaluates ABS(A) to obtain

$$ABS(-2) = 2$$

which results in the expression

$$A - B^2/C*D*2$$

    Scan 2. Evaluates $B^2 = (1/2)^2 = 0.25$; hence, the resulting expression is

$$A - 0.25 / C*D*2$$

    Scan 3. The multiplications and divisions are evaluated from left to right as follows:

$$0.25 / 4*5*2$$
$$= 0.0625*5*2$$
$$= 0.3125*2 = 0.625$$

81

The expression that results is

$$A - 0.625$$

Scan 4. $A - 0.625$ is evaluated to obtain

$$-2 - 0.625 = -2.625$$

Hence, the desired result is

$$A - B^2/C*D*ABS(A) = -2.625$$

Example 4.3-4   Evaluate the following expression:

$$(X^Y)^E$$

Solution:

    Scan 1.   n.o.

    Scan 2.   The subexpression X^Y in the arithmetic function $(X^Y)^E$ is evaluated first as follows:
        Scan 2.1   n.o.
        Scan 2.2 $X^Y = 3^2 = 9$
        Scan 2.3   n.o.
        Scan 2.4   n.o.
    Continuing with Scan 2 we have the reduced expression 9^E which is evaluated to obtain

$$(9)^E = 9^3 = 729$$

    Scan 3.   n.o.

    Scan 4.   n.o.

    That is, $(X^Y)^E = 729$

Example 4.3-5   Evaluate the expression

$$X^{Y^E}$$

Solution:

    Scan 1.   n.o.

    Scan 2.   Here, the exponentiations indicated in the given expressions are evaluated, working from right to left, in accordance with the order of evaluation given in Table 4.3-1 for Precedence Level 2. Thus, we have

$$Y^E = 2^3 = 8$$

    and

$$X^8 = 3^8 = 6561$$

    Scan 3.   n.o.

    Scan 4.   n.o.

Thus, the desired result is

$$X^{Y^E} = 6561$$

Comment: We note that although the expressions $(X^Y)^E$ and X^{Y^E} in Examples 4.3-4 and 4.3-5 may appear to be the same, they are far from equal, as is evident from their values which are 729 and 6561, respectively.

STRING EXPRESSIONS   4.4

A **string expression** is a sequence of string variables and string constants that is formed using **string functional operators.** We will now introduce five string functional operators and discuss some of their aspects.

1. **Concatenation** (or **catenation**) **operator:** In information processing problems, it is often necessary to link one string to another. This can be done using the concatenation operator, the syntax of which is given by

    CONCAT(A,B)

    where A and B are character string constants or variables. For example, if ABC has the value

    '$_\wedge$BLOW$_\wedge$'

    then

    D ← CONCAT(ABC, 'WIND')

    yields

    '$_\wedge$BLOW$_\wedge$WIND'

    which is stored in the location assigned to the character string variable D. Again, from the definition of CONCAT, it follows that

    CONCAT('JOE',ABC) yields 'JOE$_\wedge$BLOW$_\wedge$'

2. **Substring operator:** Sometimes we may want to extract a substring, such as a keyword or phrase, from a string expression. The desired extraction can be carried out by using a substring operator. Its syntax is given by

    SUBSTR(A, m, n)

    where A is a character string constant or variable, and m and n are constants or variables with positive integer values. The character position of the string A from which a substring is to be extracted is given by m. The number of characters that are to be extracted is specified by n. If n is omitted, then the rest of the string starting at character m will be extracted. For example,

    (a)   SUBSTR('A$_\wedge$PURPLE$_\wedge$HAT', 3,6) = 'PURPLE'
    (b)   SUBSTR('A$_\wedge$PURPLE$_\wedge$HAT', 3) = 'PURPLE$_\wedge$HAT'
    (c)   SUBTR('WIL''S$_\wedge$PRIDE', 7,5) = 'PRIDE'

83

3. **Put-string operator:** While processing string expressions, it may be necessary to remove a substring and replace it with another substring. This is achieved by "overlaying" the substring to be inserted over part of the original string. This can be achieved by the put-string operator, the syntax of which is given by

$$PSUBST(A,B,n)$$

where A, B are character string constants or variables, and n is a numeric integer constant or variable. A is the character string onto which the string B is to be overlayed, and n indicates the point at which the overlay is to begin. For example,

$$PSUBST('FUNNIER_{\wedge}CLOWN', 'CIRCUS_{\wedge}', 1)$$

results in

$$'CIRCUS_{\wedge\wedge}CLOWN'$$

We observe that the operator PSUBST causes B, which has the value 'CIRCUS$_{\wedge}$', to be written over part of A, which has the value 'FUNNIER$_{\wedge}$CLOWN'. The overlay commenced at the first character of A, since n = 1.

4. **Index operator.** This operator enables the location of the *first* occurrence of a certain substring in a given string. It is useful for editing textual material to determine, for instance, whether common misspellings of a word are present. Its syntax is

$$INDEX(A,B)$$

where A and B are character string constants or variables. A is the given character string in which we search for the substring B. For example, we have

(a) $INDEX('BLUE_{\wedge}SKY_{\wedge}COUNTRY', 'SKY') = 6$, and
(b) $INDEX('BLUE_{\wedge}SKY_{\wedge}SKY_{\wedge}COUNTRY', 'SKY') = 6$

since the index operator searches only for the first occurrence of the substring 'SKY' in the given character string 'BLUE$_{\wedge}$SKY$_{\wedge}$SKY$_{\wedge}$COUNTRY'.

Comment: If B is not a substring of A, then INDEX(A,B) results in zero, as follows:

$$INDEX('BLUE_{\wedge}SKY_{\wedge}COUNTRY', 'SKI') = 0$$

5. **Length operator:** The phrase *length of a character string* refers to the number of characters in the string. It is possible to find the length of a character string using this operator, the syntax of which is given by

$$LENGTH(A)$$

where A is the string with length to be determined. For example,

$$LENGTH('THE') = 3$$

Additional illustrative examples are presented in tabular form (see Table 4.4-1). The reader is encouraged to go through these examples, and insert the missing steps that lead to the corresponding answers.

4.5
STRING COMPARISON

TABLE 4.4-1. Examples of string functions

The character string expressions pertinent to the examples given below are

$$A \leftarrow \text{'UNITED}_\wedge\text{WORLD}_\wedge\text{UNIVERSITY'}$$
$$B \leftarrow \text{'}_\wedge\text{IS}_\wedge\text{AN}_\wedge\text{INSTITUTION'}$$
$$C \leftarrow \text{'THE}_\wedge\text{GOLDEN}_\wedge\text{EAGLE}_\wedge\text{TEAM'}$$

(a) CONCAT (A, B)
     = 'UNITED$_\wedge$WORLD$_\wedge$UNIVERSITY$_\wedge$IS$_\wedge$AN$_\wedge$INSTITUTION'

(b) SUBSTR(C, 4, 13) = '$_\wedge$GOLDEN$_\wedge$EAGLE'

(c) PSUBST (A, 'BASKETBALL', 14)
     = 'UNITED$_\wedge$WORLD$_\wedge$BASKETBALL'

(d) INDEX (A, 'I') = 3

(e) INDEX (SUBSTRA(A, 4), 'I') = 13

(f) SUBSTR (C, INDEX(C, 'G'), 12)
     = 'GOLDEN$_\wedge$EAGLE'

(g) CONCAT (SUBSTRA(A, 1, 13), CONCAT(SUBSTR(C, INDEX(C, 'G')), 'S')) = 'UNITED$_\wedge$WORLD$_\wedge$GOLDEN$_\wedge$EAGLE$_\wedge$TEAMS'

(h) PSUBST (C, SUBSTR(B, 8), INDEX(C, 'T'))
     = 'INSTITUTIONEAGLE$_\wedge$TEAM'

(i) PSUBST (C, SUBSTR(B, 8), INDEX(SUBSTR(C, 2), 'T'))
     = 'THE$_\wedge$GOLDEN$_\wedge$EAGLEINSTITUTION'

(j) PSUBST (C, SUBSTR(B, 8), INDEX(C, 'TE'))
     = 'THE$_\wedge$GOLDEN$_\wedge$EAGLE$_\wedge$INSTITUTION'

(k) LENGTH(SUBSTR(C, 4, 13)) = 13

(l) LENGTH (CONCAT(C, SUBSTR(B, 8))) = 32

STRING COMPARISON 4.5

Computers often work with string information. Text editing, for example, is a very common application in which editorial changes are made or errors are corrected. Other applications include creating reports, musical scores, artwork, etc.

In this section, we consider the problem of comparing character strings. Specifically we ask the following question: What does it mean for a character string constant to be less than ($<$), greater than ($>$), or equal to ($=$) another? The answer plays an important role in the development of sorting routines that enable lists of character strings to be rearranged in lexicographic order.

String comparison can be realized by carrying out the following

85

three steps, which are illustrated with the strings A = 'BOOK', and B = 'BOOTS':

1. The shorter string is extended by adding blanks so that the lengths of the two strings are equal. Thus, A is modified to obtain

    $$\text{'BOOK}_\wedge\text{'}$$

2. The strings 'BOOK$_\wedge$' and 'BOOTS' are compared character by character from the left to the right in accordance with a collating sequence. In our case, we will consider a specific collating sequence, as follows:

    $$\wedge < \text{special characters} < \text{a through z} < \text{A through Z} < \text{0 through 9}$$

    We can make four observations with respect to this collating sequence:
    (a) The symbol $\wedge$ (which denotes a blank) is less than any other character.
    (b) Special characters are less than lowercase alphabetic characters.
    (c) Lowercase characters are less than capitals.
    (d) Capitals are less than numeric characters.

3. Record the *first time* the character-by-character comparison process in Step 2 results in two *unequal* characters, say A_k and B_k, which belong to A and B, respectively. Then, use the following decision rule:

    $$A_k < B_k \text{ implies that } A < B$$

    and

    $$A_k > B_k \text{ implies that } A > B$$

Thus, for the case A = 'BOOK$_\wedge$' and B = 'BOOTS', the related comparison can be illustrated as follows:

$$
\begin{array}{ccccc}
B & O & O & K & \wedge \\
\updownarrow & \updownarrow & \updownarrow & \updownarrow & \\
B & O & O & T & S
\end{array}
$$

We observe that the first time the comparison process results in unequal characters is when K is compared with T. That is, $A_k = K$ and $B_k = T$. Now, from the collating sequence it follows that $K < T$; hence, the preceding decision rule implies

$$\text{'BOOK'} < \text{'BOOTS'}$$

Thus, if an alphabetic sort were being carried out using the approach described, then BOOK would sort before BOOTS.[3] The programmer does not have to specify the three-step process. The use of comparison

---

[3] Note that if every character in string A equals the corresponding character in string B, then A is said to be equal to B; i.e., A = B.

operators with two character strings will cause the process to take place. We now present an additional example to emphasize this point.

4.5
STRING COMPARISON

Example 4.5-1

(a) Given three strings A, B, and C, develop a structured algorithm that picks the string with the smallest value and prints it out. Draw the corresponding flowchart.

(b) Use the flowchart developed in (a) to obtain the trace table for the case when the input is given by

> 'UWU'
> 'UNITED''S$_\wedge$WORLD'
> 'UNITED$_\wedge$WORLD'

Solution:

(a) We consider the following initial solution:

   1. Read values for the variables A, B, and C.
   2. Find the minimum of the values.
   3. Print the minimum value.

   A corresponding refined algorithm is

   1. Read values for the variables A, B, and C.
   2. Find the minimum of the values.
       2.1 Assign the value of A to a variable MIN.
       2.2 If B is less than MIN, then assign the value of B to MIN.
       2.3 If C is less than MIN, then assign the value of C to MIN.
   3. Print the minimum value.
       3.1 Print 'MIN=', MIN.
       3.2 END.

   Note that this algorithm is the same as that devised to find the largest number in a given set of numbers (see Example 3.6-4), except for the variable declarations. For the corresponding flowchart, see Figure 4.5-1.

(b) The trace table obtained via the flowchart in Fig. 4.5-1 with

> 'UWU'
> 'UNITED''S$_\wedge$WORLD'
> 'UNITED$_\wedge$WORLD'

is as follows:

Symbol #	A	B	C	MIN
1	UWU	UNITED S$_\wedge$WORLD	UNITED$_\wedge$WORLD	
2				UWU
3				
4				UNITED'S$_\wedge$WORLD
5				
6				UNITED$_\wedge$WORLD
7				

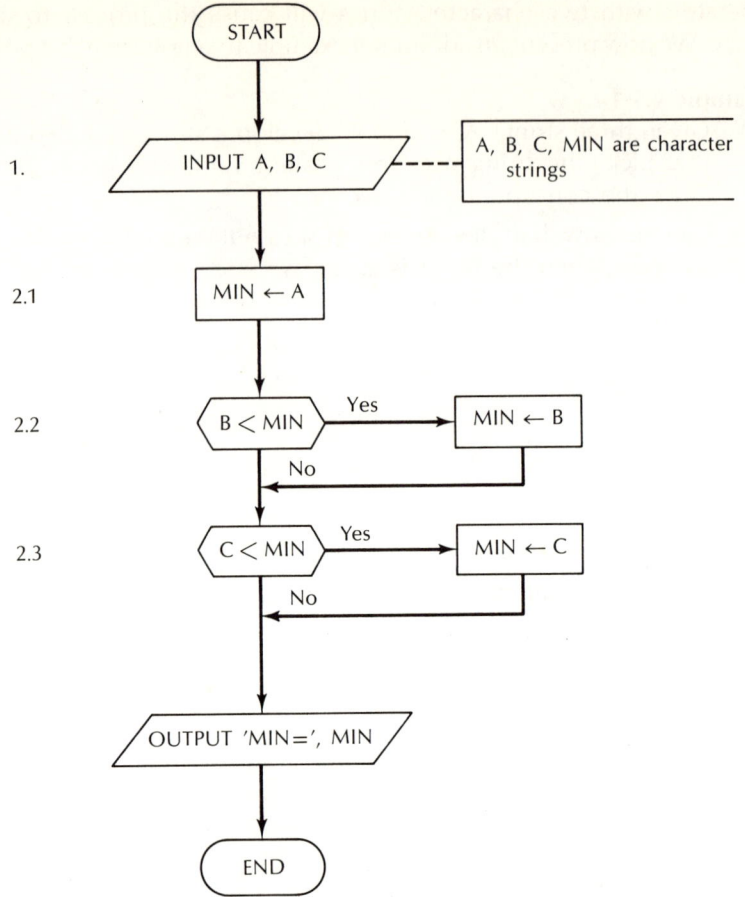

```
                    START
                      │
                      ▼
1.        /  INPUT A, B, C  / ─ ─ ─   A, B, C, MIN are character
                                      strings
                      │
                      ▼
2.1           ┌──────────────┐
              │   MIN ← A    │
              └──────────────┘
                      │
                      ▼
2.2         < B < MIN >  ──Yes──►  [ MIN ← B ]
                │ No                      │
                ▼◄───────────────────────┘
2.3         < C < MIN >  ──Yes──►  [ MIN ← C ]
                │ No                      │
                ▼◄───────────────────────┘
              /  OUTPUT 'MIN=', MIN  /
                      │
                      ▼
                    END
```

FIG. 4.5-1. Flowchart pertaining to Example 4.5-1.

4.6 SUMMARY

We introduced a number of numeric and string operators in this chapter and discussed their role in representing arithmetic and string expressions, respectively. The intent was not to introduce *all* the operators and functional operators available in programming languages, but to acquaint the reader with the basic rules for some of the operators that are commonly encountered. Next, we outlined a systematic procedure for evaluating arithmetic expressions via a set of precedence rules. The procedure was illustrated by means of several examples. In addition, some aspects of comparing character strings with the operations $=$, $<$, and $>$ were included.

4-1. Evaluate the following arithmetic expressions, with values for the variables: $A = 2$, $B = 4$, and $C = 3$.

(a) $(A + B)/(B - C)$ (b) $A + B/A - C$

(c) $(B)^{(A*A)}$ (d) $B/A*B$

(e) $(A + B)/A - B$ (f) $A + B/(A - B)$

(g) $A + B/A - B$ (h) $B/(A*B)$

(i) $(B/A)*B$

4-2. If $A = 3$, $Y = 3$, $Z = 4$, $P = 5$, $Q = 2$, $R = -2$, and $S = 4$, evaluate the following arithmetic expressions:

(a) $(A + Y)^{-SQRT(Z)}$ (b) $(A - R)/Z^Q*S$

(c) $A - R/Z^Q*S$ (d) $SQRT(2*LOG10(P^2*Z))$

(e) $ABS(R)$ (f) $SQRT(ABS(R) + Z/2)$

4-3. If $A = 2$, $B = 3$, $C = 2$, $D = 1$, and $G = 2$, evaluate the following expressions:

(a) $SQRT(D + B^A + B*G)$ (b) $A + B/C*D$

(c) $A - B/C + D*G$ (d) $ABS(A + B)$

(e) $SQRT(B^A)$ (f) $(A - B)/C + D*G$

4-4. Given a = 'BIG∧BAD∧WOLF', b = 'BAD∧', c = 'LITTLE∧', evaluate the following expressions:

(a) PSUBST(a, c, 1)   (b) INDEX(a, b)

(c) CONCAT(b, SUBSTR(a, 9, 4))   (d) LENGTH(a)

(e) SUBSTR(a, INDEX (a, 'BA'), 3)   (f) INDEX(a, 'B')

4-5. Given: a = 'MOPPY', b = 'RAP', c = 'P'. Evaluate

(a) INDEX(a, c)   (b) SUBSTR(a, 2, 2)

(c) CONCAT(b, PSUBST(b, 'D∧IT',2))   (d) INDEX(a, b)

(e) SUBSTR(a, INDEX(a, '0'))   (f) CONCAT(c, SUBSTR(a, 2))

The general learning objectives for this set of problems are

☐ to gain problem-solving experience with decision and branching constructs;

☐ to gain experience with arithmetic expressions and precedence of operators;

☐ to appreciate the usefulness of trace tables for checking and debugging; and

☐ to learn to develop internal program documentation.

Additional objectives pertaining to a specific problem are stated as a preface to that problem. The reader is expected to develop a structured algorithm, flowchart, and trace table for some sample data, and finally, a debugged computer program.

89

The problems that follow have been used in various computer language classes with students from different disciplines. The languages and student disciplines for which a problem has been successfully used are given at the end of each problem.

CS 4-1. *Additional objective for APL users:* to create an APL-invoked procedure.

Create a program which, if given sets of three numbers, will determine whether they represent the lengths of the sides of a triangle. It is to output one or more values according to the following table to indicate the result:

Triangle type	Output value
Not a triangle	0
Equilateral	1
Isosceles	2
Right	3
None of the above	4

Recall that an equilateral triangle has three sides of equal length, an isosceles triangle has two sides of equal length, and a right triangle has the square of the longest side equal to the sum of the squares of the other two sides. For three sides to form a triangle, the longest side must be shorter than the sum of the other two sides.

Have the problem terminate the processing when there are no more data you wish analyzed.

Echo your input.

Trace sets of data that satisfy each of the five conditions listed above.

(APL, BASIC, FORTRAN, PL/I)

CS 4-2. You are a programmer for Sam's Yacht Sales, Inc. The accountant comes to you and tells you to design a program to take care of computing sales tax. He wants each sales transaction to be a record (i.e., punched on a card) with the state of the sale (Sam sells in Kansas, Missouri, and Nebraska) and the amount of the sale. The output is to include a printed report with the states listed, along with the total sales and total sales tax for that state. Grand totals should appear at the bottom of the table. The tax rates are Missouri 4 percent, Kansas 3 percent, and Nebraska $3\frac{1}{2}$ percent.

(COBOL, FORTRAN, PL/I)

CS 4-3. Write a procedure that will compute a four-value moving average of Russian wheat production in metric tons (that is, the mean of values 1-4, then the mean of values 2-5, then 3-6, etc., working with a sliding sequence of four). Input will consist of an undetermined number of values. Include in the output

(a)   metric tons from each value, and
(b)   each calculated average.

(COBOL, PL/I)

CS 4-4.   You are given the resistor network shown below. The problem is to find the equivalent resistance for this network.

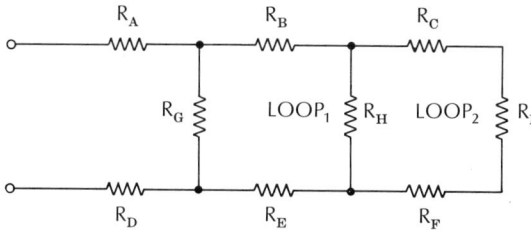

Recall that for resistances in series, the equivalent resistance is $R_T = R_1 + R_2 + \ldots R_n$; for resistances in parallel, the equivalent resistance is

$$R_p = \cfrac{1}{\cfrac{1}{R_1} + \cfrac{1}{R_2} + \ldots + \cfrac{1}{R_m}}$$

To solve the problem notice that R_H is parallel with R_C, R_I, R_F, which are in series; therefore,

$$R_{\text{Loop 2}} = \cfrac{1}{\cfrac{1}{R_H} + \cfrac{1}{(R_C + R_I + R_F)}}$$

Solve the problem and print out the intermediate equivalent resistances and the equivalent resistance of the network.

Input at least four sets of the nine resistances and calculate the equivalent network resistance.

Label your output appropriately and trace the solution with a sample set of data.

(FORTRAN[scientific, engineering], PL/I[scientific])

CS 4-5.   Evaluate

$$F(x,y) = (g(x)/h(y)) \cdot f(x,y)$$

where

$$g(x) = x^3 - 2.5x^2 + 2x + 3.1$$
$$h(y) = y^5 - y^2 + 6$$
$$f(x,y) = x(y - x)$$

91

Using a nested solution is more efficient for many programming languages. Nesting allows multiplication as opposed to exponentiation; the former is faster than the latter, in general. Nesting g(x) gives

$$g(x) = x(x(x - 2.5) + 2) + 3.1$$

Nest h(y) and solve this problem for a series of x's and y's.

Output your answers neatly labeled and have your program stop with the message 'NORMAL TERMINATION'.

Trace your flowchart solution for $x = 2$, $y = 1$.

(FORTRAN [engineering, scientific], PL/I [scientific])

CS 4-6. For purposes of this problem, we will define four types of quadrilaterals (four-sided figures):

TYPE 1: Four sides equal   TYPE 2: Two pairs of sides equal

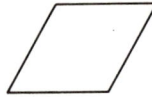

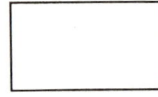

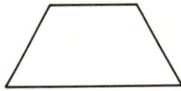

TYPE 3: Three sides equal   TYPE 4: No pair of sides equal

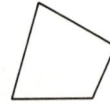

The data will consist of the side lengths of an indeterminate number of four-sided figures. For each figure, the length of the sides will appear in clockwise order:

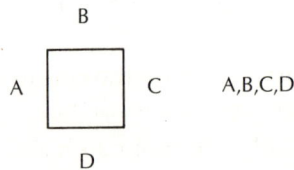

A,B,C,D

Create a program to determine which type each figure represents. Output the following for each figure:
(a)   the length of each of the four sides, and
(b)   the type of quadrilateral.

Note: This is a difficult problem.

FORTRAN [architecture, engineering], PL/I [scientific])

CS 4-7.   There are 36 465 days in the 20th century, if one starts with March 1, 1900. The sequence number of a day from 3/1/1900 can be calculated as follows:

Y = the last two digits of the four-digit
year representation

M = month

D = day

If M is greater than 2, replace M by M minus 3; otherwise, replace M by M plus 9 and Y by Y minus 1. The sequence number of a day is then given by

$$K = \left[\frac{1461 * Y}{4}\right] + \left[\frac{153 * M + 2}{5}\right] + D$$

where [ ] indicates that the computations should result in integers. Therefore, they should be done as intermediate calculations and the result assigned to an integer.

Input a date. Output the sequence number since January 1, 1900. Be sure to add the 59 days from January 1, 1900 to March 1, 1900.

(FORTRAN [general], PL/I)

CS 4-8. For many years the surgeon general has warned people of hazards smoking poses to health. A heart attack risk factor may be determined from the following table:

	Exercise	
	Above norm	**Below norm**
Smoker	3	4
Nonsmoker	1	2

The information consists of whether each subject smokes (coded '1') or does not smoke (coded '2') and whether the subject's exercise factor is below the norm (coded '2') or equal to or above the norm (coded '1').

For each subject you classify, output his or her exercise code, smoking status code, and heart attack risk factor.

(FORTRAN [general], PL/I [general])

CS 4-9. Evaluate F(x) * G(x) if x takes the values 1, 2, 3, 4. The polynomials F and G are defined as

$$F(x) = x^4 + 6x^3 + 3x^2 - 5x + 3$$
$$G(x) = x^5 - 4x^4 + 2x^3 - x^2$$

Nested evaluation of polynomials is more efficient for many programming languages, as nesting allows multiplication as opposed to exponentiation. For instance, F(x) can be written as $F(x) = x(x(x(x + 6) + 3) - 5) + 3$. Devise a solution to this problem using decisions and branching, and nested multiplication.

93

Output your answers, neatly labeled, and have your program stop with the message NORMAL TERMINATION.

Trace the algorithm for x = 1 and x = 2.

(FORTRAN [engineering and scientific], PL/I [scientific])

CS 4-10. The advent of rock music has brought with it concern about levels of environmental sound. In the current problem, you will label each of an indefinite number of sound sources according to their decibel (db) level code.

Decibel-level code	Rating label
1	Low db level
2	Moderate db level
3	High db level
4	Unsafe db level

Each record you read will contain a sound source and a decibel level code. Your output should include
(a) the sound source;
(b) the decibel level code; and
(c) the rating level.

(FORTRAN [scientific], PL/I [general, scientific])

CS 4-11. In 1626 Peter Minuit is said to have purchased Manhattan Island from a tribe of the Algonquin Indian nation with trinkets valued at $24. Manhattan Island was assessed at a value of $8 billion in 1971. Discover whether Minuit would have been wiser to invest for his descendents the $24 at 3 percent interest compounded annually commencing January 1, 1626. The value of the investment can be calculated from

$$\text{Amount} = \text{Principal} * (1 + \text{rate\_of\_interest})^{\text{\#\_of\_periods}}$$

Output the following:
(a) principal;
(b) amount on January 1, 1979; and
(c) an evaluation of his original investment strategy. (For instance output "GOOD STRATEGY" if $8 billion is greater than Amount.)

Now repeat this for interest rates of 3.5 percent, 4 percent, 4.5 percent, and 5 percent.

(COBOL, PL/I [business, general])

CS 4-12. Develop a procedure that will inform a sales clerk how to make change with a minimum number of coins. Assume all items for sale cost at least a penny and never more than a dollar. A clerk can make change if told how many coins of each denomination should be returned to the customer.

Echo your input. Output should consist of the number of half-dollars, quarters, dimes, nickels, and pennies the customer should receive. Input is a single purchase price. Terminate the program if the purchase cost is greater than a dollar and when there are no more data.

(PL/I [business, scientific])

CS 4-13. A student has recently decided to provide a bank-balance computing service for his friends. At the end of each month, those interested in the service bring a list of their account transactions for the month and their old balance.

The student has hired you to develop a program which will process the monthly transactions for each of his customers. You will want to read in the amount of the old balance as the current balance value and then process each transaction, one at a time. Enter a negative value (−) when a check is written and a positive value (+) when a deposit is made. Only the last transaction card for an individual should contain a zero.

Output the following:
(a)   old balance as read in;
(b)   each transaction; and
(c)   new balance.

Repeat the process until the monthly statements are produced for all customers.

(PL/I [business])

CS 4-14. A nearby drugstore was recently informed of a new luxury tax to be imposed on all items classified as cosmetics. They have computerized cash registers and have hired you to program the new tax calculation. Each item subject to the new tax is to be coded with a one, while all others are to be coded with a zero. The program is to total all the items purchased by each customer and add the 10 percent luxury tax on the taxable items. A tax code of minus one signals the end of each customer's purchases.

Output the following:
(a)   tax code of item;
(b)   amount of tax if taxable item; and
(c)   total amount of customer's purchases.

Now repeat until all customers' purchases are totaled.

(PL/I [business, general])

CS 4-15. The problem is to program a version of Euclid's algorithm for finding the greatest common divisor (gcd) of two integers using the subtraction process. Essentially, what Euclid developed is as follows:

Given two integers, determine the larger of the two. (*Note:* if they are equal—what is the answer?) Subtract the

95

smaller from the larger and consider these two values as the old smaller and the difference. If the old smaller is zero, then the difference is the gcd; if not, consider the new pair as before. Have your program terminate with the message 'NORMAL TERMINATION' when there are no more pairs of numbers.

(PL/I [computer science])

CS 4-16. *Additional learning objective:* to introduce the use of built-in functions.

An interesting concept applicable to computer science is the notion of palindrome. A **palindrome** is a word or phrase that reads the same forward as backward, ignoring punctuation and blank symbols. Two examples are MADAM, and A MAN A PLAN A CANAL PANAMA. Write a program that detects palindromes. Since blanks, punctuation marks, and capitalizations make no difference, you may enter only pertinent characters.

Output the following.
(a) the input phrase, and
(b) either 'is a palindrome' or 'is not a palindrome', whichever is appropriate.

Note: This is a difficult problem.

(PL/I [computer science])

CS 4-17. *Additional learning objective:* to introduce the use of built-in functions.

Develop a computer program that, given a date such as 7/4/1973, will calculate the day of the week—in this case, Wednesday. In 7/4/1973, the character 7 indicates the month July, 4 indicates the fourth day of the month, and 1973 indicates the year 1973.

The problem may be solved for the months March–December by the following formula:

$$I = \text{Day} + 2 \cdot \text{Month} + \left[\frac{3 \cdot \text{Month} + 3}{5}\right]$$
$$+ \left(\text{Year} + \frac{\text{Year}}{4} - \frac{\text{Year}}{100} + \frac{\text{Year}}{400} + 1\right)$$

Then the number of the day of the week, J, is calculated

$$J = \text{MOD}(I,7) + 1$$

where MOD is a built-in function that will take the remainder of the division of I by 7.

Example for July 4, 1973 = 7/4/73:

$$\text{Day} = 4 \qquad \text{Month} = 7 \qquad \text{Year} = 1973$$

so

$$I = 4 + 14 + \frac{24}{5} + 1975 + \frac{1973}{4} - \frac{1973}{100} + \frac{1973}{400} + 1$$

$$= 4 + 14 + 4 + 1973 + 493 - 19 - 19 + 4 + 1 = 2472$$

Then

$$J = \left(\frac{2474}{7}\right) + 1 = 3 + 1 = 4$$

So July 4, 1973 was a Wednesday; i.e., it was the fourth day of the week.

Read a series of dates including July 4, 1973; April 22, 1961; and March 10, 1959. Print out the day of the week these dates occurred. If a January or February day is input, output the message that the program does not handle this date.

(PL/I [computer science, general, scientific])

CS 4-18.   An interesting task involved in the study of writing styles is to determine the number of occurrences of the articles, *the, a,* and *an,* and the connective, *and.* Write a program that will read a short paragraph as a series of words and determine the occurrences of the previously named words.

The input should be written as 'THE', 'HOUSE', etc. for ease in manipulating it. The output should contain an echo of the input and a count of the number of all words and of each of the listed words. Make sure your output is well labeled.

(PL/I [general])

CS 4-19.   Another interesting task in studying writing styles is to determine the number of occurrences in a manuscript of each of the vowels. That is, how many times does the letter *A* appear in the manuscript, how many times does the letter *E* appear, and so on for the letters *I, O,* and *U.* All we would need to do to perform this task manually is scan the manuscript from beginning to end, tallying each *A, E, I, O,* or *U.* The input is to be read as a series of sentences. Each sentence is a character string, which is assumed to be limited to 80 characters.

The output from your program should contain a count of the total number of letters in the manuscript, and a count for each of the vowels. Make sure that your output is well labeled.

(PL/I [general])

CS 4-20.   Four neighboring cities, Northville, Southville, Eastville, and Westville, have sales taxes which are 4 percent, 2 percent, 3 percent, and 3.5 percent, respectively. You have been hired by a firm doing business in these four cities to program tax that they owe to each city from individual daily transactions. The data consist of the name of the city followed by the amount of the sale (in dollars and cents).

Write a program that will read an unspecified number of data cards and print the city, amount of sale, tax, and total cost of the sale. Additionally, after all transactions have been read, print the total tax owed to each city.

(PL/I [general])

CS 4-21. Find all the two-digit numbers between 0 and 100 that are equal to the absolute value of the difference of the square of their digits. That is, if given a two-digit number, D_1D_2, it must be true that

$$10D_1 + D_2 = |D_1^2 - D_2^2|$$

Print out the following values in three lists:

$$10D_1 + D_2 \qquad |D_1^2 - D_2^2| \qquad EQUAL$$

where EQUAL has the value 0 if $10D_1 + D_2 \neq |D_1^2 - D_2^2|$; otherwise, EQUAL has the value 1.

(FORTRAN [scientific], PL/I [scientific])

CS 4-22. Find the real roots of a general quadratic equation

$$Ax^2 + Bx + C = 0$$

where the real roots are given by

$$X_1, X_2 = \frac{-B \pm \sqrt{B^2 - 4AC}}{2A}$$

Complex roots occur if $B^2 - 4AC < 0$; calculate these if you know how.

Terminate your program with the message 'NORMAL TERMINATION' when there are no more values of A, B, C to read.

The output should appear as follows:

SOLUTIONS TO QUADRATIC EQUATIONS					
EQUATION	EQUATION VALUES				
NUMBER	A	B	C	ROOT 1	ROOT 2
1	X	X	X	X	X

(FORTRAN [scientific], PL/I [scientific])

CS 4-23. The Fibonacci sequence of numbers is well known among scientists. The sequence begins 0, 1, 1, 2, 3, 5, 8, 13 . . . . After the first two numbers have been read or assigned, each successive number is the sum of the previous two numbers in the sequence. Write a program to compute and list all numbers below 1000 that are in the sequence.

(FORTRAN [general], PL/I [general])

KEYWORDS

field
file
format
record
round
stream
truncate

5

Input/Output and Format Considerations

GIGO: Garbage In–Garbage Out

It is important that communication between computer systems and their users be as accurate and convenient as possible. Two major factors involved are (a) the hardware that is available and (b) how effectively a programmer uses available hardware.

There are a variety of input/output (I/O) media; for example, typewriter, terminals, cards, printer paper, line drawings on paper, pictures on televisionlike screens, human voice, bank checks, and product codes. Two examples are shown in Figures 5.0-1 and 5.0-2. It

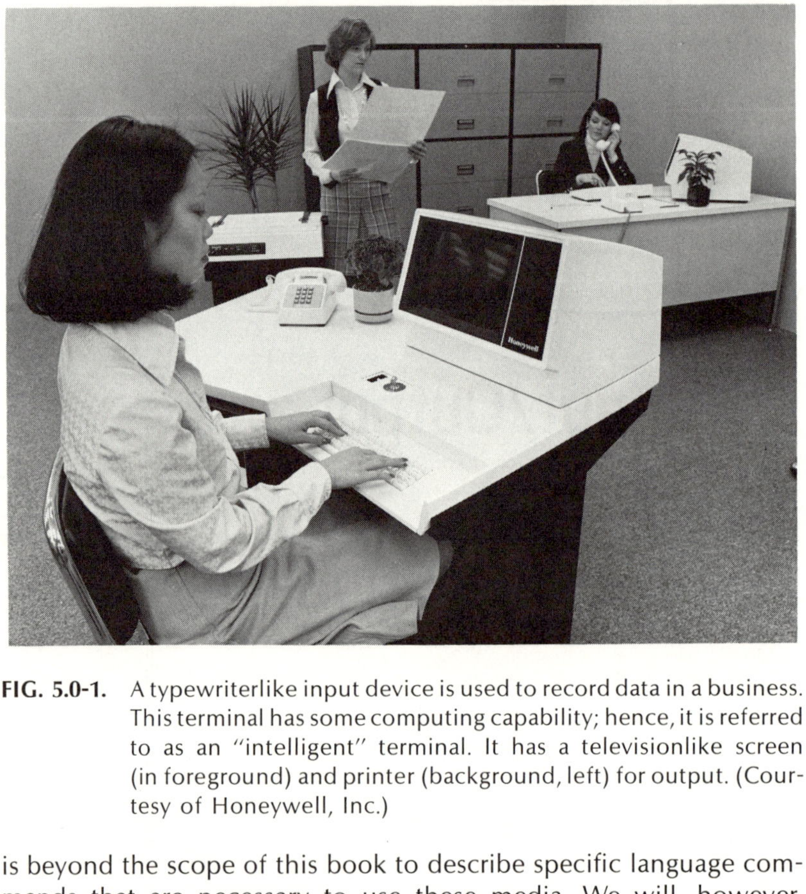

FIG. 5.0-1. A typewriterlike input device is used to record data in a business. This terminal has some computing capability; hence, it is referred to as an "intelligent" terminal. It has a televisionlike screen (in foreground) and printer (background, left) for output. (Courtesy of Honeywell, Inc.)

is beyond the scope of this book to describe specific language commands that are necessary to use these media. We will, however, provide some background material that is pertinent to effectively using the more common I/O media.

In the previous chapters, we have assumed that there is no need to specify the manner in which data are to be organized on our input medium (say, cards) before being read in via an INPUT statement. A similar assumption has been made with respect to the organizational aspects of the data to be stored as output on a medium (say, a line printer) via an OUTPUT statement. In other words, we have assumed an input-output situation that is **format-free** or **stream-oriented.**

100

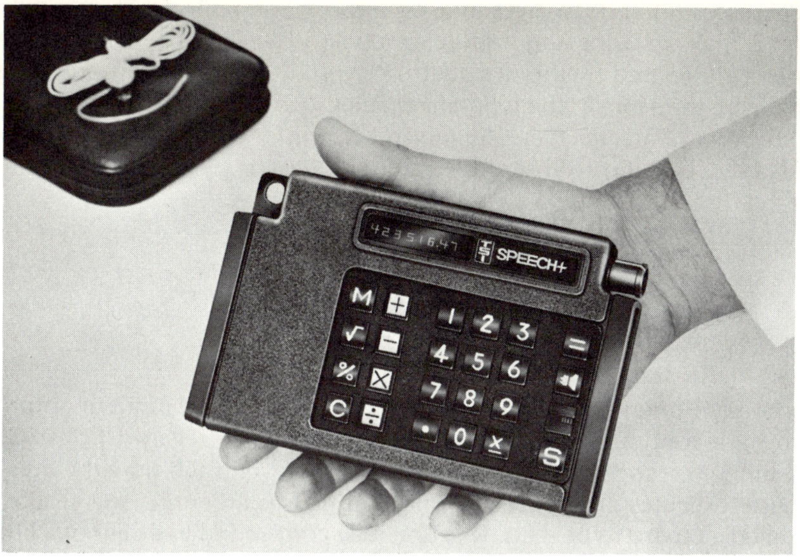

FIG. 5.0-2. A calculator with I/O used by the blind. The input is numerical information via the keyboard; the output is the corresponding result in the form of a spoken word. (Courtesy of Telesensory Systems, Inc.)

The use of a free format for I/O operations is quite acceptable, provided

1. the feature is available in the language being used, and
2. the I/O format desired is that which is produced by the free-format convention.

If either of these conditions is not satisfied, it is necessary to resort to other forms of I/O that use **format statements** to provide a means of explicitly describing the form of the I/O data. In essence, format statements deal with specifications pertaining to

1. the position of data on input or output, and
2. the characteristics of numbers and strings that appear in the I/O data.

In this chapter we will discuss the logical model of I/O that allows a format-free situation to exist. Then we will turn to a **record-oriented model,** which is the one used when the programmer wishes to describe I/O data through formats.

At the outset, a clear distinction must be made between the following processes:

(a) describing input data as they are stored on an input medium such as cards;

(b) describing output data as they are to be stored on an output medium such as printer paper; and

(c) describing memory locations used to store the data in the machine.

101

Processes (a) and (b) are achieved through format statements, while (c) is realized using annotation symbols (declarations). Such descriptions are related by the fact that the type of incoming and outgoing data are the same as the type of variables in which the data are stored. No other characteristics—length, precision, or magnitude—are necessarily common to these descriptions.

5.1 STREAM-ORIENTED I/O

In stream-oriented I/O, the programmer considers the input or output to be a continuous stream of characters with no regard to physical boundaries, such as the end of a card. However, each time the computer executes an INPUT statement, it causes an entire unit of data called a **record** to be read. This unit may correspond to a card or a line of input from a terminal. In order to have the input or output appear as a stream, a special section of memory is used for temporary storage purposes. This section of memory is generally referred to as a **buffer.** To illustrate the use of a buffer with card input, consider reading a card on which four values have been punched. Then, execution of the statement

INPUT A,B

causes the card to be read into the buffer and the first two values subsequently stored into the variables A and B. The other two values remain in the buffer until the next INPUT statement is encountered. For example, suppose this statement is

INPUT X,Y,Z

Execution of this statement causes the two values remaining in the buffer to be stored into the variables X and Y. Then the next card is read into the buffer to provide a value for the variable Z. These buffer operations are summarized in Figure 5.1-1. Buffers are used in a similar way with respect to the output of stream-oriented data.

To recapitulate:

- ☐ Stream-oriented I/O is a logical way for the programmer to consider the input or output data.
- ☐ The input values appear on the input medium separated by some punctuation mark, such as a blank(s) or comma.
- ☐ The output values appear on the output medium in a form that is predetermined by the language used. Our flowchart language outputs with no punctuation for separation of values.

Input:

(a) Buffer contents after INPUT A,B is executed

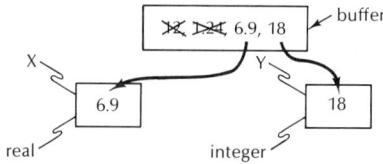

(b) Buffer contents after INPUT X,Y,Z is begun

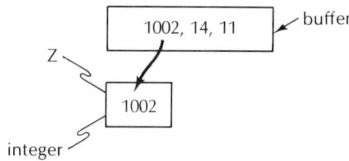

(c) Buffer contents after second card is read to get value for Z

(d) Buffer contents after INPUT X,Y,Z is executed

FIG. 5.1-1. Buffer operations.

RECORD-ORIENTED I/O 5.2

The study of record-oriented I/O is based upon the concept of a record, which is a unit of data. It may contain one value or, more commonly, a logically related set of values (for example, a student record at a university, which may include name, local address, home address, identification number, grade-point average, etc.) Some input media may physically constrain the length of a record. For instance, a

103

punched card is always a record that may be no longer or no shorter than the card capacity. When magnetic tape is used for record storage, the record length is essentially unlimited. A common magnetic tape system may store 800 characters per inch, and the tape could be 732 meters (2400 feet) long.

Record-oriented I/O requires that the programmer describe the contents of the records from the first column of the record through the last column to be used. In addition, the following three entities must also be specified:

1. the location of each value;
2. the type of each value; and
3. the size of the location where each value is to be found.[1] Each time a read command (e.g., INPUT) is given, an entire record is read. If the programmer fails to use part of the record, it is lost.

5.3 FIELDS

The notion of a field is fundamental to the development of formats. A **field** is simply a sequence of one or more characters, as it appears on either the input or the output. For example, the real number -258.117 and the string constant 'ANSWER' represent two fields.

An important attribute is **field width,** which is defined as the number of characters in the field. Thus, for example, the fields

$-$	2	5	8	.	1	1	7

and

$-$	A	1	3	.	$	+

have field widths of 8 and 7, respectively.

Fields can be classified into two major categories; namely,

1. **numeric fields,** which contain numeric constants, and
2. **character string fields,** which contain character string constants.

One of the basic distinctions between these fields is that string values are not (in general) used in arithmetic operations because there is a basic difference in the way in which the respective data are stored.

---

[1] If the memory location is not large enough to hold the information it is to store, the value is **truncated** or **rounded** before it is stored. To truncate means to drop digits or characters that cannot be accommodated. To round means to store a value closest to the one we desire to store. In this book we will assume that information is stored by resorting to truncation.

The manner in which a given field is represented in an I/O medium depends upon whether it is a numeric field or a character string field, as illustrated in the following examples:

Ex. 1: The integer 13 is punched in a field width of 4 on a card, and printed in a field width of 4 as

∧	∧	1	3

or sometimes as

0	0	1	3

Ex. 2: The real number −7.32 is punched in a field width of 7 on a card, and printed in a field width of 7 as

∧	∧	−	7	.	3	2

or sometimes as

−	0	0	7	.	3	2

Ex. 3: The character string constant '13' is punched in a field width of 4 on a card, and printed in a field width of 4 as

1	3	∧	∧

Ex. 4: The character string constant 'HELLO' is punched in a field width of 7 on a card, and printed in a field width of 7 as

H	E	L	L	O	∧	∧

In Examples 3 and 4, we note that the apostrophes needed to represent character string constants in flowcharts and stream-oriented input do not appear in the field. The following two remarks pertain to the manner in which numeric and character string fields are stored on input and output media:

1. Numeric fields are **right-justified,** as illustrated in Examples 1 and 2. That is, the rightmost character of a set of characters is moved to the right end of the field, and all unoccupied locations in the field are filled with blanks or zeros.[2]

2. Similarly, character string fields are said to be **left-justified** on I/O media, as illustrated in Examples 3 and 4.

---

[2] It is convenient to use only one of these. We will assume that blanks (∧) are used to fill unoccupied locations in a field.

5.4 RECORDS AND FIELDS

A natural extension of the concept of a field is a record, which now can be defined as a contiguous sequence of fields. The number of characters in a record depends on the I/O medium and the logical length of the information. For example, a record stored on a standard 80-column punched card always consists of 80 characters, although many of the characters may be blanks. The number of characters contained in a record on printer paper varies with the width of the printer, but 120 to 132 characters are common. Records may also be stored on other media such as magnetic tapes, magnetic drums, or magnetic disks. Record lengths associated with such media are variable, and are under the control of the programmer.

Next, we define a **file** as a collection of related records; so it follows that we can think immediately in terms of input and output files. It is necessary to have a condition that may be used to determine whether or not the end of an input data file or stream has been reached. We refer to such a condition by the term **end-of-file.** This term was introduced earlier, in Chapter 3, in connection with conditional branching (see Figure 3.5-2). It is used in a flowchart decision symbol as follows:

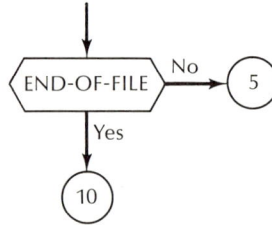

The notion of a record leads in a logical manner to that of a **record-oriented file,** in which I/O data is regarded as being made up of a set of **discrete** (noncontinuous) records. Conversely, when the I/O data is regarded as a *continuous* stream of characters, we refer to a **stream-oriented file.** Format-free I/O is stream-oriented.

5.5 DIFFERENT TYPES OF FIELDS

There are several types of fields that can be used to describe data in records used for I/O operations. Information pertaining to these fields is included in format statements which appear in annotation symbols of flowcharts. We will now enumerate the different types of fields and indicate the manner in which they are used by means of illustrative examples.

☐ *I-field:* Has the form Iw, where w denotes the field width. It is used to describe integer fields.

An illustration of the I-field is given in Figure 5.5-1. Since the

L   INPUT A ----- A integer, FORMAT(I5)

2   A ← A+1121

3   OUTPUT A ----- FORMAT(I7)

Input record:

| ^ | ^ | − | 2 | 1 |

Columns: |←———— 5 ————→|

TRACE TABLE

Symbol #	A
1	−21
2	1100

Output:

ʌʌʌ1100

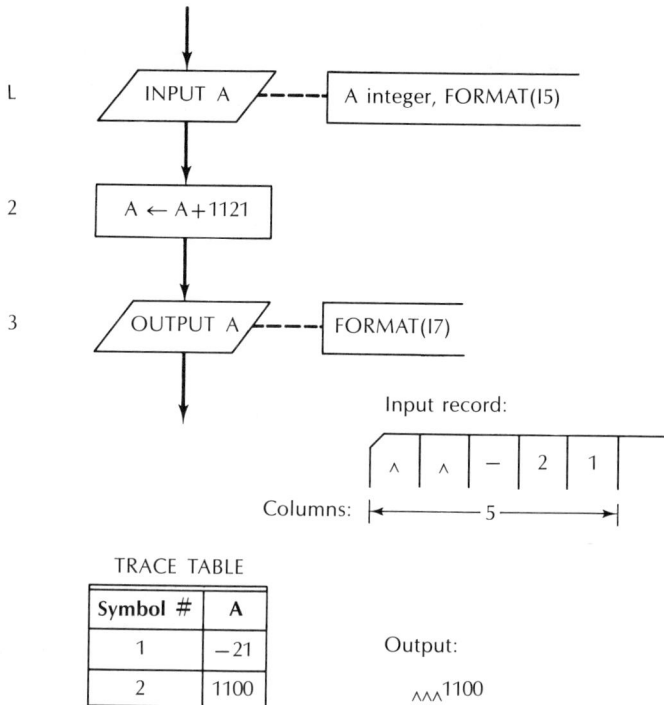

FIG. 5.5-1.   Example of the I-field.

integer variable A is read in with FORMAT (I5), the first five columns of the input record are scanned, and the number that appears in these columns is assigned to A. From the input card of Figure 5.5-1, it follows that A = −21. Thus when Symbol 2 is executed, this value of A is replaced by the value of A + 1121, which yields A = 1100.

Now, consider the OUTPUT statement in Symbol 3. The I7 specification in the statement FORMAT (I7) implies that seven columns are to be allotted for the output of A. Thus, the corresponding output is

ʌʌʌ1100

☐   *X-field:* Has the form wX, where w is the field width. It enables us to skip over a specific field width while executing INPUT and OUTPUT statements. The X-field has the form X(w) in some languages.

Figure 5.5-2 illustrates the use of the X-field. As a consequence of the format statement associated with Symbol 1, the following action is taken:

(a)   I3 causes the integer variable A to be assigned the value 987 which is in the first three columns of the input:

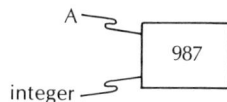

A ⟍
          987
integer ⟍

107

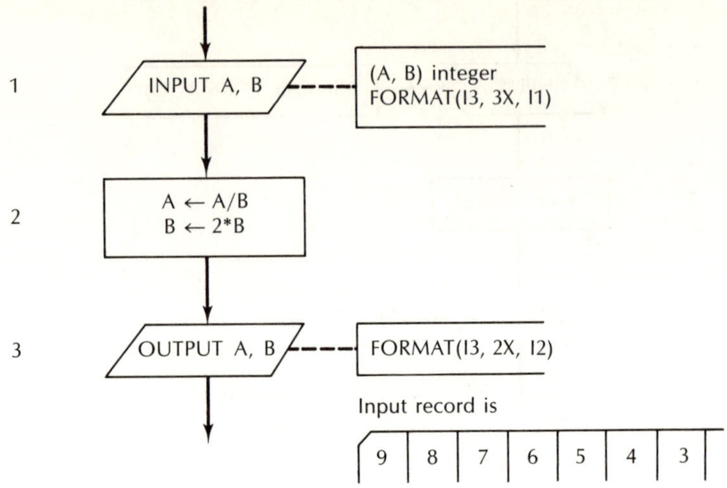

FIG. 5.5-2. Example of the X-field.

(b)   3X causes the next three columns to be skipped.
(c)   I1 causes the integer variable B to be assigned the value in the column that follows the columns that were skipped in (b):

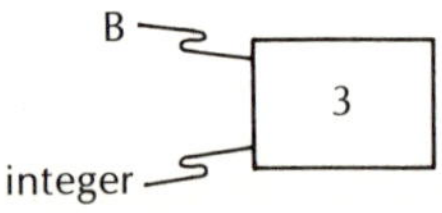

Next, Symbol 2 is executed; that is

$$A \leftarrow 987/3 = 329$$
$$B \leftarrow 2*3$$

which results in A = 329 and B = 6. Finally, when OUTPUT A, B in Symbol 3 is executed, the output will appear as follows:

$$329_{\wedge\wedge\wedge}6$$

We observe that the two blanks that separate 329 and $_\wedge 6$ are due to the 2X specification in the format statement associated with the preceding OUTPUT statement. The blank associated with $_\wedge 6$ is due to the fact that 6 is right-justified in the I2 field.

☐   *A-field:* Has the form Aw, where w is the field width. It is used in the I/O operations of character string constants. The A-field also has the form A(2).

   We make the following additional remarks with respect to the A-field:
(a)   Whenever the length of the string is less than the field width, then the rightmost positions in the field width are filled with blanks.
(b)   If the length of the string is greater than the field width, the leftmost characters that cannot be accommodated are lost (truncated).

(c) No arithmetic operations can be done with values that are read in using the A-field.

(d) This field is also referred to as the **alphameric** or **alphanumeric field;** that is, it consists of letters of the English alphabet, blanks, decimal digits, and special characters.

As an illustration of the A-field, we consider the example shown below. The annotation (COLOR, SIZE) character string (4) implies that COLOR and SIZE are character string variables that are allocated memory locations holding exactly four characters each.

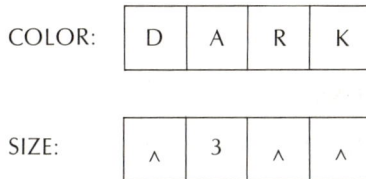

INPUT
COLOR, SIZE

(COLOR, SIZE) character string (4)
FORMAT(A4, 2X, A2)

The input record is

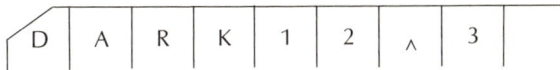

| D | A | R | K | 1 | 2 | ∧ | 3 |

From the FORMAT statement, it is apparent that the characters in the first four columns are assigned to the variable COLOR. The next two columns are skipped as a consequence of the 2X specification. The A2 specification then causes the rightmost two positions of a field width of 4 to be filled with blanks. Thus, COLOR and SIZE are stored as follows:

COLOR:

| D | A | R | K |

SIZE:

| ∧ | 3 | ∧ | ∧ |

Character string constants can be used in the output as they were in stream I/O. In implementing programming languages, this is accomplished in one of two ways. Either they may appear in the format, itself, or may be part of the output list, with an A-field specified for them in the format. Figure 5.5-3 illustrates the use of each of these ways of using character string constants in input/output.

☐ *F-field:* Has the form Fw.d, where w is the total field width and d is the number of digits to the right of the decimal. It is used to facilitate the input/output of real constants. In some languages this may appear as F(w,d), while F(w) is used in certain languages if an integer field is desired.

We will discuss the use of the F-field in input and output operations, respectively.

109

Input record:

| J | D | B | R | O | W | N | E | L | L | Y | ∧ |

↓

INPUT NAME ----- Name character string (8)
COST integer
X integer
FORMAT(2X, A10)

↓

Comment: The above input command results in the storage of 'BROWNELL' in the location assigned to NAME:

NAME

character
string (8)

BROWNELL

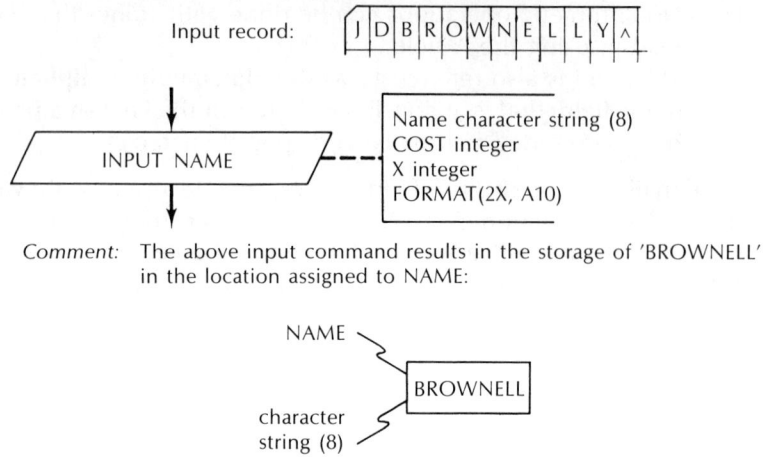

This is because NAME is allocated 8 characters. A10 causes the next 10 characters to be transmitted, and truncation of the last 2 characters in the input record occurs.

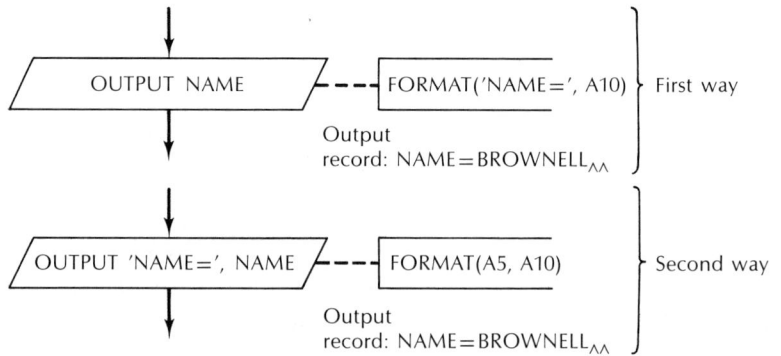

↓

OUTPUT NAME ---- FORMAT('NAME=', A10) } First way

↓

Output
record: NAME=BROWNELL∧∧

↓

OUTPUT 'NAME=', NAME ---- FORMAT(A5, A10) } Second way

↓

Output
record: NAME=BROWNELL∧∧

FIG. 5.5-3. Output of character string constants.

INPUT OPERATIONS

These are explained by means of the examples in Table 5.5-1.

Ex. 1: We observe, in Table 5.5-1, that there is no decimal in the first six columns of the input. A decimal point is therefore *inserted* in accordance with the F6.1 specification to obtain (in memory)

A — | 25.9 | — real

Thus, the following rule can be stated:

Rule 1: *When a decimal point is not present in the input field, it is assumed to occur d digits from the right end of the field, and the entire field width can be used for the digits and the sign bit. The sign need not appear if the number is positive.*

110

Ex. 2: In this case, a decimal point is present in the first six

TABLE 5.5-1. F-field examples related to input operations

Example #		Input data:
1.	INPUT A — — — FORMAT(F6.1)	∧ ∧ ∧ 2 5 9
2.	INPUT A — — — FORMAT(F6.1)	− 2 5 . 9 3
3.	INPUT B — — — FORMAT(F5.2)	2 6 3 8 7
4.	INPUT C — — — FORMAT(F8.3)	∧ 2 6 3 . 8 7 0 ⟵——8 columns——⟶

columns of the input. The decimal number that appears in these columns is directly assigned to the variable. Thus,

$$ A \longleftarrow \boxed{-25.93} \quad \text{real} $$

which leads to the following rule:

Rule 2: *The presence of a decimal point in the input field over-rides the "d" portion of the Fw.d specification.*

Ex. 3: Since no decimal point appears in the first five columns of the input, one is inserted in accordance with the F5.2 specifications to obtain

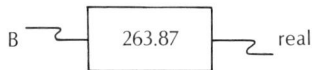

$$ B \longleftarrow \boxed{263.87} \quad \text{real} $$

Ex. 4: Because there is a decimal in the first eight columns of the input, the d ($= 3$) portion of the F8.3 specification is ignored. Thus we have

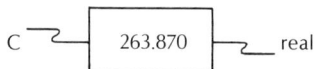

$$ C \longleftarrow \boxed{263.870} \quad \text{real} $$

In the interest of consolidating the concepts presented in the previous sections, let us now consider an additional example.

111

Example 5.5-1 Given the following input record:

| ∧ | ∧ | 1 | 6 | · | 2 | ∧ | ∧ | 2 | 3 | ∧ | A | B | $ | 1 | 2 | 3 | ∧ | 9 | $ | 1 | |

what values are assigned to the variables A, B, and C if the statement

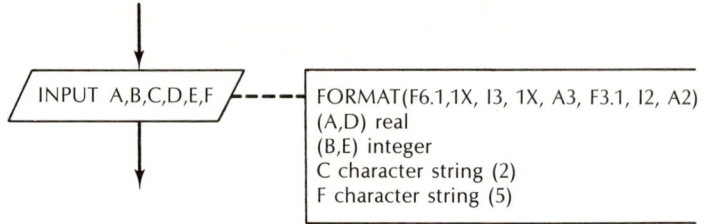

INPUT A,B,C,D,E,F ----- FORMAT(F6.1,1X, I3, 1X, A3, F3.1, I2, A2)
(A,D) real
(B,E) integer
C character string (2)
F character string (5)

is executed?

Solution: The variable A is associated with the F6.1 portion of the format statement. Thus we obtain

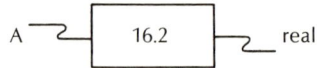

A ⊃ | 16.2 | ⊂ real

Next, one column is skipped and B is assigned the integer value in the next three columns due to the I3 specification. Hence,

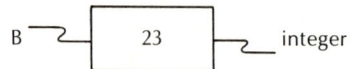

B ⊃ | 23 | ⊂ integer

One more column is skipped. Then A3 inputs 'AB$'. Since C has length 2, we obtain

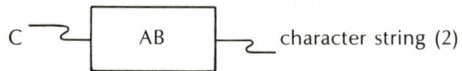

C ⊃ | AB | ⊂ character string (2)

The specification associated with D assigns it a number with one digit to the right of the decimal. Thus,

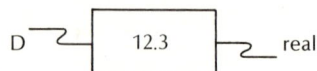

D ⊃ | 12.3 | ⊂ real

Next, the integer field of width 2 results in E being assigned the integer 9:

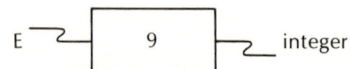

E ⊃ | 9 | ⊂ integer

Finally, the A2 specification associated with F results in

F ⊃ | $1∧∧ | ⊂ character string (5)

112

The use of the F-field specification for output operations is similar to that for input operations. We illustrate these by four examples.

Ex. 1:

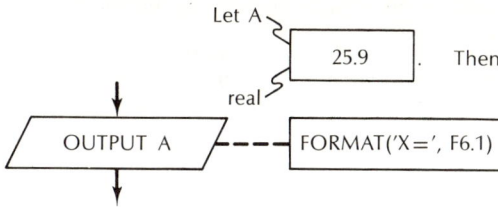

yields the output

$$X = _{\wedge\wedge}25.9$$

since a field width of 6 is provided by the format specification for A.

Ex. 2:

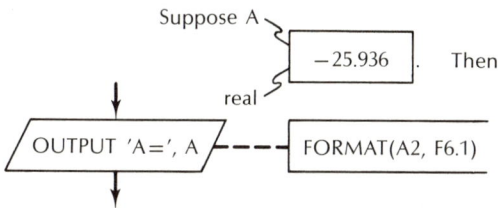

results in the output

$$A = _{\wedge}-25.9$$

which consists of 8 characters in accordance with the preceding format. However, we observe that since the format allows for only one digit to the right of the decimal in the output field, the digits 3 and 6 in −25.936 do not appear in the output, although they are still available in memory.

Ex. 3:

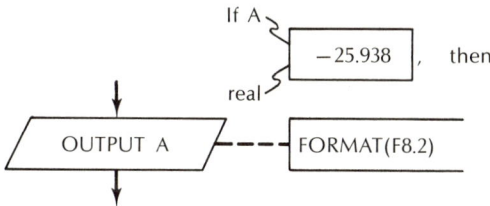

leads to the output $_{\wedge\wedge}-25.93$. We remind the reader that truncation is assumed when there are more decimal digits in memory than output positions.

Ex. 4:

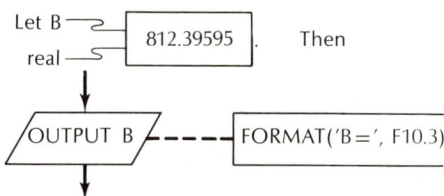

113

yields the following output:

$$B = {}_{\wedge\wedge\wedge}812.395$$

☐ *E-field:* As was the case with the F-field, this field is also used for I/O operations on real numbers. Its general form is as follows:

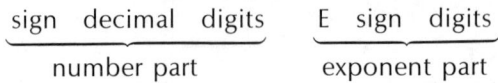

sign decimal digits	E sign digits
number part	exponent part

In output operations, the sign in the number part is not printed if the number is positive.

Next, this type of representation is illustrated by three examples.

Ex. 1: Since $345.67 = .34567 \times 10^3$, its E-field representation is given by .34567E+3.

Ex. 2: The number .00034567 can be equivalently written as $.34567 \times 10^{-3}$. Thus, the E-field representation is given by .34567E−3.

Ex. 3: Since $-1.7 = -.17 \times 10^1$, its E-field representation is given by −.17E+1.

FIXED DECIMAL REPRESENTATION

The notion of fixed decimal representation was introduced in Section 3.3. It is a way of storing numeric values in memory, and is used with I/O operations in a free-format environment, or with E- and F-field formats. The following examples illustrate the manner in which fixed decimal representation may be used.

Ex. 1: If the input is given by

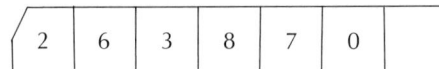

2	6	3	8	7	0

and the statement

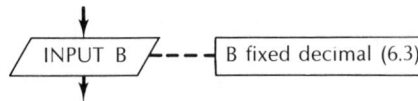

INPUT B ---- B fixed decimal (6.3)

is executed, then we obtain

B ⟶ | 2 | 6 | 3 | 8 | 7 | 0 | ⟵ fixed decimal (6.3)

Ex. 2: If the input is given by

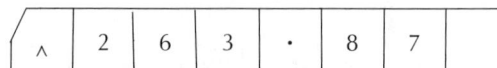

∧	2	6	3	.	8	7

and the statement

is executed, then the value assigned to C is as follows:

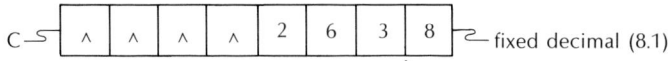

It is observed that the (8.1) specification for the size of memory causes the digit 7 to be lost (truncated) after it is read from the card.

Ex. 3: Let

Then execution of the statement

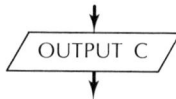

with the formats listed below, results in the corresponding outputs shown alongside the formats.

FORMAT (F8.3) -100.021
FORMAT (F8.2) $_\wedge-100.02$
FORMAT (F8.0) $_{\wedge\wedge\wedge\wedge}-100$
FORMAT (F9.3) $_\wedge-100.021$
FORMAT (F7.3)   100.021 (may cause an error due to loss of sign)

CONCLUDING REMARKS

(a) Suppose A, B, C, and D are four numeric variables, the values of which are to be read in using an F8.3 specification. This can be accomplished as follows:

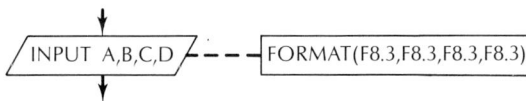

A convenient feature of flowchart language is that the F8.3 specification need not be repeated. This is because we can use the following statement, which is equivalent to the one given above:

115

In the three additional examples that illustrate this feature, the symbol $\equiv$ denotes that the statements that appear on either side of it are equivalent:

$$\text{FORMAT}(2\text{I}3) \equiv \text{FORMAT}(\text{I}3,\text{I}3)$$
$$\text{FORMAT}(3\text{A}2,2\text{X},\text{I}4) \equiv \text{FORMAT}(\text{A}2,\text{A}2,\text{A}2,1\text{X},1\text{X},\text{I}4)$$
$$\text{FORMAT}(2\text{F}6.1,3\text{I}3) \equiv \text{FORMAT}(\text{F}6.1,\text{F}6.1,\text{I}3,\text{I}3,\text{I}3)$$

(b) Parentheses may be used to group a series of format phrases (field specifications) as follows:

$$\text{FORMAT}(2(\text{I}3,1\text{X},\text{F}6.1)) \equiv \text{FORMAT}(\text{I}3,1\text{X},\text{F}6.1,\text{I}3,1\text{X},\text{F}6.1)$$
$$\text{FORMAT}(2(\text{I}3,1\text{X}),\ 3(\text{A}4)) \equiv \text{FORMAT}(\text{I}3,1\text{X},\text{I}3,1\text{X},\text{A}4,\text{A}4,\text{A}4)$$

5.6 SUMMARY

This chapter focused on format considerations in I/O operations. Stream and record I/O were described and the notions of fields, records, and files were introduced. A variety of field types were defined, and their applications with respect to I/O of numeric and string variables were illustrated by means of examples.

problems

5-1. Select the output value for X from the choices listed. Given: X = 14.5548.

$$\fbox{OUTPUT X} - - - - \fbox{FORMAT('ANSWER =', F7.3)}$$

(a) ANSWER = $_\wedge$14.554   (b) $_\wedge$14.554
(c) ANSWER =   (d) ANSWER = 14.5548
(e) None of these

5-2. Suppose I = 'DAYS', J = 6, K = 5.1, and A = 10.118 have been stored in memory. Select the correct form of the output from the choices listed.

$$\fbox{OUTPUT I,J,K,A} - - - - \fbox{FORMAT(2X, A6, I2, F4.1, F7.2)}$$

(a) $_{\wedge\wedge}$ DAYS $_{\wedge\wedge}6_\wedge5.1_{\wedge\wedge}10.11$
(b) $_{\wedge\wedge\wedge}$DAYS$_\wedge6_\wedge5.1_{\wedge\wedge}10.11$
(c) $_{\wedge\wedge}$ DAYS $_{\wedge\wedge\wedge}6_\wedge5.1_{\wedge\wedge}10.11$
(d) $_{\wedge\wedge}$ DAYS $_{\wedge\wedge\wedge}6_\wedge5.1_{\wedge\wedge}10.12$
(e) None of these

116

5-3. Given:

INPUT P,Q ----- FORMAT(2X, I3, 3X, I2)
(P, Q) integer

The input record is 14871930−2. From the choices listed, select the values stored for P and Q.
(a) P = 148   Q = 71   (b) P = 871   Q = 20
(c) P = 871   Q = −2   (d) P = 148   Q = 30
(e) None of these

5-4. Given:

INPUT P,Q ----- FORMAT(F5.2, F5.1)
(P, Q) real

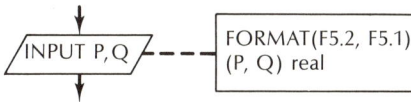

The input record is ∧∧416−1.24. From the choices listed, select the values stored for P and Q.
(a) P = 4.16   Q = −1.24   (b) P = 4.16   Q = −124
(c) P = 4.16   Q = −1.2   (d) P = 416   Q = 1.24
(e) None of these

5-5. Given:

INPUT X, Y ----- (X, Y) character string
FORMAT(1X, A2, 1X, A2)

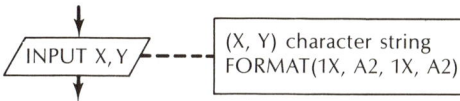

The input record is HI68HO. From the choices listed, select the values stored for X and Y.
(a) X = 'I6'   Y = 'HO'   (b) X = 'HI',   Y = 'HO'
(c) X = 'HI∧∧'   Y = '32∧∧'   (d) X = 'I6∧∧'   Y = 'HO∧∧'
(e) None of these

5-6. Choose the numerical value of .1632E−2 from the following:
(a) 16.32   (b) .001632
(c) .01632   (d) −.001632
(e) None of these

5-7. Given:

INPUT A,B,C ----- FORMAT(2X,I3,3X,I2,F6.2)
(A,B) integer, C fixed
decimal (4.2)

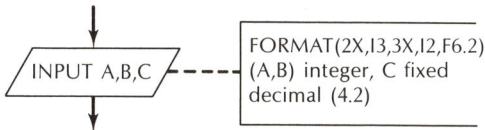

The input record is 76129310∧3∧−1021. What values are stored for A, B, and C?

5-8. Given:

The input record is $_{\wedge\wedge\wedge}123-25.94-1.92_{\wedge\wedge}$. What values are sorted for A, B, and C?

5-9. Given:

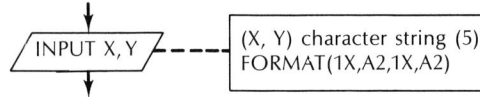

The input record is WAIT37. What values are stored for X and Y?

5-10. What is the numerical value of $.1234E-2$?

5-11. Given:

If A = 123.9321, what is the output?

5-12. Suppose I = 'QUIZ', J = 2, C = 3.0, and A = 16.736 are in memory. Depict the output if the following statement is executed:

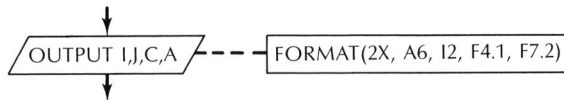

5-13. Given:

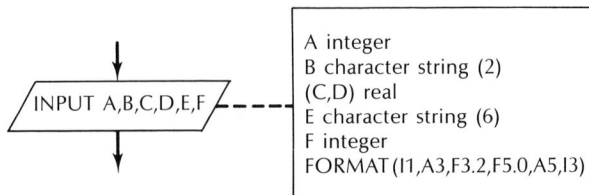

If the input record is 9ABC12691.91NUTTS678, what are the values stored for the variables A through F?

5-14. Given:

INPUT X,Y,Z

(X, Y, Z) fixed decimal (4.2)
FORMAT(F6.2,F5.2,F9.1)

If the input record is $+14.21-1241_{\wedge\wedge}10.23_{\wedge\wedge}$, what are the values stored for X, Y, and Z?

6

Data Structures

Once a person has understood the way in which variables are used in programming, he has understood the quintessence of programming. (E. W. Dijkstra, 1972)

KEYWORDS

array
deque
list
lower bound
queue
stack
subscript
tree
upper bound

Data processed by computers may be classified into two major categories: (a) numeric data and (b) nonnumeric data. By the term **numeric data,** we mean data stored in the form of numbers, subject to arithmetic operations. Such data may be encountered in a variety of disciplines, such as physics, mathematics, banking, engineering, psychology, biology, medicine, business, and statistics. On the other hand, by **nonnumeric** (or **alphameric**) **data** we mean data in the form of character strings. These are made up of letters of the English alphabet, blanks, decimal digits, and special characters. We encounter nonnumeric data in such diverse applications as the following:

- ☐ Common business problems.
- ☐ Military logistics.
- ☐ Inventory control in factories and warehouses.
- ☐ Management of airline reservations.
- ☐ Maintenance of student/patient records in academic/medical institutions.
- ☐ Production scheduling in factories.

There is a fundamental difference in the use of computers for processing the two categories of data. Nonnumeric data processing generally involves the storage, organization, and maintenance of large volumes of data, while performing a relatively small number of calculations on individual **items** (i.e., pieces of information) of that data. In contrast, computers used for processing primarily numeric data are generally required to perform a large number of calculations on relatively small volumes of numeric data. Hence, it is common to associate the terms **number crunching** and **data processing** when computing with numeric and nonnumeric data, respectively.

The problem of organizing data for efficient storage and retrieval in each of the preceding types of computing has received a great deal of attention in the field of computer science. In this connection, the term **data structure** is used—it refers to the method of organizing data and the resulting interrelations between data items and their addresses (or variable names), so that the organization is useful to the programmer.[1] Examples of data structures that are used with numeric and nonnumeric data include the following:

1. linear lists;
2. two-dimensional arrays;
3. trees; and
4. files.

In this chapter we will restrict our attention to the first three data structures in the preceding list. Some aspects related to files will be

---

[1] By *addresses* is meant data locations in memory.

considered in Chapter 9, which concerns business data processing. We will study the problem of processing some of these data structures using flowchart language in Chapter 7.

It is often necessary to process collections of related data items. If the relationship among these items is a linear ordering, then the collection of items is called a **linear list.** We will now consider four types of linear lists: (a) one-dimensional arrays, (b) stacks, (c) deques, and (d) queues.

ONE-DIMENSIONAL ARRAYS

The simplest manner in which a group of related data can be organized in computer memory is by arranging the given items in the form of a **one-dimensional array.** This type of an array is a list with a *single name,* consisting of elements that are of the *same type;* the elements are stored in contiguous memory locations in a computer.[2]

We consider an array of eight elements, and let JIM denote its name. Then the elements of the array are given by $JIM_1, JIM_2, \ldots, JIM_8$. If the memory of the computer is organized so that each data item can be accommodated in a single memory location, then the location or address of a typical item can be readily obtained. For example, if the starting address of the array JIM is 3000, the address of the element JIM_7 is given by 3006 (see Figure 6.1-1).

Address in memory	Contents are values of
...	...
3000	JIM_1
3001	JIM_2
3002	JIM_3
3003	JIM_4
3004	JIM_5
3005	JIM_6
3006	JIM_7
3007	JIM_8

FIG. 6.1-1. Organization of a one-dimensional array in memory.

When dealing with arrays, two factors are of concern, as follows:

1. Creation of the array in memory so that one can use it in a program. This is done via declarative statements that appear in annotation boxes.

2. Referencing of the array information, which can be achieved in two ways: (a) by referencing individual array elements, and (b) by referencing the entire array.

---

[2]A one-dimensional array is also referred to as a *vector.*

We use **subscript variables** or **constants** to select an element in a one-dimensional array. A subscripted variable represents an array that is stored in contiguous memory locations. Each memory location is uniquely identified by giving it a number, or **subscript,** indicating its position in the array. For example, if we wish to store our ending savings account balances for each of the first six months of a year, we could refer to them by the name BALANCE of type fixed decimal (5.2), as indicated in Figure 6.1-2.

FIG. 6.1-2. A one-dimensional array called BALANCE.

The preceding array, BALANCE, can be created in memory by means of the following declarative statement:

BALANCE(6) fixed decimal (5.2)

Thus, BALANCE(6) implies that six contiguous locations are set aside in memory to store the six ending balances. The portion of the statement "fixed decimal (5.2)" defines the type of the array.

To be able to reference an element in BALANCE, we must specify exactly which element is desired by providing a subscript for each element. Such a subscript is enclosed in parentheses following the name of the array. Furthermore, the subscript must be an integer constant, variable, or expression. For example,

$$BALANCE(1) \leftarrow 100.02$$

will store 100.02 in BALANCE(1). And

$$BALANCE(2) \leftarrow BALANCE(1) + 100.21$$

will reference the value of BALANCE(1), add 100.21 to it, and then store the resulting sum in BALANCE(2). Thus, array elements are used in a manner very similar to simple variables.

The preceding references to BALANCE have been made with constant values for the subscript. However, an integer variable can also be used. For example, the statements

124

will cause the output to be 753.46, since the current value of BALANCE(6) is 753.46 (see Figure 6.1-2). As a second example, let us consider the statements

$$I \leftarrow 1$$
$$BALANCE(I+2) \leftarrow BALANCE(I+1) + 100.00$$

These statements are first expressed as

$$I \leftarrow 1$$
$$BALANCE(3) \leftarrow BALANCE(2) + 100.00$$

by using the current value of I, which equals 1. Next, the value of BALANCE(2) is referenced, 100.00 is added to it, and the result is stored in BALANCE(3). Since the current value of BALANCE(2) is 210.23 (see Figure 6.1-2), it follows that

$$BALANCE(3) = 310.23$$

From this discussion, it is apparent that we are now in a position to reference individual array elements by making use of subscripts. In several programming languages, however, it is also possible to reference entire arrays without resorting to subscripts. For instance

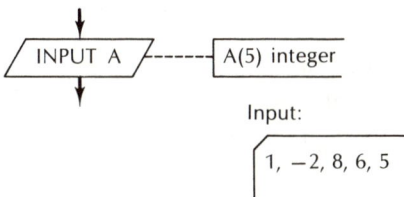

reads five values to fill the five contiguous locations in memory assigned to the array A, which results in

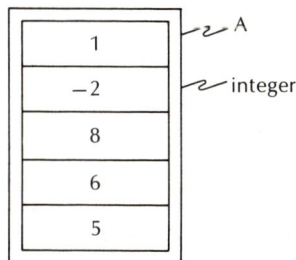

125

Similarly, the statement

$$A \leftarrow A + 1$$

causes 1 to be added to each element of A to yield

As an additional illustration, consider an array TAX, consisting of the five elements

$$TAX(1), \; TAX(2), \; TAX(3), \; TAX(4), \; TAX(5)$$

Let us assume that the values associated with these array elements are as follows:

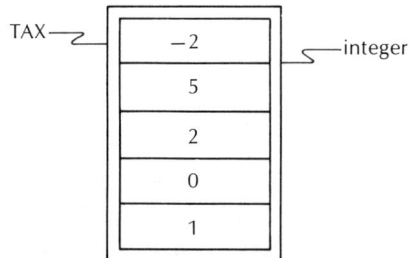

We define a subscript variable I which takes the *values* 1, 2, 3, 4, and 5. Then any element of TAX can be represented in terms of I as TAX(I), $I = 1, 2, \ldots, 5$; (i.e., I equals 1 or 2 . . . or 5). Now, let Y be an integer variable. Then it is straightforward to verify that execution of each of the following statements results in $Y = 2$:

(a) $Y \leftarrow TAX(3)$, yields $Y = 2$

(b) $K \leftarrow 3$
     $Y \leftarrow TAX(K)$, yields $Y = 2$

(c) $K \leftarrow 1$
     $Y \leftarrow TAX(K+1) - 3$, yields $Y = 5-3=2$

(d) $K1 \leftarrow 1$
     $K2 \leftarrow 2$
     $K3 \leftarrow 5$
     $Y \leftarrow TAX(K1) + TAX(K2) - TAX(K3)$,
     yields $Y = -2+5-1=2$

It is important to remember that if an expression appears in the parentheses following TAX, then *it must be evaluated first*—before attempting to use it. For example, in (c), where $TAX(K+1) = TAX(2)$,

$$Y \leftarrow TAX(2) - 3$$

which yields

$$Y \leftarrow 2$$

since the current value of TAX(2) is 5. It is also worthwhile noting that there is a fundamental difference between two statements of the form

$$W \leftarrow TAX(K+1) \qquad \textbf{(6.1-1)}$$

and

$$Z \leftarrow TAX(K) + 1 \qquad \textbf{(6.1-2)}$$

From Equations (6.1-1) and (6.1-2) it follows that

W has the value of $\left\{\begin{matrix} \text{"contents of memory location assigned to} \\ \text{the array element TAX(K+1)"} \end{matrix}\right\}$

Z has the value of $\left\{\begin{matrix} \text{"contents of memory location assigned to} \\ \text{the array element TAX(K) + 1"} \end{matrix}\right\}$

Clearly, W and Z are different, since they refer to two distinct locations in memory. Therefore, in general, the values assigned to W and Z will be quite different upon execution of the statements in Equations (6.1-1) and (6.1-2). For the case K = 3,

$$W \leftarrow TAX(4) = 0$$

and

$$Z \leftarrow TAX(3) + 1 = 2 + 1 = 3$$

To recapitulate, in the case of scalar variables, such as A, X, Y, subscripted variables can be declared, as illustrated in Figure 6.1-3.

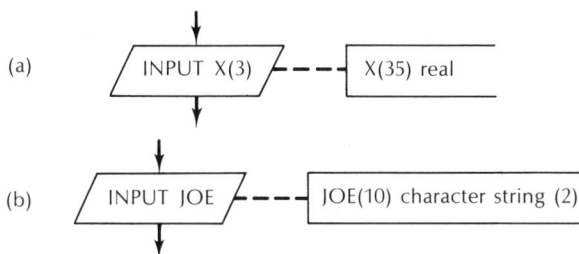

(a) INPUT X(3) ---- X(35) real

(b) INPUT JOE ---- JOE(10) character string (2)

FIG. 6.1-3. On declaring subscripted variables.

Declaration of the array X causes 35 locations in memory to be reserved for the elements X(1), X(2), X(3), . . . , X(35). Since X is declared as a real array, numeric constants would be stored in these locations. Thus the INPUT X(3) statement will cause a numeric constant to be read into the location in memory assigned to the array element X(3). However, when the array name appears *without* a subscript, a value is read for each element as shown in Figure 6.1-3(b), where a total of 10 values are read from the input data stream. They are then placed in the corresponding memory locations assigned to the array JOE. These values would then be treated as character string constants, since JOE is declared as a character string array.

127

STACKS, DEQUES, AND QUEUES

A **stack** is a data structure in which data items are arranged in computer memory in a manner resembling a stack of plates in a cafeteria; i.e., items are stacked one on the other (see Figure 6.1-4). For convenience, each item in the stack is considered to be a single character. In general, stack items could be numbers, words, records, etc. The term **pointer** implies a word that contains the address of another word. In Figure 6.1-4, for example, the pointer contains the address of the "top of the stack."

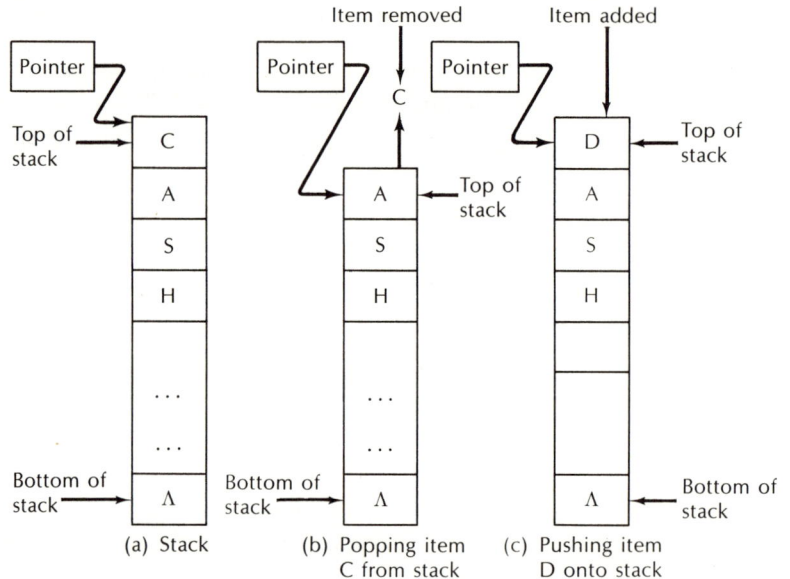

FIG. 6.1-4.   Illustration of a stack, and its related POP and PUSH operations.

When adding an item to the stack, it can be placed only on the top, and only the topmost item can be removed. The process of removing an item from a stack is referred to as **popping the stack.** Conversely, the term **pushing** refers to the process of adding to the stack. Thus, items can be either removed or added via the POP and PUSH operations, respectively, as illustrated in Parts (b) and (c) of Figure 6.1-4.

From the preceding discussion, it is apparent that a stack works on a "last-in–first-out" basis. A one-dimensional array may be used to store a stack.

A **queue** is a data structure that works on a "first-in–first-out" basis (see Figure 6.1-5). It is similar to a stack, in that new items are added

FIG. 6.1-5.   A queue.

only to the top; but items are removed only from the bottom. A data structure in which it is possible to add or remove items from both ends is a **deque,** which stands for "double-ended queue" (see Figure 6.1-6).

FIG. 6.1-6. A deque.

In languages having the capability to store list data structures, a stack can be stored using a pointer, as illustrated in Figure 6.1-7. We observe that a pointer in each element "points to" the next element in the stack. The last element in the stack does not have a pointer, and this is indicated by the symbol Λ (see also Figure 6.1-4). This data structure is called a **linked list.**

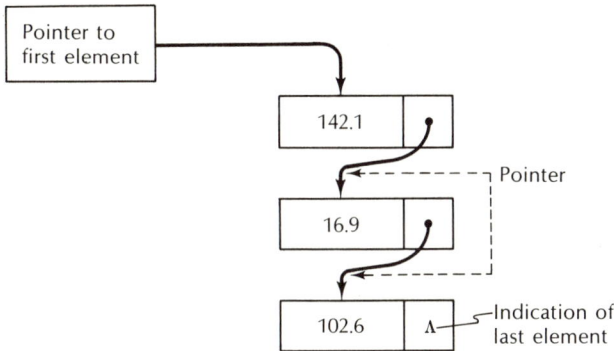

FIG. 6.1-7. A three-element stack.

The linked list has an advantage over the array for list storage, in that the former allows the list to grow without a predetermination of maximum list size. Linked lists have the disadvantage that the pointers have to be stored in addition to the element value.

TWO-DIMENSIONAL ARRAYS 6.2

The notion of placing data items in a one-dimensional array, the elements of which occupy consecutive addresses in memory, can be extended to multidimensional arrays. In this regard, we will restrict our attention to two-dimensional arrays, an example of which is given below.[3] Here, TOM is the name of a real array, and TOM (I, J) denotes its element in Row I and Column J:

TOM (1,1)	TOM (1,2)	TOM (1,3)
TOM (2,1)	TOM (2,2)	TOM (2,3)
TOM (3,1)	TOM (3,2)	TOM (3,3)
TOM (4,1)	TOM (4,2)	TOM (4,3)

---

[3] A two-dimensional array can be used to store a *matrix* or *table*—e.g., a map mileage table.

Inspection of this two-dimensional array shows that it consists of 4 rows and 3 columns. Note that, by convention, the row subscript precedes the column subscript. To illustrate the manner in which this array is stored in memory, we assume that the address allotted to the element TOM(1, 1) is 200, and that the remaining elements appear in memory in the following order:

TOM(1, 2), TOM(1, 3), TOM(2, 1), TOM(2, 2), TOM(2, 3)

and so on, until

TOM(4, 1), TOM(4, 2), TOM(4, 3)

In other words, we assume that the corresponding organization in memory (see Figure 6.2-1) is obtained by reading the array TOM on a *row-by-row* basis (row-major order).[4]

Address in memory	Contents are values of
...	...
200	TOM (1,1)
201	TOM (1,2)
202	TOM (1,3)
203	TOM (2,1)
204	TOM (2,2)
205	TOM (2,3)
206	TOM (3,1)
207	TOM (3,2)
208	TOM (3,3)
209	TOM (4,1)
210	TOM (4,2)
211	TOM (4,3)
...	...

FIG. 6.2-1. Storage of a two-dimensional array with 4 rows and 3 columns, on a row-by-row basis.

The idea of handling one-dimensional arrays via subscripted variables is readily extended to multidimensional arrays. Procedures for declaring and referencing multidimensional arrays are similar to those associated with the one-dimensional case. These are summarized as follows:

1. Memory locations must be reserved through the use of declarations in annotation symbols. Such declarations must give both the number of dimensions and the bound on each dimension. For example, the declaration

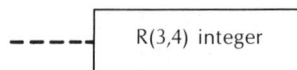

- - - - - | R(3,4) integer |

---

[4]An array can also be read on a column-by-column basis (column-major order). The specific mode depends on the programming language. For example, FORTRAN-IV and PL/1 use the column and row modes, respectively.

implies that R is a two-dimensional array consisting of 3 rows and 4 columns, and can be thought of as:

R(3,4)
integer

5	0	2	4
6	5	−1	7
−3	4	2	0

Thus a total of 12 locations would be assigned in memory for the 12 elements of R. Since this array is declared as type integer, only integer values would be expected to be stored in these locations.

2. Array elements can be referenced using integer constants, integer variables, or expressions that evaluate to integers as subscripts. Consider the following examples:

	Reference	**Value**
(a)	$R(2,2)$	5
(b)	$R(1,4)$	4
(c)	$I \leftarrow 2$	
	$J \leftarrow 4$	
	$R(I,J)$	7
	$R(I-1, J-1)$	2

In these examples, we observe that I and J are variables used as subscripts.

3. In some programming languages, it is possible to reference entire arrays. For instance, the statement

$$R \leftarrow R + 1$$

results in:

R
integer

6	1	3	5
7	6	0	8
−2	5	3	1

4. Multidimensional arrays of dimension 3 or more can be formed by using an appropriate number of subscripts. For example, in order to store a book of 100 tables, in which each table consists of 30 rows and 4 columns, we would create a three-dimensional array via the declaration

TABLE (100,30,4)

where TABLE is the name of the array.

131

In concluding our discussion of arrays, we note that in some languages it is possible to explicitly specify the **upper** and **lower bounds** on a subscript. Up to this point the upper bound has been explicitly specified, as in the declaration

```
- - - -[  A(4) real          ]
```

where the upper bound is 4; the lower bound is taken to be 1, although it is not specified. However, in such declarations as

```
- - - -[  A(0:3) real         ]
```

both the lower and upper bounds are explicitly specified as 0 and 3, respectively, and separated by a colon. Thus we have

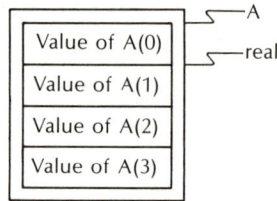

```
        ┌──────────────┐        ⌐──A
        │ Value of A(0) │        ⌐──real
        ├──────────────┤
        │ Value of A(1) │
        ├──────────────┤
        │ Value of A(2) │
        ├──────────────┤
        │ Value of A(3) │
        └──────────────┘
```

As an additional example, the declaration

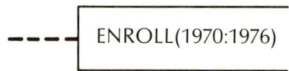

```
- - - -[  ENROLL(1970:1976)    ]
```

could represent the enrollments at an institution during the 1970–1976 period. As a result, the following array would be created:

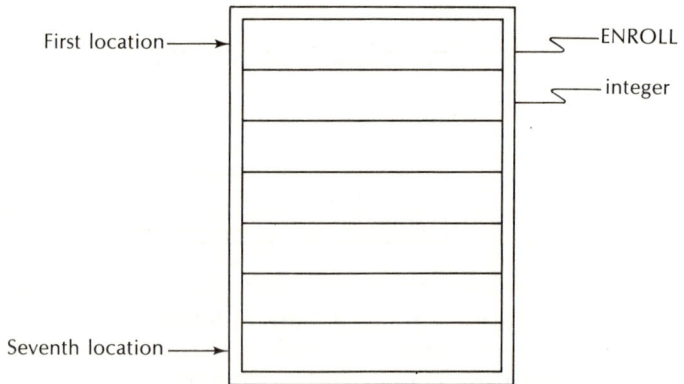

```
First location ──→  ┌──────────────┐   ⌐──ENROLL
                    │              │   ⌐──integer
                    ├──────────────┤
                    │              │
                    ├──────────────┤
                    │              │
                    ├──────────────┤
                    │              │
                    ├──────────────┤
                    │              │
                    ├──────────────┤
                    │              │
                    ├──────────────┤
Seventh location ──→│              │
                    └──────────────┘
```

In the first location the value of ENROLL (1970) is stored, while the rest of the enrollment figures related to the 1971–1976 period are stored in the second through seventh locations, respectively.

The preceding representation can also be extended to multidimensional arrays. For example, for the two-dimensional case that follows,

9 elements for the array X are set aside as a consequence of the declaration

---- X(−1:1, 18:20) integer

Value of X(−1,18)	Value of X(−1,19)	Value of X(−1,20)
Value of X(0,18)	Value of X(0,19)	Value of X(0,20)
Value of X(1,18)	Value of X(1,19)	Value of X(1,20)

X

integer

TREES **6.3**

A **tree** is composed of a hierarchy of elements called **nodes,** the uppermost level of which has only one node, called a **root.** The set of nodes constituting a tree are linked by **branches** (see Figure 6.3-1),

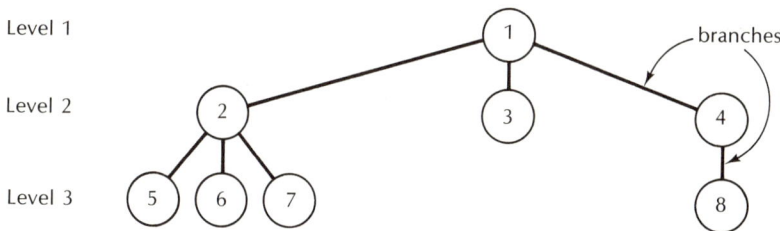

Level 1

Level 2

Level 3

branches

FIG. 6.3-1. Example of a tree.

where the symbol ◯ denotes a node. With the exception of the root, every node has exactly one node related to it at a higher level, and this is called its **parent.** A parent node can have one or more related nodes at a lower level, called **children.** Nodes with no children are referred to as **leaves.** For example, Node 3 and Nodes 5 through 8 in Figure 6.3-1 are leaves.

It is important to note that the notion of hierarchy is inherent in a tree. As a consequence, tree structures are frequently used to represent organizational charts. A partial tree structure (which could conceivably represent a small-scale company) is shown in Figure 6.3-2.

Trees are very useful as logical structures of data, and can be stored efficiently in computer memory. A general discussion of the procedures used for storing trees is beyond the scope of this book. However, we will introduce the basic idea of storing trees by restricting our attention to a specific class called **binary trees**—those in which each parent can have at most two children. Two examples of binary trees are shown in Figure 6.3-3.

Let us assume that we are to store a number of data items in memory using a binary tree structure, by storing each item at a node of the tree.

133

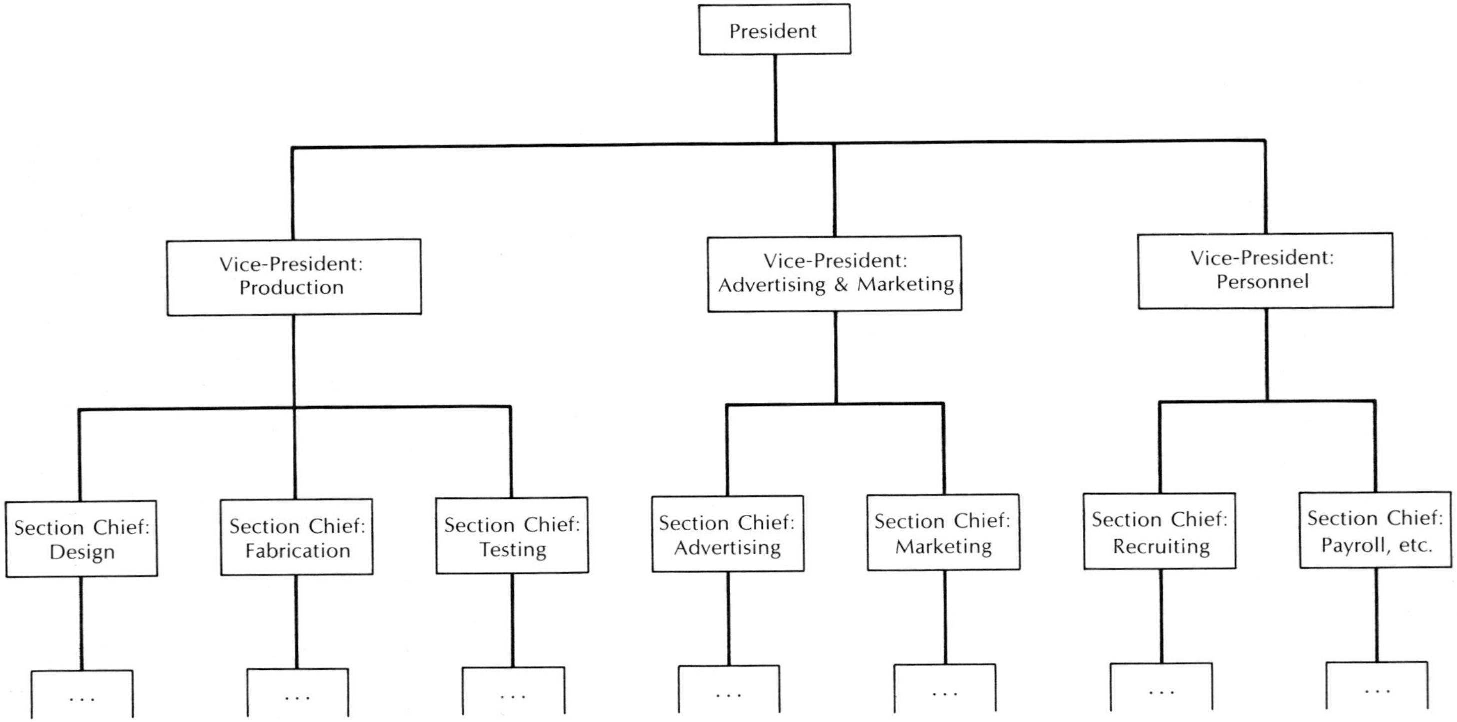

FIG. 6.3-2. The notion of a tree structure.

FIG. 6.3-3. Examples of binary trees.

For convenience of description, let us further assume that the data items are simply a set of English letters, as shown in Figure 6.3-3. Then it follows that this tree can be stored in memory by associating each of its nodes with several words containing the following information: (a) the data item at the node and (b) two pointers, one to the leftmost child (or node) and the other to the rightmost child (or node). We note that a START pointer giving the address of the root is required in order to enter the tree. An example of this arrangement is shown in Figure 6.3-4.

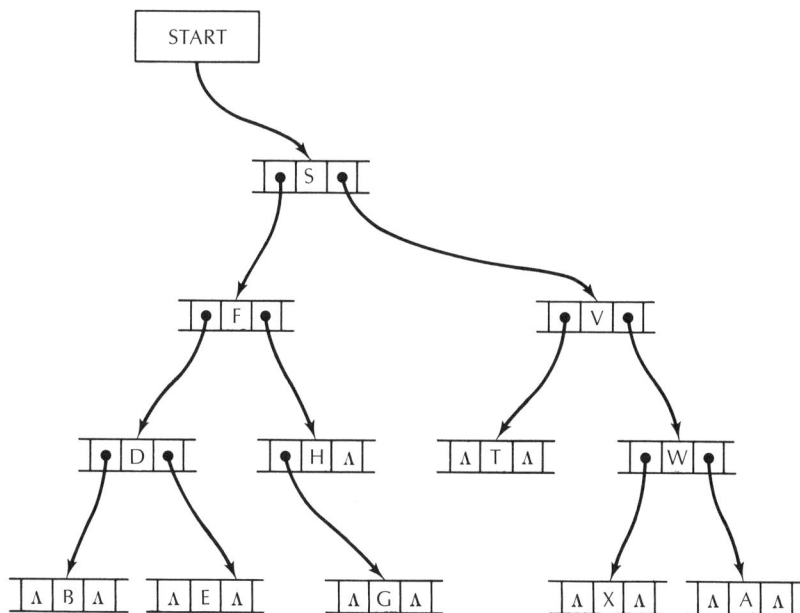

FIG. 6.3-4. Data structure for the binary tree in Figure 6.3-3(a).

SUMMARY 6.4

In this chapter we introduced the notion of data structures that provide more conceptually logical ways of utilizing computer memory. The list was the first type of data structure considered. Three forms of the list were discussed: (a) stacks, (b) queues, and (c) deques. In

135

addition, two data structures called arrays and trees were introduced.

There are additional types of data and data structures available in programming languages that we did not discuss in this chapter. Some allow the user to create new types of data and then declare related variables. For example, the type DAYS with a range of values 'SUN', 'MON', . . . , 'SAT' could be created. A variable WEEKDAYS could be declared with the subrange 'MON, . . . , 'FRI' by Declare WEEKDAYS type DAYS ('MON', . . . , 'FRI'). If the user tried to assign to WEEKDAYS a value not in the subrange, an error would occur. The extension of the language to allow the programmer to create new data structures is also a powerful feature of some languages, such as PASCAL.

problems

6-1. How many elements are there in the following arrays?
(a) I(10)   (b) A(2,3)
(c) Y(11)   (d) X(11,10)
(e) I(10), character string (3)   (f) A(2,3) integer
(g) Y(11) integer   (h) X(11,10,3)

6-2. How many elements are there in the following arrays?
(a) A(5) character string
(b) B(4,5) integer
(c) C(3,2,3) real

6-3. Declare a single array to hold the student number and three exam scores for every student in a class. Assume there are 600 students in the class.

6-4. Declare an array to hold the numbers and names of textbooks available in the bookstore. Assume there are 1224 different textbooks. How will the book numbers have to be stored?

6-5. Declare a two-dimensional array with Rows 1–130, Columns 6–9. Let the name of this array be WAGES and let its type be fixed decimal (8.2).

problems for computer solution

The general learning objectives for this set of problems are

☐ to gain a problem-solving experience involving iterative cal-culations;

☐ to gain experience with one-dimensional arrays; and

☐ to learn the use of an anticipated end-of-data condition.

Additional objectives pertaining to a specific problem are stated as a preface to that problem. The languages and student disciplines for which a problem has been successfully used are given at the end of each problem.

136

CS 6-1. Develop a table that illustrates the effect of both simple interest and compound interest upon a savings account. For simple interest use the equation $b = p(1 + ni)$, where p is the value of the principal on deposit, n is the number of years that principal remains on deposit, i is the simple yearly interest rate, and b is the final balance. The compound interest upon a savings account is $b = p(1 + i)^n$, where the letters have the same meaning as given for simple interest.

Create an output table showing the account balance at the end of each year for each interest calculation scheme up to n years. Your input is p, i, and n.

The program should terminate a calculation if n exceeds 10, or for bad data; i.e., $i \leq 0$, $n < 1$. Lack of data should also terminate the program.

(COBOL, FORTRAN [general], PL/I [scientific])

CS 6-2. This problem involves reading data from cards, checking one of the data fields for accuracy, then writing accepted card images onto a transaction file that could serve as input to other programs. The input includes the personnel identification number—a 10-digit number. For convenience in updating the personnel file in a later program, the transaction file is to be created in increasing order by ID number. Your program should reject any cards that are not already in increasing order by ID number. A validity check of the ID should be performed. This can be done using a check-digit procedure in which one digit present is calculated by a predetermined rule from the values of the other digits. This procedure is called a **hash total.** Such a procedure will *not* detect all errors, but will reduce the chances of transcription error. There is *no* scheme that will detect *every* mistake that might occur.

For checking (creating) the tenth digit of the ID number, use the following method:
(a) Form a sum of the digits one, four, and seven.
(b) Form a sum of digits two, five, and eight; multiply by three.
(c) Form a sum of digits three, six, and nine; multiply by nine.
(d) Form a single-digit hash total of these three sums by summing them and retaining only the units portion of this new sum.
(e) The hash total digit should be the tenth digit of the ID.

This procedure should detect all single incorrect digit errors, all transposition errors, and most multiple-digit errors.

If the card record survives all of these tests, then enter it in the transaction file.

(COBOL [business], PL/I [business])

CS 6-3. An approximation to the value of the area under a curve (definite integral) can be obtained using the rectangular rule. As shown in the following diagram, the shaded area is used as an approximation to the area under the curve c between the points X_1 and X_4:

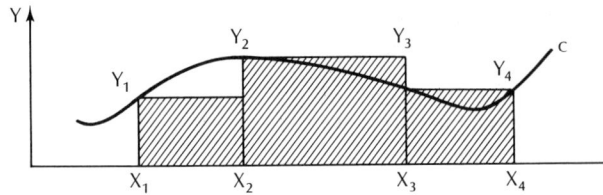

$$AREA_R = Y_1(X_2 - X_1) + Y_2(X_3 - X_2)$$
$$+ Y_3(X_4 - X_3) + \cdots + Y_{n-1}(X_n - X_{n-1})$$

Develop an algorithm that will read two lists of numbers $X_1 \ldots X_n$ and $Y_1 \ldots Y_n$ and store that in two arrays. This algorithm should be limited to 30 points. Form the area approximation using the rectangular rule.

(FORTRAN [architecture, engineering], PL/I [scientific])

CS 6-4. As Americans have become more aware of the need for energy conservation, more attention is being given to home insulation. In this problem you will be given information regarding an indefinite number of houses. It will consist of (a) the number of rooms in the house, (b) the measurement in perimeter (linear) feet of outside wall, and (c) the measurement in perimeter (linear) feet of inside wall for each room. You must then produce a report for the house reflecting (a) the amount of insulation required for the walls of the house, (b) the cost of all the insulation, and (c) the percent of the insulation of the house required for the outside wall (in perimeter feet). *Note:* The measurement for the inside walls will have been included twice. Use $2.10 per perimeter foot as the cost of insulation for the outside wall, and $1.70 per perimeter foot as the cost of insulation for the inside wall.

(FORTRAN [architecture])

CS 6-5. Calculate the value of the sine of x using a Maclaurin expansion for the values of the angle from 0° to 10° in steps of 0.5°. Compare the answer calculated with the value resulting from use of the built-in function SIN available on your computer system. Print the output degree, sine, SIN, degrees, sine, SIN in six columns across the page. That is, have 10 values of degree in each column. The expansion of sine Y is

$$F(y) = sine\ (y) = y - \frac{y^3}{3!} + \frac{y^5}{5!} - \frac{y^7}{7!} + \cdots$$

Use the nested polynomial for sine as follows:

$$\text{sine}(y) = \left(\left(\frac{y^2}{120} - \frac{1}{6}\right) \cdot y^2 + 1 \cdot y\right)$$

This formula expresses the angle in radians, not degrees, so you must convert x in degrees to y in radians in order to use it.

Remember $y/\pi = x/360$, or $y = 0.01745329x$, where y is the angle in radians and x is the angle in degrees.

Does the built-in function SIN require the angle in radians or degrees?

(FORTRAN [engineering, scientific], PL/I [scientific])

CS 6-6. The problem here is to compute a list of values, C, where

$$C_i = (A_i - B_i)^2 \qquad 1 \le i \le K$$

Next, calculate general coefficients

$$D_k = \sum_{i=1}^{k} (A_i - B_i)^2 \qquad 1 \le k \le K$$

and find the largest C_i. Print these in a table with four columns containing values of A, B, C, and D. Finally, print the largest C. Also report the value of K; that is, the number of data items read (the number of A's).

**(FORTRAN [architecture, engineering, scientific],
PL/I [scientific])**

CS 6-7. In processing a set of empirical data, a scientist is often faced with the problem of finding a curve that "fits" the data. One fit method, called least squares, produces an equation describing a curve through the data such that the sum of the squares of the deviations of the data points from the curve is a minimum:

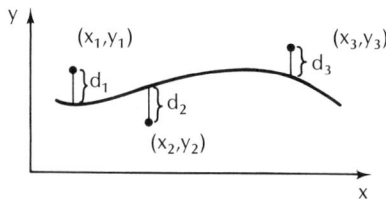

The set of points is $\{(x_1,y_1), (x_2,y_2) \cdots (x_n,y_n)\}$. If the equation is

$$f(x) = a_0 + a_1x + \cdots + a_jx^j$$

then

$$d_i = y_i - f(x_i) = y_i - (a_0 + a_1x_1 + \cdots + a_jx_i^j)$$

We wish to minimize

$$C = \sum_{i=1}^{n} d_i^2 = \sum_{i=1}^{n} [y_i - f(x_i)]^2$$

Do not be concerned with the minimization other than to note that a linear least squares fit (i.e., $f(x) = a_0 + a_1 x$) is obtained if a_0 and a_1 are calculated as follows:

$$a_0 = \frac{\Sigma y_i \Sigma x_0^2 - \Sigma(x_i y_i) \Sigma x_i}{n \Sigma x_i^2 - (\Sigma x_i)^2}$$

$$a_1 = \frac{n \Sigma(x_i y_i) - \Sigma x_i \Sigma y_i}{n \Sigma x_i^2 - (\Sigma x_i)^2}$$

Find a linear least squares fit for a set of 10 points; that is (x_i, y_i) pairs. Have your program write out a_0 and a_1.

It is not required, but it might be of interest to you to calculate each d_i. That is, calculate

$$d_i = y_i - f(x_i) = y_i - a_0 - a_1 x_i$$

(APL, BASIC, FORTRAN [engineering, scientific], PL/I [scientific])

CS 6-8.   You have been hired by a school district to write a program to analyze scores on class assignments. The program will read in a class and school identification card followed by an indeterminate number of student cards. Each student card contains the following information: student number (range 1–100) and score (range 0–100). You are to produce a report listing the student number, score, and difference from the average for each class. This implies that you must read all the data for a class and calculate an average before such a list can be printed.

(FORTRAN [general], PL/I [business, general])

CS 6-9.   Students from many high schools take college entrance examinations. The superintendent of schools in your county wishes a summary report of the scores of people from each high school in his or her district who took the examination.

The input will contain three pieces of information: a student identification number, a high school identification number, and a student's examination score.

There are three high schools and no more than 10 people in each school who took the examination. The input is presorted by high school. The superintendent wishes a listing by high school of all scores identified by student number. Following this, detailed listings are used to obtain the high score for each school, the average score for each

140

school, and the low score for each school. Create well-labeled output for the superintendent to use.

(FORTRAN [general], PL/I [general])

CS 6-10. The old Christmas carol, "A Partridge in a Pear Tree," has a very simple melody and repetitious verses. Each new verse adds just one line to the refrain (see Figure CS 6-10). The longest verse, the last, has a total of 14 lines, of which 3 have been repeated in every verse. Each new line added to the refrain is repeated until the end of the song. Notice that the first line of each verse changes. Read and store a minimum of information and then print a copy of the song. As a challenge, you may wish to print it exactly as it appears in Figure CS 6-10.

A PARTRIDGE IN A PEAR TREE

ON THE FIRST DAY OF CHRISTMAS
MY TRUE LOVE SENT TO ME
A PARTRIDGE IN A PEAR TREE.

ON THE SECOND DAY OF CHRISTMAS
MY TRUE LOVE SENT TO ME
  TWO TURTLEDOVES, AND
A PARTRIDGE IN A PEAR TREE.

ON THE THIRD DAY OF CHRISTMAS
MY TRUE LOVE SENT TO ME
  THREE FRENCH HENS,
  TWO TURTLEDOVES, AND
A PARTRIDGE IN A PEAR TREE.

ON THE FOURTH DAY OF CHRISTMAS
MY TRUE LOVE SENT TO ME
  FOUR CALLING BIRDS,
  THREE FRENCH HENS,
  TWO TURTLEDOVES, AND
A PARTRIDGE IN A PEAR TREE.

ON THE FIFTH DAY OF CHRISTMAS
MY TRUE LOVE SENT TO ME
  FIVE GOLD RINGS,
  FOUR CALLING BIRDS,
  THREE FRENCH HENS,
  TWO TURTLEDOVES, AND
A PARTRIDGE IN A PEAR TREE.

ON THE SIXTH DAY OF CHRISTMAS
MY TRUE LOVE SENT TO ME
  SIX GEESE A-LAYING,
  FIVE GOLD RINGS,
  FOUR CALLING BIRDS,
  THREE FRENCH HENS,
  TWO TURTLEDOVES, AND
A PARTRIDGE IN A PEAR TREE.

ON THE SEVENTH DAY OF CHRISTMAS
MY TRUE LOVE SENT TO ME
  SEVEN SWANS A-SWIMMING,
  SIX GEESE A-LAYING,
  FIVE GOLD RINGS,
  FOUR CALLING BIRDS,
  THREE FRENCH HENS,
  TWO TURTLEDOVES, AND
A PARTRIDGE IN A PEAR TREE.

ON THE EIGHTH DAY OF CHRISTMAS
MY TRUE LOVE SENT TO ME
  EIGHT MAIDS A-MILKING,
  SEVEN SWANS A-SWIMMING,
  SIX GEESE A-LAYING,
  FIVE GOLD RINGS,
  FOUR CALLING BIRDS,
  THREE FRENCH HENS,
  TWO TURTLEDOVES, AND
A PARTRIDGE IN A PEAR TREE.

ON THE NINTH DAY OF CHRISTMAS
MY TRUE LOVE SENT TO ME
  NINE DRUMMERS DRUMMING,
  EIGHT MAIDS A-MILKING,
  SEVEN SWANS A-SWIMMING,
  SIX GEESE A-LAYING,
  FIVE GOLD RINGS,
  FOUR CALLING BIRDS,
  THREE FRENCH HENS,
  TWO TURTLEDOVES, AND
A PARTRIDGE IN A PEAR TREE.

ON THE TENTH DAY OF CHRISTMAS
MY TRUE LOVE SENT TO ME
  TEN PIPERS PIPING,
  NINE DRUMMERS DRUMMING,
  EIGHT MAIDS A-MILKING,
  SEVEN SWANS A-SWIMMING,
  SIX GEESE A-LAYING,
  FIVE GOLD RINGS,
  FOUR CALLING BIRDS,
  THREE FRENCH HENS,
  TWO TURTLEDOVES, AND
A PARTRIDGE IN A PEAR TREE.

ON THE ELEVENTH DAY OF CHRISTMAS
MY TRUE LOVE SENT TO ME
  ELEVEN LADIES DANCING,
  TEN PIPERS PIPING,
  NINE DRUMMERS DRUMMING,
  EIGHT MAIDS A-MILKING,
  SEVEN SWANS A-SWIMMING,
  SIX GEESE A-LAYING,
  FIVE GOLD RINGS,
  FOUR CALLING BIRDS,
  THREE FRENCH HENS,
  TWO TURTLEDOVES, AND
A PARTRIDGE IN A PEAR TREE.

ON THE TWELFTH DAY OF CHRISTMAS
MY TRUE LOVE SENT TO ME
  TWELVE LORDS A-LEAPING,
  ELEVEN LADIES DANCING,
  TEN PIPERS PIPING,
  NINE DRUMMERS DRUMMING,
  EIGHT MAIDS A-MILKING,
  SEVEN SWANS A-SWIMMING,
  SIX GEESE A-LAYING,
  FIVE GOLD RINGS,
  FOUR CALLING BIRDS,
  THREE FRENCH HENS,
  TWO TURTLEDOVES, AND
A PARTRIDGE IN A PEAR TREE.

FIG. CS 6-10. Old English carol.

CS 6-11. Develop an algorithm that will determine the weighted average of a list of integers. The list is of length M. First form and print a frequency count of the occurrences of these integers. Then form the weighted average from

$$W_{AVG} = \frac{\sum_{N=1}^{M} N^2 \cdot FREQ_N}{\sum_{N=1}^{M} FREQ_N}$$

Where $FREQ_N$ stands for the frequency of occurrence of the Nth number.

(FORTRAN [scientific])

CS 6-12. Compute the summary statistics (arithmetic mean and standard deviation) for a set of data.

The mean is computed by adding up all the data values and dividing by the number of values.

The standard deviation is indicative of the spread of values from the mean. It is computed by finding the difference of each value from the mean, squaring the difference, adding up all the squared differences, dividing by n (the number of values), and then taking the square root.

Your program should read and count data values until there are no more. Assume there will never be more than 50 values (observations). After the end of data has been detected, calculate the mean and standard deviation.

(FORTRAN [scientific], PL/I [scientific])

CS 6-13. The corporation for which you currently work manufactures replacement parts for sailboats and sells only to retailers. There is need for a frequency distribution of the parts sold each month. The company is currently selling 23 parts, numbered consecutively from −1, 0, 1 to 21. Each time a sale is made, a record is created with the number of the part, the number of the parts sold, and the amount received. At the end of the month, you are to submit a report with the following information:

Part number	Number sold	Amount received	Average sale price
−1	2	10.50	5.25
0	200	200.00	1.00
⋮	⋮	⋮	⋮

The records used as input to this program are not ordered and are indeterminate in number.

(PL/I [business])

CS 6-14. You are fortunate enough to secure a summer job at the Wiggett Manufacturing Company. The head Wiggett desires a monthly sales report summary for his sales force and you are selected to program the solution.

The company employs 10 salespeople, denoted by numbers 0–9. Each time a salesperson sells some wiggets, a record is prepared showing salesperson number, number of wiggetts sold, and dollar amount received.

After reading all sales records for a month, prepare a report that gives the following information:

Person number	Wiggetts sold	Dollars volume	Average/Wiggett
0	10	100	10
1	10	50	5
⋮	⋮	⋮	⋮

Note that there are a varying number of sales records each month, and they cannot be assumed to be ordered. Check the input for validity of the values before accepting it as input.

(PL/I [business])

CS 6-15. You are to develop a procedure that summarizes a set of sales transactions. As output, your manager wants a report for each of the 10 products the company sells and the total sales for the previous day. The company has many distributors. Each day they phone the data-collection section with a report of item number and quantity sold on each item. The manager wishes a summary report giving the total sales of each item.

The data will come to you as a set of records of unknown length with each record containing item number and quantity sold for that item.

Furnish well-labeled reports to your manager to meet this requirement for data reduction. Be sure to check for validity of item number and quantity sold.

(PL/I [business])

CS 6-16. In large organizations, budgets are used for planning purposes. Often, they are used for control, as well, after they are implemented. By segmenting expense items into fixed cost and variable cost categories and providing the capability of producing a budget whenever projections of income vary from original projections, a useful management tool is created. This is often called a variable or flexible budget. Such a budget, although stated in specific dollar amounts, is based on percentage of some stated criterion (usually income); it can be adjusted as this criterion fluctuates.

Examples of fixed costs are salaries, standard telephone charges, and office supplies. Variable costs based upon sales volume might be commissions, freight, and long distance calls.

Given the following variable budget, write a program to extend it for annual sales of $100 000, $110 000, $120 000, $130 000, $140 000.

Print the budgets appropriately labeled across a printer page, as follows:

Goody Manufacturing Company Variable Budget—Sales Division Year Ended December 31, 1979		
Expense items	**Fixed per month**	**Variable per $100 sales**
Supervisory salaries	$54,000	
Office salaries	11,300	
Travel		$0.70
Telephone	520	0.07
Depreciation—Office equipment	100	
Supplies	400	0.61
Commissions		9.50
Freight	350	1.02
Advertising	4,600	
	$71,270	$11.90

(COBOL, PL/I [business])

CS 6-17. Write a program that summarizes beef price differences in a group of up to 50 cities. Each data input record contains the name of the city (maximum 25 characters) and the price per pound of hamburger at a local supermarket.

Your program should compute the average price of hamburger over all the cities and then print a table of cities, hamburger prices, and the amounts above or below the average price. (Use a positive number if price is above the average; a negative number if price is below the average.)

(PL/I [business])

CS 6-18. The general manager of a corporation producing transistor radios wants a program to summarize the productivity of his various manufacturing locations. An input record will contain the name of the company (maximum 25 characters), the branch number (a 3-digit number), and the productivity factor of the branch.

Your program should compute the average productivity rate for the corporation. Print a table showing (a) the name of the corporation, (b) its average productivity, (c) location number, (d) productivity factor for that location, and (e) the difference between the productivity of a location and the average productivity for the entire corporation.

Run your program for at least one corporation's data, but

design it so that it can process data pertaining to several.

(PL/I [business])

CS 6-19. Develop an algorithm that will determine all the prime numbers between 1 and N using the procedure known as the sieve of Eratosthenes. According to this procedure, first form a list of the numbers between 1 and N. Then eliminate all numbers evenly divisible by 2 that are greater than 2. Next eliminate all numbers evenly divisible by 3 that are greater than 3, and so on. When you have reached the number N/2 and have eliminated all those evenly divisible by N/2, the prime numbers are those left on the list. You might indicate element elimination by placing a zero in the list element and not printing zero elements. One concern you may have is how to determine if one number is evenly divisible by another. This can be accomplished using integer arithmetic. For example, A is evenly divisible by B if

$$A = I*B$$

where

$$I \leftarrow A/B$$

Having found all the primes in the list, find the twin primes—pairs of primes that differ by 2 (e.g., 11 and 13).

(PL/I [computer science])

CS 6-20. *Additional learning objective:* to use a two-dimensional array.

A mechanical method used to encode or decode messages relies on the transliteration of the plain message in a given order. You are to write a program which will either encode or decode messages. Consider messages of no more than 48 characters in length. Whether the message is to be encoded or decoded will depend upon the description; the number 1 will be used to indicate ENCODING and the number 0 will be used to indicate DECODING.

Use the following scheme:

(a) Consider the message as a matrix of 8 rows and 6 columns. For example,

THIS PROGRAM WILL EITHER ENCODE OR
DECODE MESSAGES

	1	2	3	4	5	6
1	T	H	I	S	P	R
2	O	G	R	A	M	W
3	I	L	L	E	I	T
4	H	E	R	E	N	C
5	O	D	E	O	R	D
6	E	C	O	D	E	M
7	E	S	S	A	G	E
8	S					

145

(b) The coding translation is to be performed by interchanging rows and columns in a specific order as agreed upon between the sender and receiver. In this problem assume that Column 1 and Column 4 will be interchanged, then Row 2 and Row 4, and finally Row 1 and Row 8.

The input will consist of the descriptor code and the message. The output will comprise an echo check of all input, the task (whether encode or decode), and the encoded or decoded message.

(PL/I [computer science])

CS 6-21. Write an algorithm that will estimate the population for future years. Then use this algorithm to estimate the population for 10 years into the future and print out a table containing columns of information for year, population, and deviation from mean population over the 10-year period.

To simulate population growth, start with an initial birthrate (BRATE), population (POPUL), year (YR), and deathrate (DRATE) of the previous year. This model keeps track of the population by adding the number of births and subtracting the number of deaths. The assumption is that the rates of death and birth are constant.

$$\text{Number of births in year}_i: \quad (NB_i) = \text{Population in year}_i \ (POPUL_i) \times BRATE$$
$$\text{Number of deaths in year}_i: \quad (ND_i) = POPUL_i \times DRATE$$
$$POPUL_{i+1} = POPUL_i + NB_i - ND_i$$

(PL/I, [general])

CS 6-22. You are to develop a program that accepts moves from two players in a tic-tac-toe game. This rather simpleminded approach to game playing on the computer will (a) display the progress of the game and (b) determine which player is the winner.

Create a 3 x 3 array to represent the game board. Then assign Player 1 an '0' with a point value of 1 and assign Player 2 an 'X' with a point value of −1. Print the 3 x 3 array with 'X' and 'O', but maintain the array with 0, 1, and −1. If any row, column, or diagonal has a value of 3 or −3, there is a winner—Player 1 or Player 2, respectively.

Read in a record from Player 1, indicating the move by board position; then read one in from Player 2. If a player attempts a move to an occupied square, note this and allow the other player the win.

When a player wins, note this. Then clear the board and start over. The players always alternate plays. If nine moves are accumulated, the game is a tie. Print the board after each pair of moves and at game termination.

(PL/I [general])

CS 6-23. Assume you are the registrar of a college. Each semester you want to know who all the students are, their grade-point averages, and their classes in school. The number of students varies from semester to semester. Available as input are student records in the following format:

Field contents	Record columns	Type
SSN	9	Numeric
Name	20	Alphabetic
Class	1	Numeric
(Semester info.)		
Hours ATT	(4,1)	Numeric
Hours passed	(4,1)	Numeric
Grade points	(4,1)	Numeric
(Cumulative record)		
Hours ATT	(4,1)	Numeric
Hours passed	(4,1)	Numeric
Grade point	(4,1)	Numeric

Read each student record and determine the semester and cumulative GPA, according to

$$GPA = \frac{\text{Grade points}}{\text{Hours passed and failed}}$$

Compute New Class Code:

Hours passed	Class
0–30	1
31–60	2
61–90	3
90–up	4

Print out the information for each student.

The registrar knows the president would like to have the following additional information:
(a) number of students by class;
(b) average semester GPA;
(c) average cumulative GPA; and
(d) number of hours passed and failed.
Conclude your output with this information.

(PL/I [business, general])

CS 6-24. An automobile speedometer is connected to one of the wheels of the car, converting the number of revolutions per second to a speed in miles per hour. Several factors can affect the accuracy of the speedometer reading including the tread wear, the atmospheric temperature, and the air pressure in the tire. Even if the speedometer correctly measures revolutions per second, its reading may differ from the actual speed of the car. Some highways have mileposts that allow a driver traveling at a fixed speedometer reading to determine the actual speed of the car.

Write a program that will calculate the number of seconds between mileposts for a car traveling at speeds of 10 to 95 miles in 5-mile increments. Information should be printed in four columns across a printer page as miles/hour, seconds, miles/hour, seconds. The first column should have readings for 10–55 miles per hour; the third column should contain readings for 60–95 miles per hour.

$$distance = velocity*time$$

Remember distance is one mile and time is in seconds, so velocity in miles per hour must be converted to miles per second in order to use it.

(PL/I [general])

CS 6-25. Create a program that will produce a monthly report showing the highest, lowest, and average temperature for a city. The input will consist of the name of the city, the name of the month, the number of days in the month, and the average daily temperatures for each day of the month.

The output should include monthly reports in the form of a table showing (a) the month, (b) the temperature corresponding to each day, (c) the deviation of that temperature from the monthly mean, and (d) the high, low, and mean temperatures for the month.

(PL/I [general]

CS 6-26. Create a sorting algorithm. One you might use searches a list for the smallest element, eliminates it from the list, and places it as the first element in a second list, and so forth. You are to assume any list the algorithm is to handle will contain numbers in the range −100 to 100.

Read a list of indeterminate length, but known to have fewer than 30 numbers. Sort this list. Echo your input, report the length of the list, and print the sorted list.

(PL/I [general, scientific])

CS 6-27. For our solar system, determine the following:
(a) average distance of each planet from the sun;
(b) an approximation of the number of miles each planet travels in one of its years. Approximate this by assuming they travel circular orbits of radius equal to their average distance from the sun. Then

$$circumference = 2 \times radius \times 3.1415926;$$

(c) average speed of each planet in miles per earth hour.

Display these three values for each planet in a neatly labeled table that includes the following information:

Planet	Perihelion (miles)	Aphelion (miles)	Duration of year (earth-days)
Mercury	29 000 000	43 000 000	88
Venus	67 000 000	68 000 000	225
Earth	91 000 000	95 000 000	365
Mars	128 000 000	155 000 000	687
Jupiter	460 000 000	510 000 000	4 333
Saturn	840 000 000	940 000 000	10 759
Uranus	1 700 000 000	1 870 000 000	30 685
Neptune	2 780 000 000	2 810 000 000	60 188
Pluto	2 770 000 000	4 600 000 000	90 700

The perihelion of a planet's orbit is the point at which the planet is closest to the sun; the aphelion is the point at which it is farthest from the sun. The average distance of a planet from the sun is the average of its aphelion and perihelion.

(PL/I [scientific])

CS 6-28. It is a common problem to determine if there is a correlation between different variables in an experiment. Develop a computerized procedure to calculate the coefficient of correlation R for two variables X and Y, both having N observed values. That is, there are N values of X and N values of Y. Read in N and then N records, each with a value of X and a value of Y.

The coefficient of correlation R is calculated as follows:

$$R = \frac{N*SUMXY - SUMX*SUMY}{\sqrt{(N*SUMXX - SUMX^2)\,(N*SUMYY - SUMY^2)}}$$

where

$$SUMX = \sum_{i=1}^{N} X_i$$

$$SUMY = \sum_{i=1}^{N} Y_i$$

$$SUMXX = \sum_{i=1}^{N} X_i^2$$

$$SUMYY = \sum_{i=1}^{N} Y_i^2$$

and

$$SUMXY = \sum_{i=1}^{N} X_i Y_i$$

149

Clearly label your output and allow the program to read multiple sets of data.

(FORTRAN [scientific], PL/I [general, scientific])

CS 6-29. Assume that you have been hired by the agronomy department to analyze data collected from sorghum plots this past summer. Four different varieties from 10 different locations are to be analyzed. Complete data will be provided from each location. That is, you will be given the yield per acre for each of the four varieties by location. You will want to store these data in separate arrays. After reading the data from all locations calculate the average yield for each variety.

Develop the following table as output:

ANALYSIS OF SORGHUM YIELDS				
LOCATION	VARIETY 1	VARIETY 2	VARIETY 3	VARIETY 4
1	XXX	XXX	XXX	XXX
2				
⋮				
10	XXX	XXX	XXX	XXX
AVE YIELD	XXX	XXX	XXX	XXX

(FORTRAN [agriculture, general, scientific])

CS 6-30. *Additional learning objective:* to use subscripted variables in a table-lookup situation.

Assume two files have been created. One is a salesperson file in the following format:

Record columns	Description	Format
11–14	Salesperson number	I(4)
16–17	Commission percent	F(2.2)
80	Code	I(1)

The second file is a product file in the following format:

Record columns	Description	Format
11–13	Product number	I(3)
14–17	Price (per unit)	F(4.2)
80	Code	I(1)

Next, a file in random order—a transaction file—is read and processed against tables created from these two files. The format for this file is

Record columns	Description	Format
11–14	Salesperson number	I(4)
18–20	Product number	I(3)
23–25	Quantity sold	I(3)

For each such card compute the salesperson's commission as

Quantity sold*Price*Commission percent

Output the salesperson's number, the product number, the quantity sold, the total value of the sale, and the salesperson's commission. Each transaction card represents a new output line, even though one salesperson may have several transaction cards.

There are fewer than 40 salespeople and fewer than 40 products. The salesperson numbers are to be in increasing order, but not consecutive. The product numbers are from 1 to 40, in strictly increasing order.

(COBOL, PL/I [business, general])

7

Loops

7.0 INTRODUCTION

One of the common justifications for using a computer is that it provides an economical way of performing repetitive tasks. Computers can carry out such tasks reliably and accurately. This chapter introduces some of the constructs that are used in programming languages to describe the process of repeating given tasks. Since the process of repeating a task involves a starting point to which we must return, it is natural to associate the notion of loops, or closed paths, with such processes. A series of algorithm steps that is executed repeatedly is called a **loop.** There are basically two situations in which a set of steps in an algorithm would have to be repeated:

1. when an algorithm performs the same operations on different data items; and
2. when the results of an algorithm are obtained by performing certain operations repeatedly, beginning with a given set of data and using results of a repetition(s) in further repetitions.

There are two distinct types of **loop structures** that are available in common programming languages. These are known as the indefinite and definite loops. An indefinite loop repeats a set of operations until a condition terminates the repetitions. On the other hand, a definite loop repeats a fixed number of times, the number being determined when the loop is initially encountered.

7.1 INDEFINITE LOOPS

An **indefinite loop** is one in which repetition is controlled by a condition that depends on variables internal to it. Two such loops are as follows:

Left flowchart: INPUT X → END-OF-FILE (Yes exits loop; No) → SUM ← SUM+X → loops back.

Right flowchart: X ← (X1−X2)/2. → Y1 ← A*X²+B*C+C → ABS(Y1) > 0.001 (No exits) → Yes → Y1 > 0. (Yes → X1 ← X) → No → X2 ← X → loops back.

$X \leftarrow (X1-X2)/2.$

$Y1 \leftarrow A*X^2+B*C+C$

$ABS(Y1) > 0.001$

$Y1 > 0.$

$X1 \leftarrow X$

$X2 \leftarrow X$

To represent such loops in flowchart language, we use a symbolism which closely parallels the constructs in programming languages for the indefinite form of a loop.

The indefinite loops shown in Figure 7.1-1 have the property that *before* a loop is executed, a test is conducted. If the condition (i.e., ABS(Y1) > .001) associated with this test has the "value" *true,* then the set of steps to be repeated (i.e., the loop body) is executed. Conversely, the loop body is not executed if the value of the condition is *false.* In the latter case, control is transferred to the statement that immediately follows the loop body.

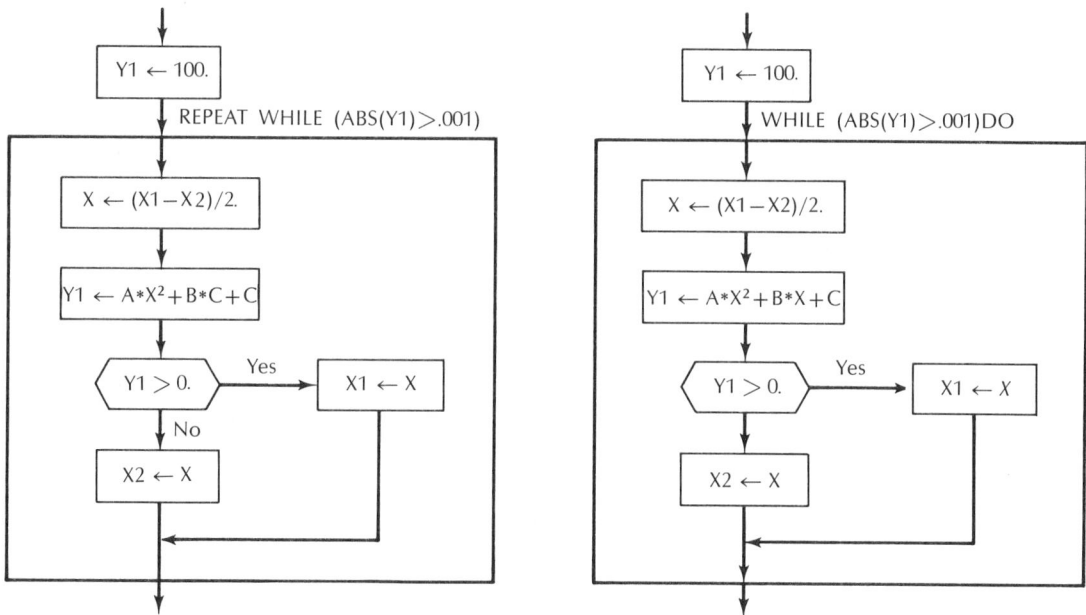

FIG. 7.1-1. Flowchart representations of indefinite loops (pretest of condition).

Another form of the indefinite loop tests for a condition for termination *after* a loop is executed (see Figure 7.1-2). Here, the loop body is executed at least once. On the other hand, the indefinite loops shown in Figure 7.1-1 are such that the loop body need not be executed at all, if so desired.

The indefinite loop that involves the END-OF-FILE condition can be represented in two ways, as summarized in Figure 7.1-3. A programming language normally allows only one of these constructions.

In closing, we remark that indefinite loops are used in the following situations:

1. where the number of times a series of statements is to be repeated cannot be predetermined; and

2. where it can be predetermined, but is simpler to check an exit condition than to compute a count.

155

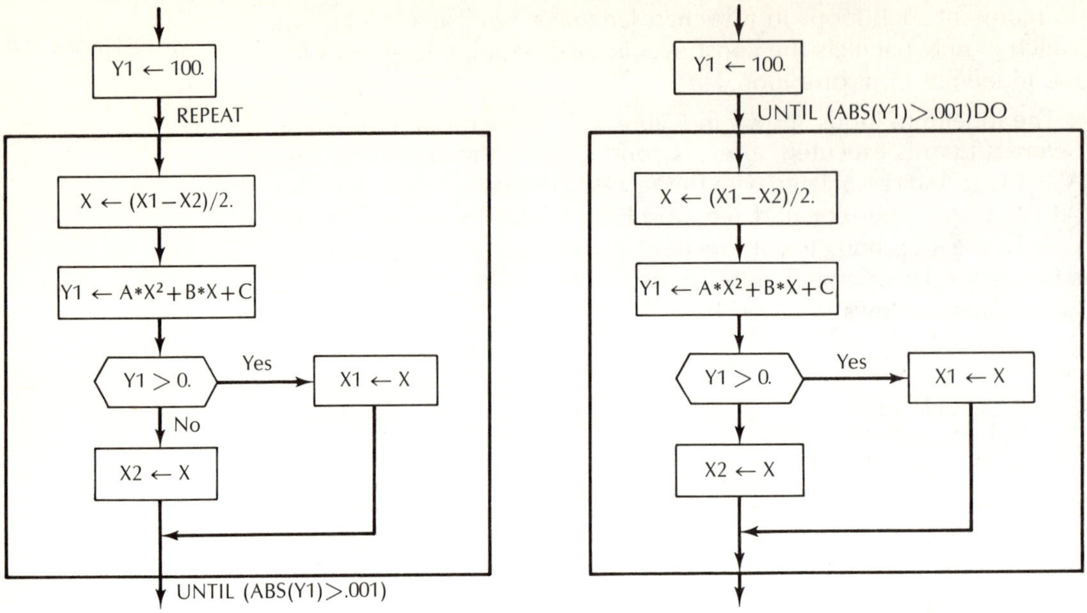

FIG. 7.1-2. Flowchart representation of indefinite loops (posttest of condition).

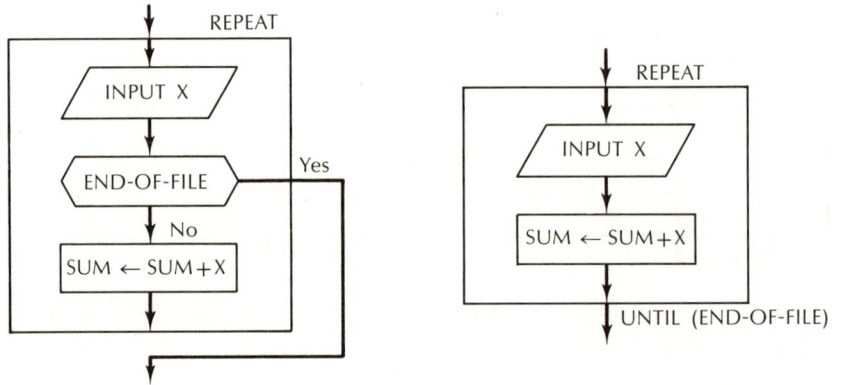

FIG. 7.1-3. Flowchart representation of indefinite loops (END-OF-FILE).

7.2 DEFINITE LOOPS

A **definite loop** is used when the number of repetitions can be predetermined, and can be conveniently used to manipulate elements of an array. There are basically two types of definite loops available in computer programming languages (see Figure 7.2-1). A programming language will normally provide one form or the other, but not both.

Before summarizing some of the advantages/disadvantages of these structures, it is instructive to identify the various **components** that constitute a loop structure. They are the following:

156

1. **Initialization:** provides starting values for the loop control variable. The term **loop control variable** refers to the variable used in a loop to control the looping process.

2. **Body:** consists of the steps to be repeated.

3. **Modification:** changes the value of the loop control variable prior to the next repetition of the loop.

4. **Test:** determines whether or not the loop is to be repeated through examination of the value of the loop control variable.

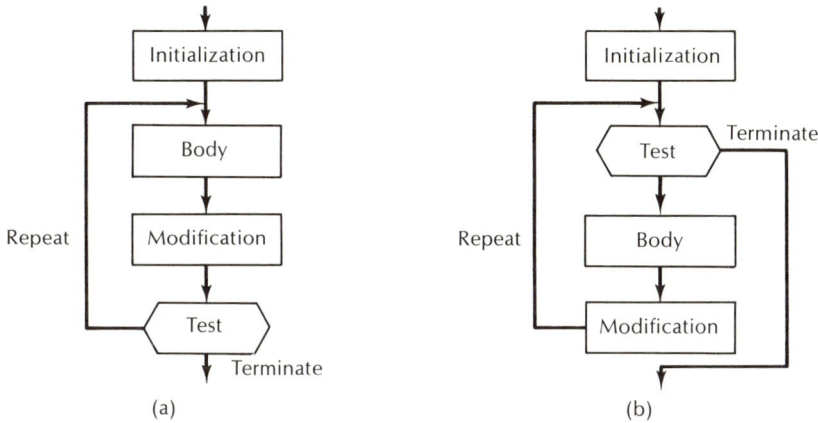

FIG. 7.2-1. Two types of definite loop structures.

In the interest of comparing the loop structures shown in Figure 7.2-1, we summarize as follows:

☐ The body of the loop structure in Figure 7.2-1(a) is always executed at least once, since the test for loop termination is not made until after the body of the loop has been executed. This may be a disadvantage in that the steps in the body cannot be skipped over. The loop structure in Figure 7.2-1(b) does not have this disadvantage, since the test component appears before the body. This loop structure, however, requires an extra test for any given number of loop repetitions.

☐ The modification step of the loop structure in Figure 7.2-1(a) is carried out even when the loop is not to be repeated.

Consider the illustrative example (Figure 7.2-2) where the four loop components have been identified. From the trace table shown in this figure it is apparent that the output that results via the OUTPUT statement in the body of the loop structure is as follows:

$$X = 1.$$
$$X = 9.$$
$$X = 25.$$

LOOP NOTATION

It is convenient to use concise notation to represent definite loop structures (see Figure 7.2-3). A general form of this notation is given by

157

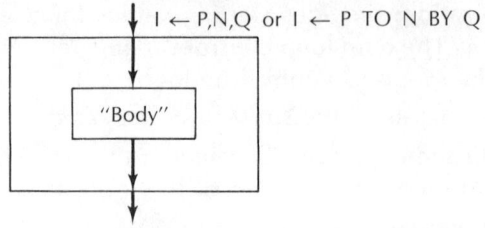

$$I \leftarrow P,N,Q \quad \text{or} \quad I \leftarrow P \text{ TO } N \text{ BY } Q$$

where P, N, and Q are positive or negative integers, constants, variables, or expressions.[1] The loop causes the value of the loop control variable, I, to be first set to the value of P. It is then incremented in steps of the value of Q, up to a value exceeding the value of N; that is,

$$I = P, (P+Q), (P+2Q), (P+3Q), \ldots, I_{max}, I_{final}$$

where

$$I_{max} \leq N \quad \text{and} \quad I_{final} > N$$

For each value of I, from P through I_{max}, the loop body is executed once. It should not be assumed that the value of I_{final} is available to the

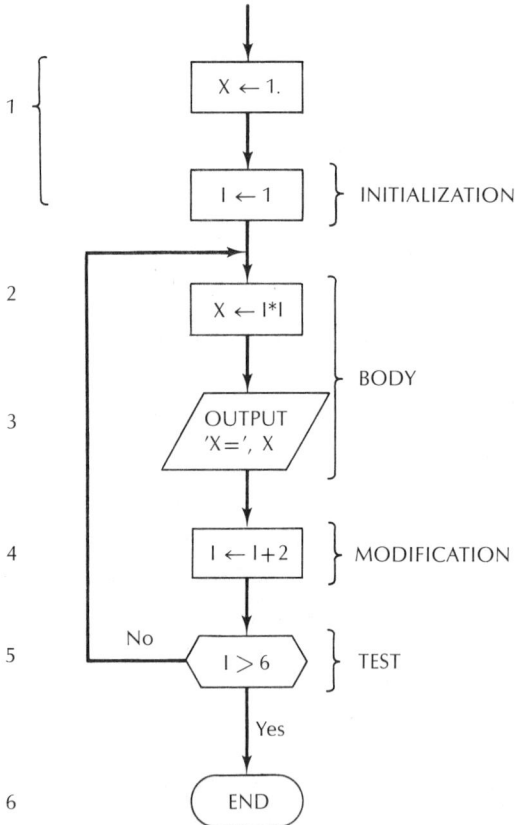

TRACE TABLE

Symbol #	I	X
1	1	1.
2		1.
3		
4	3	
5		
2		9.
3		
4	5	
5		
2		25.
3		
4	7	
5		
6		

I is the loop control variable of the loop. It takes the values 1,3,5,7

FIG. 7.2-2. A loop structure of the type shown in Figure 7.2-1(a).

[1] In certain programming languages such as FORTRAN, only positive integer values are allowed for P, N, and Q.

Note: These loop structures may be represented in flowchart language as shown below.

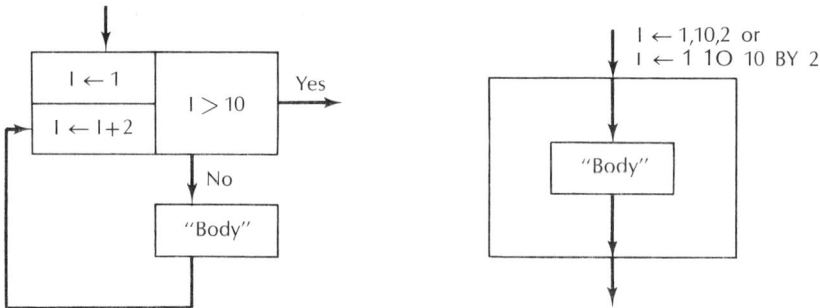

Comment: The initialization, modification, and test are all represented in one box, since only one statement is used to represent them in programming languages.

FIG. 7.2-3. Flowchart representations of definite loops.

programmer. To illustrate, a set of related examples are summarized in Table 7.2-1. From Examples 1 and 2 in this table, it is apparent that when the stepsize Q is equal to 1, there is no need to include its value in the corresponding statements I ← P,N,Q or I ← P TO N BY Q. In other words, the modification value of Q is set equal to 1 by default. Thus the statements

$$I \leftarrow P,N,1 \quad \text{and} \quad I \leftarrow P \text{ TO } N \text{ BY } 1$$

are respectively equivalent to

$$I \leftarrow P,N \quad \text{and} \quad I \leftarrow P \text{ TO } N.$$

The most common use of the definite loop is to manipulate the elements of arrays. In such cases, the loop variable takes integer values, and is treated as a subscript variable. Example 7.2-1 considers the use of a subscript variable in connection with a one-dimensional array.

159

TABLE 7.2-1. Examples related to loop notation

Example #	Notation	Values taken by the loop control variable
1.	I←1,10,1 I←1 TO 10 BY 1	1,2,3,4,5,6,7,8,9,10,11 loop body is executed for each of these
2.	I←1,10 I←1 TO 10	1,2,3,4,5,6,7,8,9,10,11 loop body is executed for each of these
3.	J←1,13,3 or J←1 TO 13 BY 3	1,4,7,10,13,16 loop body is executed for each of these
4.	J←1,14,3 or J←1 TO 14 BY 3	1,4,7,10,13,16 loop body is executed for each of these
5.	K←3,16,2 or K←3 TO 16 BY 2	3,5,7,9,11,13,15,17 loop body is executed for each of these
6.	KK← −3,3 or KK← −3 TO 3	−3,−2,−1,0,1,2,3,4 loop body is executed for each of these
7.	L← −5,4,2 or L← −5 TO 4 BY 2	−5,−3,−1,1,3,5 loop body is executed for each of these

Example 7.2-1 Let MONTH denote a one-dimensional array of size 12 which stores the values 'JAN', 'FEB', 'MAR', . . . , 'DEC'. That is, each month is represented by its first three letters.

(a) If MONTH is assumed to be in memory, what is the output if the following loop is executed?

```
                I ← 1,12,3 or I ← 1 TO 12 BY 3
┌───────────────────┼──────────────────┐
│                   │                  │
│   ┌───────────────────────────┐      │      ┌─────────────────────────────────┐
│   │ MONTH(I) ← MONTH(I+1)      │------│------│ MONTH(12) character string (4)  │
│   └───────────────────────────┘      │      └─────────────────────────────────┘
│                   │                  │
│          ┌────────────────────┐      │
│         /  OUTPUT MONTH(I)    /       │
│        └────────────────────┘         │
│                   │                  │
└───────────────────┼──────────────────┘
                    │
```

(b) What are the contents of MONTH(1), MONTH(2), . . . , MONTH(12) after the loop is executed?

Solution:

(a) 'JAN', 'FEB', 'MAR', etc. represent literal fields; hence, they are left-justified when stored in memory (see Section 5.1). Since each location in memory can store four characters (see annotation box), the elements of the array MONTH appear in memory as follows:

$$\begin{array}{ll} \text{MONTH(1):} & \text{JAN}_\wedge \\ \text{MONTH(2):} & \text{FEB}_\wedge \\ \text{MONTH(3):} & \text{MAR}_\wedge \\ & \cdots \\ \text{MONTH(10):} & \text{OCT}_\wedge \\ \text{MONTH(11):} & \text{NOV}_\wedge \\ \text{MONTH(12):} & \text{DEC}_\wedge \end{array}$$

We observe that the loop control variable, I, takes the values 1,4,7,10, and 13. Execution of the statement contained in the body of the loop yields the following results:

For I = 1: MONTH(1) ← MONTH(2) causes the contents of MONTH(1) to be replaced by the contents of MONTH(2); namely, 'FEB$_\wedge$'. Thus the statement

OUTPUT MONTH(1)

yields the output

FEB$_\wedge$

Similarly, for I = 4,7, and 10 we obtain the following results:

For I = 4: MONTH(4) ← MONTH(5)
and OUTPUT MONTH(4) yields

MAY$_\wedge$

161

For I = 7:   MONTH(7) ← MONTH(8)
               and OUTPUT MONTH(7) yields

$$AUG_\wedge$$

For I = 10:   MONTH(10) ← MONTH(11)
               and OUTPUT MONTH(10) yields

$$NOV_\wedge$$

and then I = 13 and the loop exits.

Thus, the overall output is as follows:

$$FEB_\wedge$$
$$MAY_\wedge$$
$$AUG_\wedge$$
$$NOV_\wedge$$

(b) The preceding discussion implies that the final contents of the array elements MONTH(1), MONTH(2), ..., MONTH(10), MONTH(11), and MONTH(12) are

MONTH(1):	$FEB_\wedge$
MONTH(2):	$FEB_\wedge$
MONTH(3):	$MAR_\wedge$
MONTH(4):	$MAY_\wedge$
MONTH(5):	$MAY_\wedge$
MONTH(6):	$JUN_\wedge$
MONTH(7):	$AUG_\wedge$
MONTH(8):	$AUG_\wedge$
MONTH(9):	$SEP_\wedge$
MONTH(10):	$NOV_\wedge$
MONTH(11):	$NOV_\wedge$
MONTH(12):	$DEC_\wedge$

Comment: Definite loop structures are commonly referred to as "DO loops" in some programming languages, since they are implemented via the DO statement, the general form of which is

DO loop control variable ← initial, final, modification

or

DO loop control variable ← initial TO final BY modification

where the loop control variable is also referred to as the **index** or **counter** of the loop. If we denote the loop control variable (index, counter) by I, then the preceding statement causes the following set of actions to be taken:

☐ Set the initial value of I to be the value of initial.

☐ Increment the value of I by the value of modification each time the loop body is executed.

☐ Execute the loop until I has a value greater than the value of final.

In the case of input/output operations, loop structures can be used effectively in conjunction with the INPUT and OUTPUT statements. The corresponding loop structures are referred to as **implied (or implicit) loops.** Examples of implied loops are summarized in Table 7.3-1, and are self-explanatory.

TABLE 7.3-1. Examples of implied loops

Example #	Notation	Action taken
1.	INPUT ABC - - - - - ABC(10)	Causes 10 values to be read from the input data stream and stored in the memory locations assigned to the array elements ABC(1), ABC(2),...,ABC(10).
2.	INPUT (ABC(I),I ← 1,10) - - - - - ABC(10)	Causes same action as in Example 1.
3.	OUTPUT (ABC(JJ),JJ ← 1,10,2) - - - - - ABC(10)	Causes the output of the current values associated with ABC(1), ABC(3), ABC(5), ABC(7), and ABC(9).
4.	OUTPUT ((A(K,J),I ← 1,2), J ← 1,2) - - - - - A(2,2)	Causes the output of the current values associated with A(I,J) in the following order: A(1,1) A(2,1) A(1,2) A(2,2)

Loops can be "nested" in one another. The term **nested loop** refers to a loop that is contained in another loop, as illustrated below:

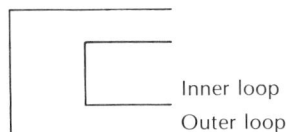

Inner loop
Outer loop

There are two important properties associated with nested loops:

1. Any number of nestings is permissible.
2. A loop nested within another loop must be completely contained within the loop in which it is nested. Thus, for example,

163

a nesting corresponding to the following illustration is not allowed:

We now present two examples of nested loops.

Example 7.4-1 Develop the trace table corresponding to the flowchart given in Figure 7.4-1(a) which consists of a nesting of two loops. Assume that the input data stream is 1,2,−3,5.

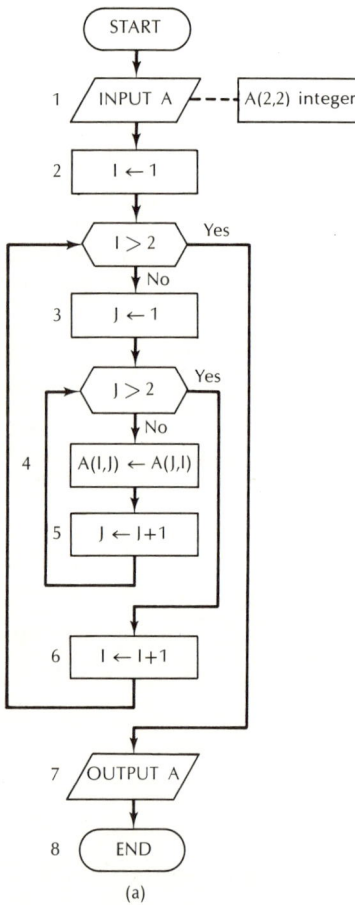

TRACE TABLE

Symbol #	I	J	A(1,1)	A(1,2)	A(2,1)	A(2,2)
1	Arbitrary	Arbitrary	1	2	−3	5
2	1					
3		1				
4			1			
5		2				
4				−3		
5		3				
6	2					
3		1				
4					−3	
5		2				
4						5
5		3				
6	3					
7						
8						

(b)

FIG. 7.4-1. Flowchart and trace table pertaining to Example 7.4-1.

Solution: The INPUT statement in Symbol 1 causes the following values to be stored in the memory locations assigned to the array A[2]:

$$A(1,1) = 1, \; A(1,2) = 2, \; A(2,1) = -3, \; A(2,2) = 5$$

Using these values for the array elements, we obtain the trace table shown in Figure 7.4-1(b), from which it is apparent that the OUTPUT A statement in Symbol 7 of the flowchart yields

$$\begin{matrix} 1 & -3 \\ -3 & 5 \end{matrix}$$

Example 7.4-2

(a) Develop the trace table corresponding to the flowchart given in Fig. 7.4-2(a) for the input data stream 1, 2, −3, 5.

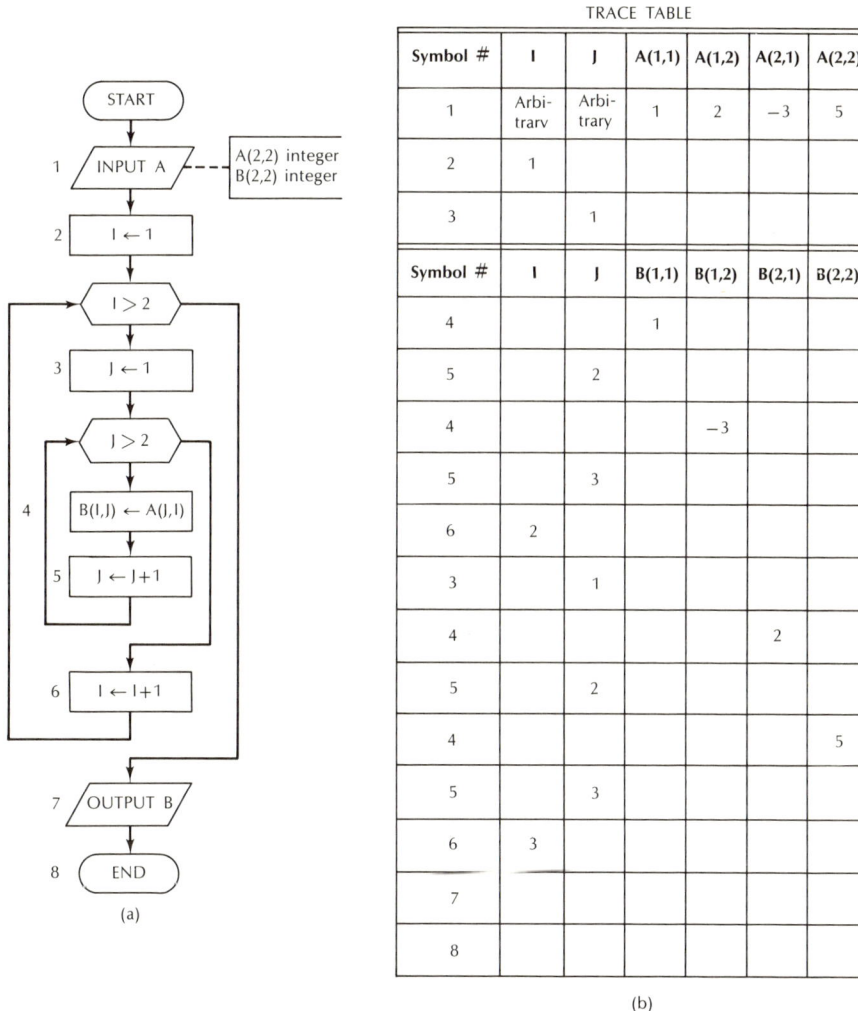

Flowchart (a):

START
1 — INPUT A — A(2,2) integer / B(2,2) integer
2 — I ← 1
I > 2
3 — J ← 1
J > 2
4 — B(I,J) ← A(J,I)
5 — J ← J+1
6 — I ← I+1
7 — OUTPUT B
8 — END

TRACE TABLE

Symbol #	I	J	A(1,1)	A(1,2)	A(2,1)	A(2,2)
1	Arbitrary	Arbitrary	1	2	−3	5
2	1					
3		1				

Symbol #	I	J	B(1,1)	B(1,2)	B(2,1)	B(2,2)
4			1			
5		2				
4				−3		
5		3				
6	2					
3		1				
4					2	
5		2				
4						5
5		3				
6	3					
7						
8						

(b)

FIG. 7.4-2. Flowchart and trace table pertaining to Example 7.4-2.

---

[2] Recall that our flowchart language associates the input data stream with the array A on a row-by-row basis.

(b) What is the relationship between the input and output arrays which are A and B, respectively?

Solution:

(a) As in the case of Example 7.4-1, the input stream causes the elements of A to take the values $A(1,1) = 1$, $A(1,2) = 2$, $A(2,1) = -3$, and $A(2,2) = 5$. The corresponding trace table is shown in Figure 7.4-2(b).

(b) As a consequence of the OUTPUT B statement in Symbol 7, we obtain the output array as

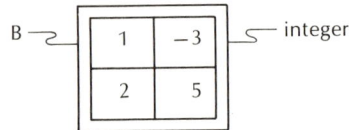

We know that the input array is

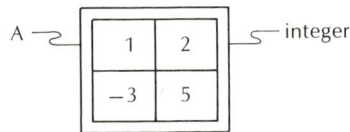

Comparing A and B, we observe that the relation between the input and output arrays is given by

$$B = A^T$$

where T denotes matrix transpose.

Comment: The algorithm in the preceding example can be used to transpose any given array A consisting of N rows and N columns by merely changing the final values of the loop variables I and J from 2 to the constant N. In addition, the declaration statement would have to be changed to read as follows:

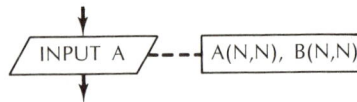

NESTED LOOP NOTATION

Nested loops, like conventional loops, can also be represented concisely using an appropriate notation. For example, suppose we wish to compute the sum of two arrays A and B and store the result in an array called C. That is, we wish to compute

$$C \leftarrow A + B \tag{7.4-1}$$

For purposes of discussion, let

Then the desired summation indicated in Equation (7.4-1) can be computed by means of a nested loop, represented by the notation

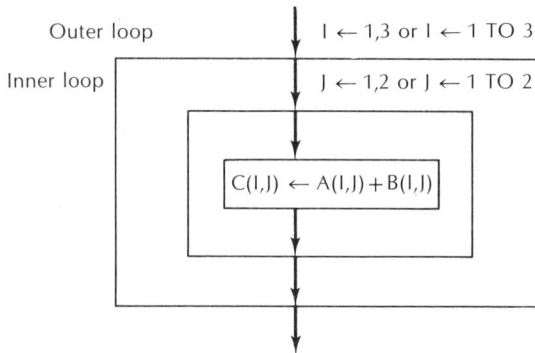

The computations implied by this notation can be listed systematically by following a simple rule; that is,

For each value taken by the outer loop control variable, the inner loop control variable takes all its values.

When we apply this rule to the case in question, we arrive at the following sets of computations:

1. $I = 1$
   — — — — —
   $J = 1$: $C(1,1) \leftarrow A(1,1) + B(1,1) \Rightarrow C(1,1) \leftarrow 0 + 6 = 6$
   $J = 2$: $C(1,2) \leftarrow A(1,2) + B(1,2) \Rightarrow C(1,2) \leftarrow 0 + 7 = 7$

2. $I = 2$
   — — — — —
   $J = 1$: $C(2,1) \leftarrow A(2,1) + B(2,1) \Rightarrow C(2,1) \leftarrow 1 + 2 = 3$
   $J = 2$: $C(2,2) \leftarrow A(2,2) + B(2,2) \Rightarrow C(2,2) \leftarrow 1 + 3 = 4$

3. $I = 3$
   — — — — —
   $J = 1$: $C(3,1) \leftarrow A(3,1) + B(3,1) \Rightarrow C(3,1) \leftarrow 1 + 4 = 5$
   $J = 2$: $C(3,2) \leftarrow A(3,2) + B(3,2) \Rightarrow C(3,2) \leftarrow -1 + 5 = 4$

where $\Rightarrow$ means "implies that."

The values of $C(I,J)$ given above can be represented in the form of an array as

167

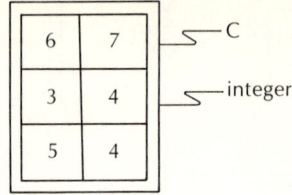

Comment: It is important to note that within the body of a loop there should be no statement that causes the value of the loop control variable (index) to change. For example, consider the following flow-chart representation of a certain algorithm that involves 3 two-dimensional arrays A, B, and C, with 4 rows and 5 columns each:

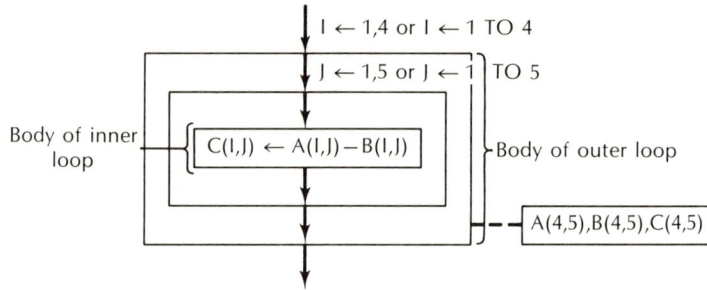

With reference to these loops, we observe that I and J are the outer and inner loop control variables, respectively. Thus, statements such as

$$I \leftarrow I + 3$$
$$I \leftarrow I * I$$
$$J \leftarrow J - 10$$

cannot be used. However, the loop variables can be used in other calculations, such as

$$K \leftarrow I + 3$$
$$A(I+1, J+3) \leftarrow 16$$

As indicated in Table 7.3-1, implied (or implicit) loops can also be used with two-dimensional arrays. For example, let us assume that the input data stream is given by

$$6,7,2,3,4,5$$

Then the statement

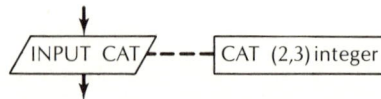

causes the following values to be assigned to the elements of CAT:

CAT(1,1) = 6, CAT(1,2) = 7, CAT(1,3) = 2; i.e., Row 1 of CAT
CAT(2,1) = 3, CAT(2,2) = 4, CAT(2,3) = 5; i.e., Row 2 of CAT

Alternatively, we can obtain the same result by means of the following statement:

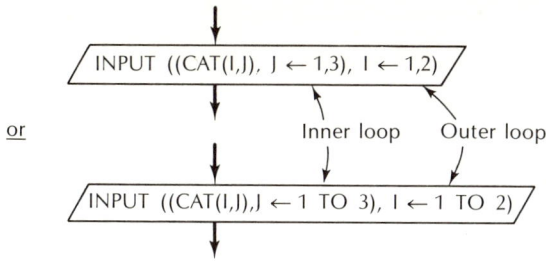

or

```
┌─────────────────────────────────┐
/ INPUT ((CAT(I,J), J ← 1,3), I ← 1,2) /
└─────────────────────────────────┘
            Inner loop    Outer loop

/ INPUT ((CAT(I,J),J ← 1 TO 3), I ← 1 TO 2) /
```

This is because the sequence of values taken by the loop control variables I and J are as follows:

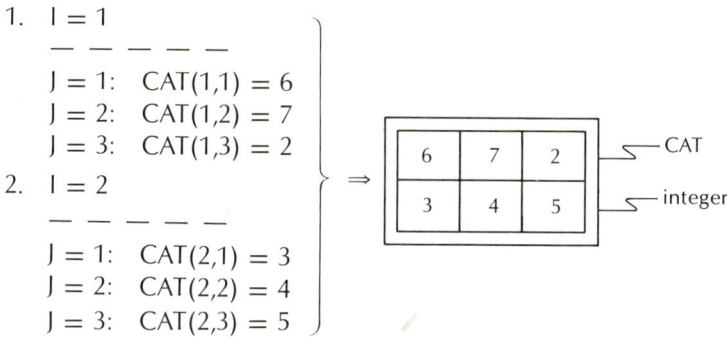

1. I = 1
    ─ ─ ─ ─ ─
    J = 1: CAT(1,1) = 6
    J = 2: CAT(1,2) = 7
    J = 3: CAT(1,3) = 2
2. I = 2
    ─ ─ ─ ─ ─
    J = 1: CAT(2,1) = 3
    J = 2: CAT(2,2) = 4
    J = 3: CAT(2,3) = 5

$\Rightarrow$

6	7	2
3	4	5

CAT
integer

ARRAY MANIPULATION USING DEFINITE LOOPS 7.5

The three additional examples in this section emphasize that definite loops can be used as a powerful tool in the manipulation of array information. The examples are restricted to one- and two-dimensional arrays, but we must keep in mind that this approach can be readily extended to arrays of higher dimensions.

Example 7.5-1 Develop a structured algorithm that enables one to read, store, and average a set of 10 test scores. Print the average score, the scores, and their difference from their average. Flowchart your solution.

Solution: The desired structured algorithm is as follows:

1. Read and sum data.
    1.1 Initialize SUM to zero.
    1.2 Repeat Steps 1.3 and 1.4 ten times.
    1.3 Read a value and store it in an array element.
    1.4 Add it to SUM.
2. Calculate and print average.

169

3. Print table of output.
    3.1   Repeat Steps 3.2 and 3.3 ten times.
    3.2   Form difference of score and average.
    3.3   Print score and difference.

See Figure 7.5-1 for a flowchart representation of this solution.

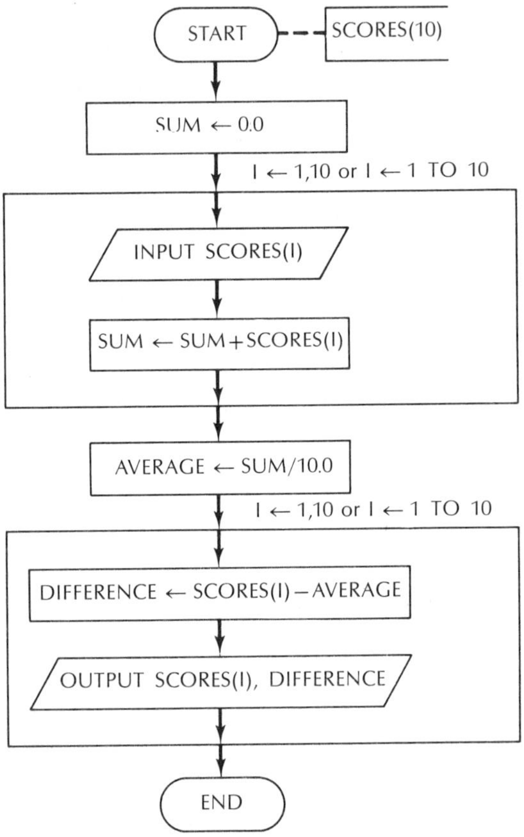

FIG. 7.5-1.   Flowchart pertaining to Example 7.5-1.

Example 7.5-2 Modify the algorithm of Example 7.5-1 for the case in which the number of students is unknown, but never exceeds 100. Flowchart your solution.

Solution: The desired modification is quite straightforward (see Figure 7.5-2). Note that this flowchart solution consists of a definite loop as well as an indefinite loop.

Example 7.5-3 Read and store the semester grades for a maximum of 100 students in a class in which five examinations were given. Print these scores along with the semester average for each student, and the difference of that average from the overall class average. Use an array SCORE (100,5). Flowchart your solution, which is to be in the form of a structured algorithm.

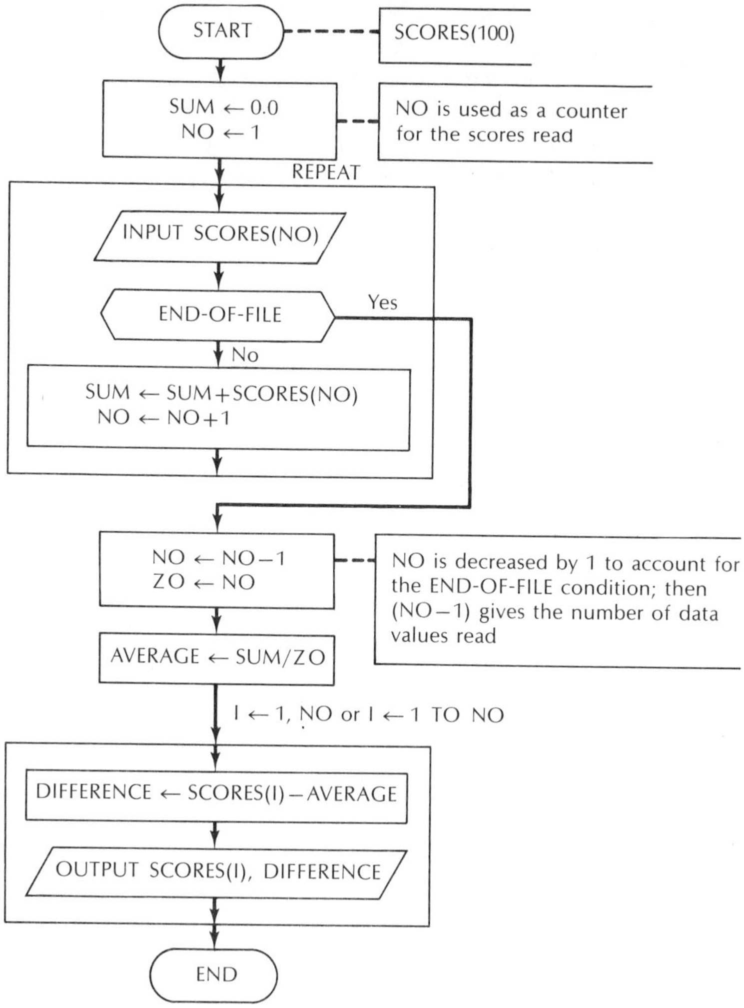

```
        ( START )------    SCORES(100)

       ┌──────────────┐    ┌─────────────────────────┐
       │  SUM ← 0.0   │----│ NO is used as a counter │
       │  NO ← 1      │    │ for the scores read     │
       └──────────────┘    └─────────────────────────┘
                 │      REPEAT
       ┌─────────┼────────────────────────────┐
       │  / INPUT SCORES(NO) /                 │
       │         │                             │
       │   < END-OF-FILE >──────Yes            │
       │         │                             │
       │        No                             │
       │  ┌──────────────────────────┐         │
       │  │ SUM ← SUM+SCORES(NO)     │         │
       │  │ NO ← NO+1                │         │
       │  └──────────────────────────┘         │
       └──────────────────────────────────────┘

       ┌──────────────┐    ┌─────────────────────────────┐
       │  NO ← NO−1   │----│ NO is decreased by 1 to     │
       │  ZO ← NO     │    │ account for the END-OF-FILE │
       └──────────────┘    │ condition; then (NO−1)      │
                 │          │ gives the number of data    │
       ┌──────────────────┐│ values read                 │
       │ AVERAGE ← SUM/ZO ││                             │
       └──────────────────┘└─────────────────────────────┘
                 │
                 │  I ← 1, NO or I ← 1 TO NO
       ┌─────────┼──────────────────────────────┐
       │ ┌──────────────────────────────────┐   │
       │ │ DIFFERENCE ← SCORES(I) − AVERAGE │   │
       │ └──────────────────────────────────┘   │
       │ / OUTPUT SCORES(I), DIFFERENCE /        │
       └────────────────────────────────────────┘
                 │
             ( END )
```

FIG. 7.5-2. Flowchart pertaining to Example 7.5-2.

Solution: It can be verified that the following structured algorithm provides a solution:

1. Read and store the five scores for each student in the array SCORE(100,5). Sum all scores to calculate the overall average.
    1.1 Repeat Step 1.2 until there is no more data; count the number of students in COUNT.
    1.2 Read and sum scores on a student.
        1.2.1 Repeat Steps 1.2.2 and 1.2.3 for a loop control variable, I, taking values 1 through 5.
        1.2.2 Read and store the score, SCORE(COUNT,I).
        1.2.3 Add SCORE(COUNT,I) to a sum.
2. Calculate average.
3. Calculate student semester average and its difference with respect to overall class average, and print.

171

3.1 Repeat Steps 3.2 through 3.6 for each student.
3.2 Initialize an accumulator SSUM to zero.
3.3 Repeat Step 3.4 for each score of a student.
3.4 Add a score to a student's SSUM.
3.5 Form student average.
3.6 Print student scores, student average, and difference from average.

See Figure 7.5-3 for a flowchart representation of this solution.

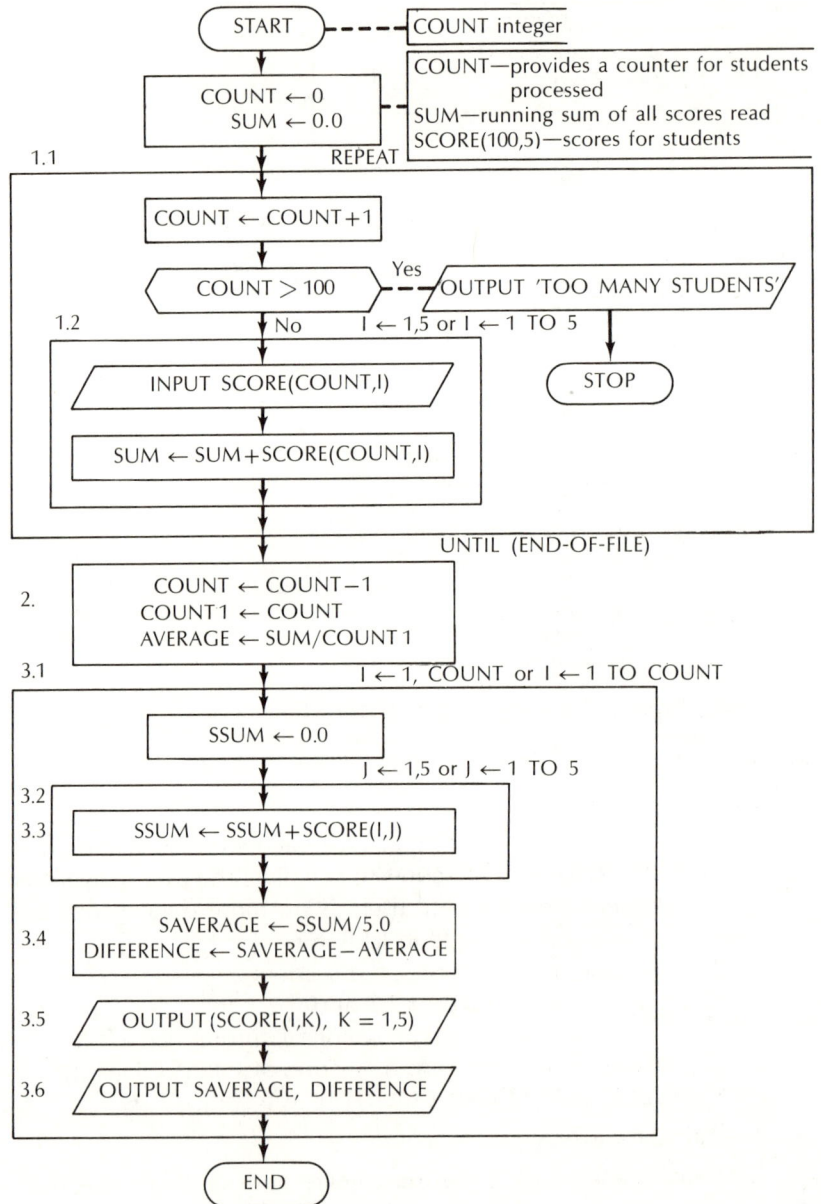

FIG. 7.5-3. Flowchart pertaining to Example 7.5-3.

Example 7.5-4 Given an array A consisting of four rows and six columns, show the set of instructions (in flowchart form) that will accomplish the following tasks:

(a) Set all the array elements in Rows 1 through 4 to zero.

(b) Set the array elements labeled with the symbol "X"as shown below to the numeric constant 99.9.

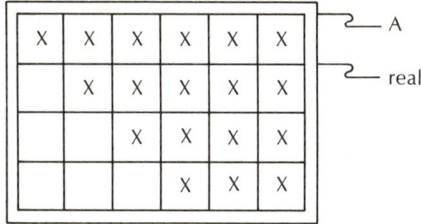

(c) Exchange the values of the array elements in Column 2 with those of the elements in Column 6.

Solution: See Figure 7.5-4 for the desired flowcharts. The reader should verify that these flowcharts do indeed accomplish the tasks as stated.

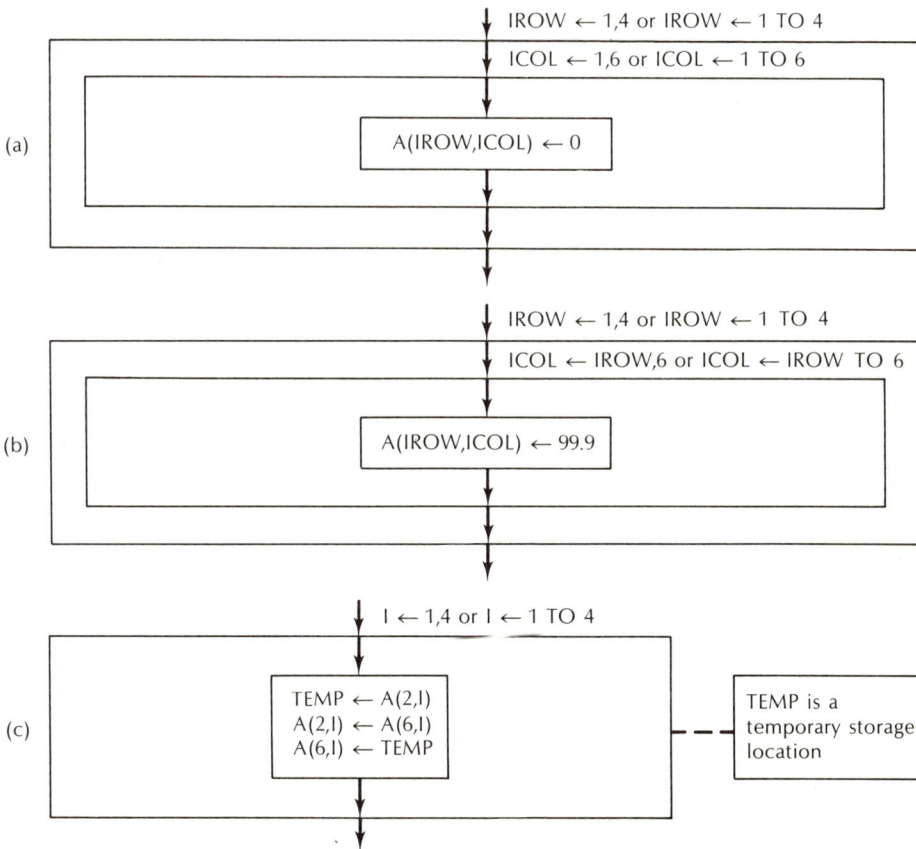

FIG. 7.5-4. Flowcharts pertaining to Example 7.5-4.

7.6 SUMMARY

In this chapter we discussed the process of repeatedly executing a sequence of statements by means of a class of constructs called loops. Two types of loops were considered; namely, definite and indefinite loops. Indefinite loops are used when the number of repetitions cannot be predetermined; definite loops are applicable when the number of repetitions is known before the loop is to be executed. It was demonstrated by means of examples that definite loops can be used effectively to manipulate array information.

problems

7-1. Consider the loop in Figure P 7-1 in which K is the loop control variable and the contents of the array ITEM are as follows:

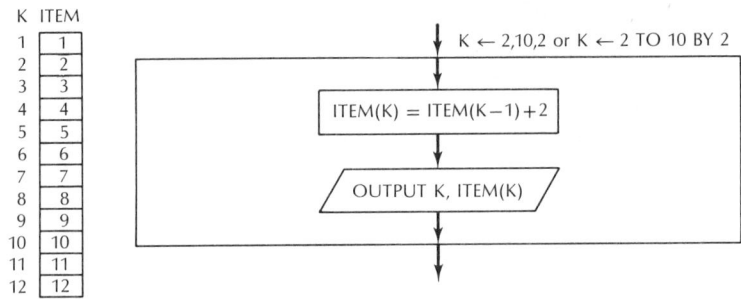

K	ITEM
1	1
2	2
3	3
4	4
5	5
6	6
7	7
8	8
9	9
10	10
11	11
12	12

$K \leftarrow 2,10,2$ or $K \leftarrow 2$ TO 10 BY 2

ITEM(K) = ITEM(K$-$1)$+$2

OUTPUT K, ITEM(K)

FIG. P 7-1.

(a) How many times is the preceding loop executed?
(b) What is the output?
(c) What are the final contents of ITEM?

7-2. Given:

$I \leftarrow 1,2$ or $I \leftarrow 1$ TO 2

$J \leftarrow 1,3$ or $J \leftarrow 1$ TO 3

OUTPUT I,J

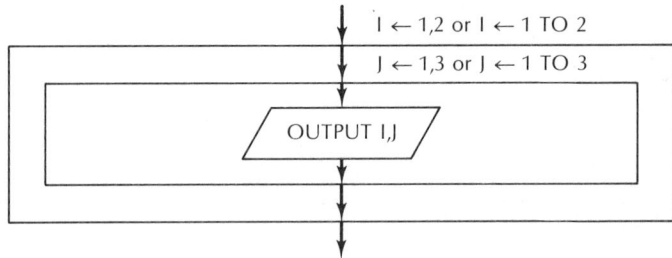

Assuming that each execution of the OUTPUT statement causes a new line to be written, find the output that results from the preceding nested loop.

7-3. Suppose MONTH is a one-dimensional array of size 12 which consists of the elements 'JAN', 'FEB', 'MAR', . . . , 'DEC'. That is, each MONTH is represented by its first three letters. If MONTH is assumed to be in memory, what is the output from the following loop?

174

$$I \leftarrow 1,11,4 \text{ or } I \leftarrow 1 \text{ TO } 11 \text{ BY } 4$$

MONTH(I) ← MONTH(I+1)

OUTPUT MONTH(I)

7-4.   Develop an algorithm that reads N test scores into an array called SCORE, and subsequently finds the average of these scores. Include an OUTPUT statement that outputs N and the average score. Assume that N does not exceed 300.

7-5.   Develop a flowchart using a definite loop and an array to read in five test scores, sum them, and print the sum. Include a decision to determine whether any score is greater than 100. Bonus questions allowed for more than 100 points per score, and the instructor wishes to set 100 as a maximum.

7-6.   Develop a structured algorithm and a flowchart that uses a definite loop and a two-dimensional array to read and store the scores on four tests for a class of 50 students.

7-7.   Assume you are given an array SCORES(100,5) consisting of scores from five different tests for 100 students in a class. Find the average on each test by taking the following condition into account: if a score of −1 is present, then it should not be considered in the averaging process. This is because a score of −1 implies that the corresponding student failed to take the exam.

7-8.   The flowchart shown in Figure P 7-8 represents an algorithm that finds the maximum value in a list of numbers called LIST. The length of the list is denoted by N, which is assumed not to exceed 10 000. Construct the trace table corresponding to the input shown below, and hence verify that the OUTPUT statement (see Figure P 7-8) results in the following output: MAXIMUM VALUE = 7.

INPUT:

card #2   1,   −3,   5,   0,   7,   3,   −5,   2,   1

card #1   9

7-9.   Change the flowchart in Figure P 7-8 such that the corresponding algorithm finds the minimum value in a list of numbers called LIST.

175

```
                          START

                        1
         INPUT N  - - - -   BIG,integer

                        2                    8
         N ≤ 10000   --No-->  OUTPUT 'ERROR'  -->  STOP

          Yes
                    I ← 1,N or I ← 1 TO N

                        3
              INPUT LIST(I)  - - - -  LIST(10000)

                        4
              BIG ← LIST(I)

                    I ← 2,N or I ← 2 TO N

              5                        6
          BIG < LIST(I)  --Yes-->  BIG ← LIST(I)

              No

                        7
         OUTPUT 'MAXIMUM VALUE =', BIG

                         END
```

FIG. P 7-8.

7-10. Assume the robot that we introduced in Chapter 3 is placed somewhere in the maze below. Write a program using the instructions in the Command List that enables the robot to get to an edge of the maze in the simplest way possible. Assume that the robot is not crosswise in the path.

Command list

☐ MOVE—moves robot forward one cell.

☐ TURN—turns robot 90° to right.

☐ IF "condition" THEN "command"—"condition" can be edge or not edge.

☐ REPEAT "statements" UNTIL "condition"— repeats all statements until "condition" is true. *Note:* This is an indefinite loop.

☐ WHILE ("condition") DO "command"

☐ BEGIN "commands" END can be used in place of a command.

☐ STOP

☐ WAIT—causes the robot to wait for 30 seconds before proceeding.

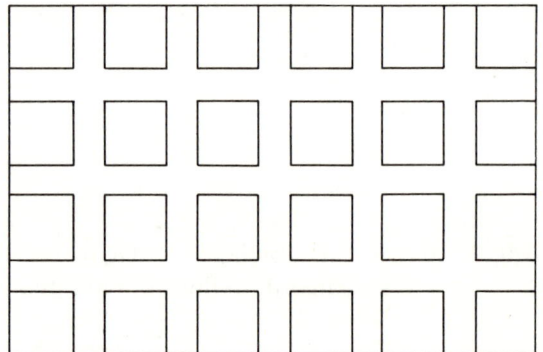

For example, the statements

```
WHILE ("not at edge")
DO BEGIN
      MOVE
      TURN
      END
```

will forever cause the robot to traverse a square in the maze, unless it bumps an edge the first time around the square.

7-11. If there are stoplights at each intersection, have the robot accomplish the task in Problem 7-10. The robot is to sense the current color of the stoplight, and, if it is red, execute one WAIT before proceeding. The command for the robot to sense the stoplight color is INPUT COLOR. The value of COLOR can then be tested to see if it is red or green.

7-12. Conditions are the same as in Problem 7-10, except that all the northern exits are open and the robot must get out of the maze. Do not let it "run away." Stop it as soon as it exits.

7-13. Work Problem 7-10, but assume only one exit is open. Have the robot find the exit and leave.

7-14. Work Problem 7-10 assuming that the robot is to find an eastern exit that is open. Assume that initially the robot is facing either north or east.

7-15. Assume the robot is placed somewhere in the following maze and is facing either north or east. It will respond only to the commands listed. You may further assume the robot does not begin at an exit and will never be crosswise in the path. Write a set of instructions that will get the robot out of the maze as easily as possible, no matter where it starts its journey. *Note:* You should be able to accomplish this task with 12 or fewer statements.

Command list

☐ Do n TIMES command—repeats the command n (an integer) times.

☐ IF condition THEN command conditions are *wall* reached or maze *exit* reached. *Note:* Negatives of commands are allowed.

☐ STOP

☐ TURN

☐ MOVE

☐ BEGIN "commands" END may be substituted for a single command.

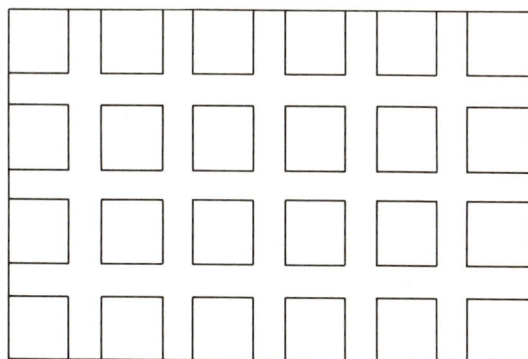

7-16. If the robot is placed in the maze given in Problem 7-15, with no known orientation except that it is not crosswise in a path, write a set of instructions that will enable it to get out of the maze.

7-17. Suppose the robot is placed in the maze given in Problem 7-15, but each intersection has a stoplight. The robot has to sense the stoplight color using INPUT COLOR, and execute the command WAIT once if the color of the stoplight is red. Write a set of instructions that will enable the robot to leave the maze.

7-18. Work Problem 7-17 with the condition that the robot must execute exactly 10 WAIT commands when the light is found to be red.

7-19. Work Problem 7-17 with the conditions that the robot must wait up to 10 times when the light is red and must move when the light is green.

8

Subalgorithms

KEYWORDS

actual parameter
cryptography
formal parameter
function subalgorithm
general subalgorithm
global variable
invoke by reference or location
invoke by value
invoked
invoking
local variable
point of invocation
return
side effect

INTRODUCTION

In studies of programming activity, researchers have reported that an individual can comprehend a limited number of program statements at any given time, in terms of the dynamic action they cause as the computer executes them. This number lies between 50 and 60—roughly one page of program statements. As we increase the number of statements an individual is required to understand, the difficulty of debugging them, proving they are correct, and maintaining them becomes increasingly difficult and time-consuming.[1] Most programs are much larger than the ideal one-page case in terms of number of statements to be executed; in fact, for many applications programs consist of anywhere from 1000 to 10 000 or more statements. We may ask, then, How can we create such programs and yet provide reasonable assurance that they are correct? A very acceptable approach is to divide the solution into a number of subtasks, each of which is small enough to be comprehended. To this end, a **subalgorithm** is a language construct that provides the desired subtask facility. It also provides the programmer with the opportunity to reduce the number of statements to be dealt with at any one time to the vicinity of the ideal case—that is, one page.

The term **subalgorithm** is an acronym for the two words subordinate algorithm. It is an algorithm that is subordinate to another algorithm. As such, subalgorithms are procedures that provide solutions to classes of subproblems.

INVOKING CONSIDERATIONS

Subalgorithms can be used in the following three cases:

1. when the steps of a subalgorithm are to be performed in *two or more* places in an algorithm to which the subalgorithm is subordinate;

2. when the steps of a subalgorithm are to be performed in *two or more algorithms*; and

3. when a large or complex algorithm must be designed.

The third case is the most important, since it enables us to express a complex problem in terms of a series of subproblems that are easier to solve using subalgorithms. These subalgorithms are then coordinated via an **invoking algorithm**.[2] In Figure 8.1-1, for example, the arrows

---

[1] Proving that programs are correct on a rigorous basis is beyond the scope of this book.

Maintaining programs refers to the process of altering them to reflect changes in a problem being solved (e.g., changes due to legislative actions on tax computation), or changes in the computer system being used.

[2] This algorithm invokes a subalgorithm to perform its task. Such algorithms may be main routines (i.e., routines to which the computer turns control when execution begins), or may be subalgorithms, themselves.

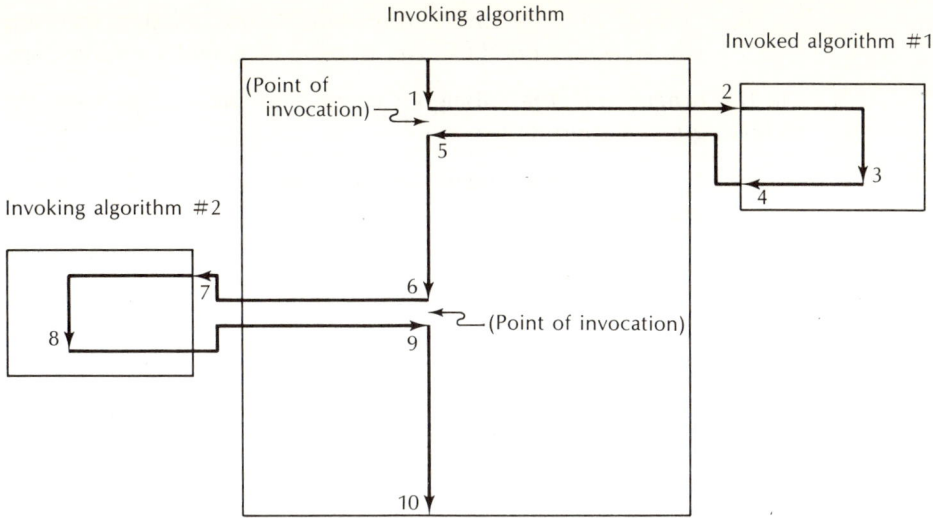

FIG. 8.1-1. Illustration of invoking algorithms, invoked algorithms, and points of invocation.

(labeled as 1-2-3-. . .-8-9-10) indicate the order in which the statements are executed. We note that once a subalgorithm is invoked, the statements in the subalgorithm are executed before control is **returned** to the invoking algorithm. From this illustration, it is apparent that an invoking algorithm is one that calls on a subalgorithm to solve a subproblem. Conversely, the subalgorithm that is called upon is referred to as the **invoked subalgorithm.** The points in an invoking algorithm at which subalgorithms are invoked or called upon are referred to as **points of invocation.**

Various subalgorithms have already been used in some of the earlier chapters, although they were not referred to as such. For example, the statement

$$Y \leftarrow SQRT(B)$$

in Figure 4.2-1 involves a subalgorithm, since each time the square root of a number (a value of B) was required, a subalgorithm by the name of SQRT was invoked. For instance

$$Y \leftarrow SQRT(9)$$

results in SQRT returning the numeric constant 3 (the square root of 9), which is then assigned to the real variable Y.

The subalgorithm approach enables a group or team of people to simultaneously work on a large or complex problem by assigning subproblems to individual members of the programming team. A carefully managed programming team can realize a substantial reduction in the time required to design, code, document, and debug algorithms by using subalgorithms.

In order to use subalgorithms in the solution of a problem, the invoking algorithm must have the following attributes:

☐ It must be able to invoke the subalgorithm.

181

☐ It must be able to communicate to the subalgorithm any values it may need to carry out the pertinent computations.

On the other hand, the subalgorithm must be able to perform the following tasks upon its execution:

☐ It must return the results of the computations to the invoking algorithm.

☐ It must return control to the invoking algorithm at a known place.

For instance, in the preceding square root example—i.e., $Y \leftarrow \text{SQRT}(9)$, the result returned is 3.0. Control is then returned to the invoking program.

A subalgorithm is identified by assigning it a unique name. Each time this name appears in an invocation, the subalgorithm's statements are executed. Values needed by the resulting computations are communicated to a subalgorithm by means of **parameters.** Parameters are also used to return the corresponding results from the subalgorithm to the invoking algorithm. Details pertaining to such parameters will be discussed in Sections 8.2 and 8.3.

As mentioned earlier, a subalgorithm must always return control to the invoking algorithm once the subalgorithm has been executed. This is accomplished through the use of a RETURN statement. The point of return will vary in accordance with the *type* of subalgorithm being used. We will consider two types of subalgorithms in this chapter: (a) function subalgorithms and (b) general subalgorithms. The interaction between such subalgorithms and their related invoking algorithms will be illustrated by means of a variety of examples.

8.2 FUNCTION SUBALGORITHM

The first type of subalgorithm we will study is called a **function subalgorithm,** and will be denoted by the acronym, **FSA.** The FSA has the property that it returns only *one* value each time it is invoked, and this value replaces its name in the expression that caused the FSA to be invoked. The general form of the FSA is defined as

Function name (parameter list)

where the **function name** identifies the FSA and the **parameter list** defines the input to the FSA in the form of a series of one or more variables. It is important to note that every FSA must be given a unique name, since this name must uniquely identify the group of statements to be executed. The execution of the FSA begins at the **entry point** identified by the name. An entry point is identified by a terminal box containing the FSA name and a parameter list, For example, the flowchart sequence is the entry point to an FSA with the name of SUM, the parameter list of which consists of two real variables, A and B.

Every FSA also contains a RETURN statement that causes *one value* (a numeric or character string constant) *to be returned to the point of invocation,* as illustrated in Figure 8.2-1. Control is also transferred

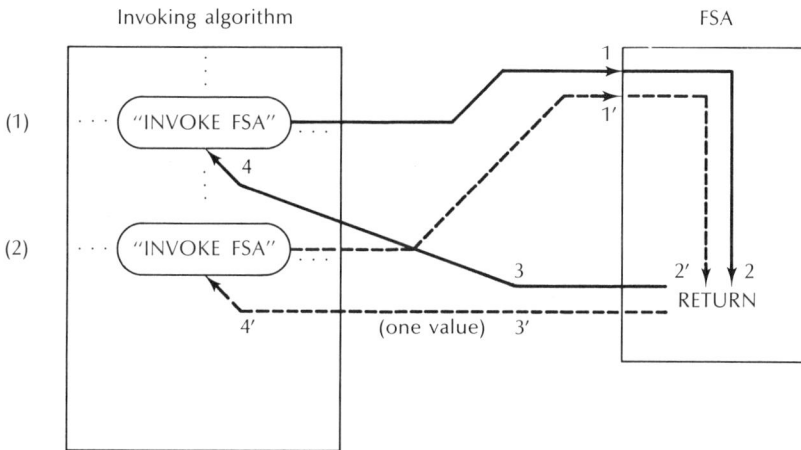

FIG. 8.2-1. Flow of execution for invoking an FSA.

back to this point. In Figure 8.2-1 we observe that the FSA is invoked twice; the points of invocation are denoted by (1) and (2). The corresponding arrows are labeled 1-2-3-4 and 1'-2'-3'-4', respectively. In each case only one value is returned.

We will now introduce two examples pertaining to FSAs, the first of which is shown in Figure 8.2-2. From this figure it is clear that there is one FSA, and its name is SQUARE. We also note that the parameter lists associated with the FSA SQUARE use the two variables X and A, one in the invoking statement, and the other in the FSA definition. To distinguish between these parameter lists, appropriate terminology must be adopted. To this end, the variable X is called an **actual parameter,** while the variable A is called a **formal parameter,** or **dummy variable.** The motivation for introducing the term *dummy variable* is to emphasize that the variable A is merely used to define the calculations to be carried out in SQUARE. When execution of the FSA commences, the value taken by the formal parameter A is equal to that of the actual parameter X. We illustrate this fact by attaching the formal parameter name A to the location of the actual parameter, as shown in Figure 8.2-2(d).

To illustrate, consider the specific input stream shown in Figure 8.2-2(a). The INPUT X statement in Symbol 2 causes X to take the value

183

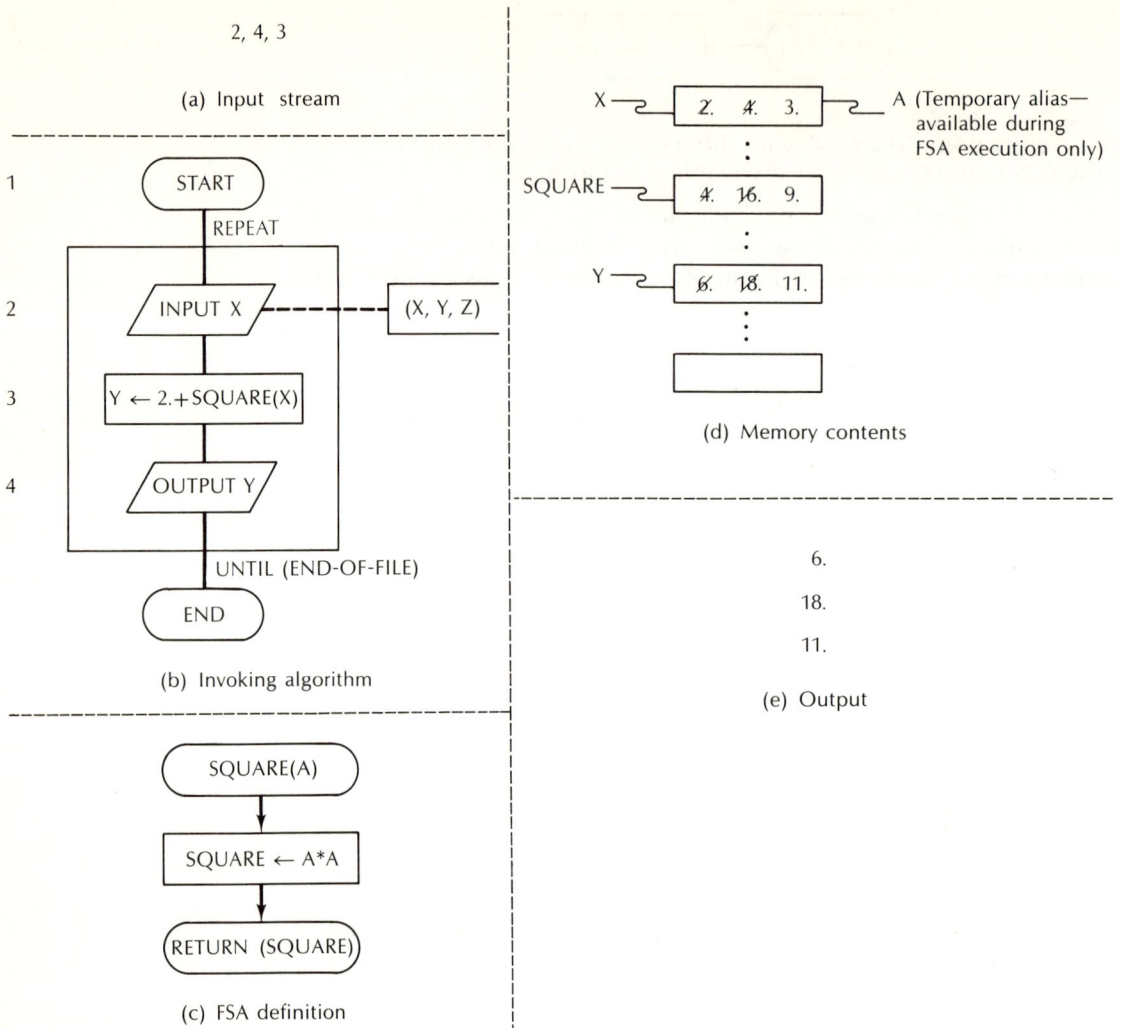

FIG. 8.2-2. Illustration of FSA SQUARE(A) which evaluates A*A.

2.0. Thus when Symbol 3 is to be executed, the SQUARE(X) portion of the statement

$$Y \leftarrow 2.0 + SQUARE(X)$$

is computed via the FSA SQUARE. Clearly, since the dummy variable A takes the same value as X, the statement

$$SQUARE \leftarrow A*A$$

yields SQUARE = 4.0, as shown in the first entry of the memory contents in Figure 8.2-2(d). This is the one value to be returned to the invoking algorithm. Its *return* is achieved by assigning it to the name of the FSA, as is the case in Figure 8.2-2(c), or by a RETURN statement that indicates the variable, constant, or expression to be returned—

184

e.g., RETURN (A*A). This returned value is substituted for the term SQUARE(X) in Symbol 3 to yield Y = 6.0. The OUTPUT statement in Symbol 4 then yields the output shown in the first line of Figure 8.2-2(e). The default type of the name of an FSA is the type of the value returned by the FSA. For example, if ITEM is the name of an FSA, then the value returned would be an integer.

Next, X takes the value 4.0, which then causes SQUARE to take the value 16.0. This in turn yields Y = 18.0, leading to the output shown in the second line of Figure 8.2-2(e). Similarly, it follows that the third value of Y that appears as output is 11.0, and the corresponding value of SQUARE is 9.0, as shown in Parts (e) and (d), respectively, of Figure 8.2-2.

As a second example, let us consider the problem of computing the value of a numeric variable R, given by

$$R \leftarrow P + Q + 3.0 \tag{8.2-1}$$

where P and Q are also numeric variables. For *each pair* of values assigned to P and Q, R must be evaluated as indicated in Equation (8.2-1). This type of computation can be conveniently carried out in terms of an FSA by rewriting the equation as

$$R \leftarrow (P + Q) + 3.0 \quad \text{or} \quad R \leftarrow SUM(P,Q) + 3.0 \tag{8.2-2}$$

where SUM(P,Q) denotes an FSA, the output of which is the value of (P + Q). Thus it follows that an appropriate invoking algorithm is as shown in Figure 8.2-3(b), while Figure 8.2-3(c) shows the corresponding definition of the FSA SUM(D,E). We note that P and Q in SUM(P,Q) are the actual parameters, while D and E in SUM(D,E) are the corresponding formal parameters or dummy variables. Thus, D is matched with P, and E is matched with Q. It is important that the type and size of D and E be identical to the type and size of P and Q, respectively.[3]

To illustrate, consider the input stream shown in Figure 8.2-3(a). It is straightforward to verify that the memory contents and the corresponding output (due to the OUTPUT statement) are as shown in Parts (d) and (e), respectively, of Figure 8.2-3.

It should be noted that the FSA SUM we have discussed can also be invoked from more than one point. For example, it may be invoked twice, as illustrated in Figure 8.2-4. It is left as an exercise for the reader to verify that the memory contents and output corresponding to the input in Figure 8.2-4(a) are as shown in Parts (d) and (e), respectively, of Figure 8.2-4.

Our discussion of function subalgorithms can be summarized by noting that their formulation requires a unique name to be associated with each FSA. It is also necessary to provide formal parameters to communicate values to be used in their calculations, and provide a

[3] *Size* refers to dimensionality, in the event P, Q, D, and E are arrays.

1, 2, 7, 9, – 1, 2

(a) Input stream

REPEAT

INPUT P,Q ----- (P,Q,R) integer

R ← SUM(P,Q)+3

OUTPUT R

(UNTIL END-OF-FILE)

(b) Invoking algorithm

SUM(D,E) --- (D,E,SUM) integer

SUM ← D+E

RETURN (SUM)

(c) FSA definition

P — 1̸ 7 –1 — D   (Temporary aliases—
                   available during
Q — 2̸ 9̸ 2 — E   FSA execution)

SUM — 3̸ 1̸6 1

R — 6̸ 1̸9 4

(d) Memory contents

6
19
4

(e) Output

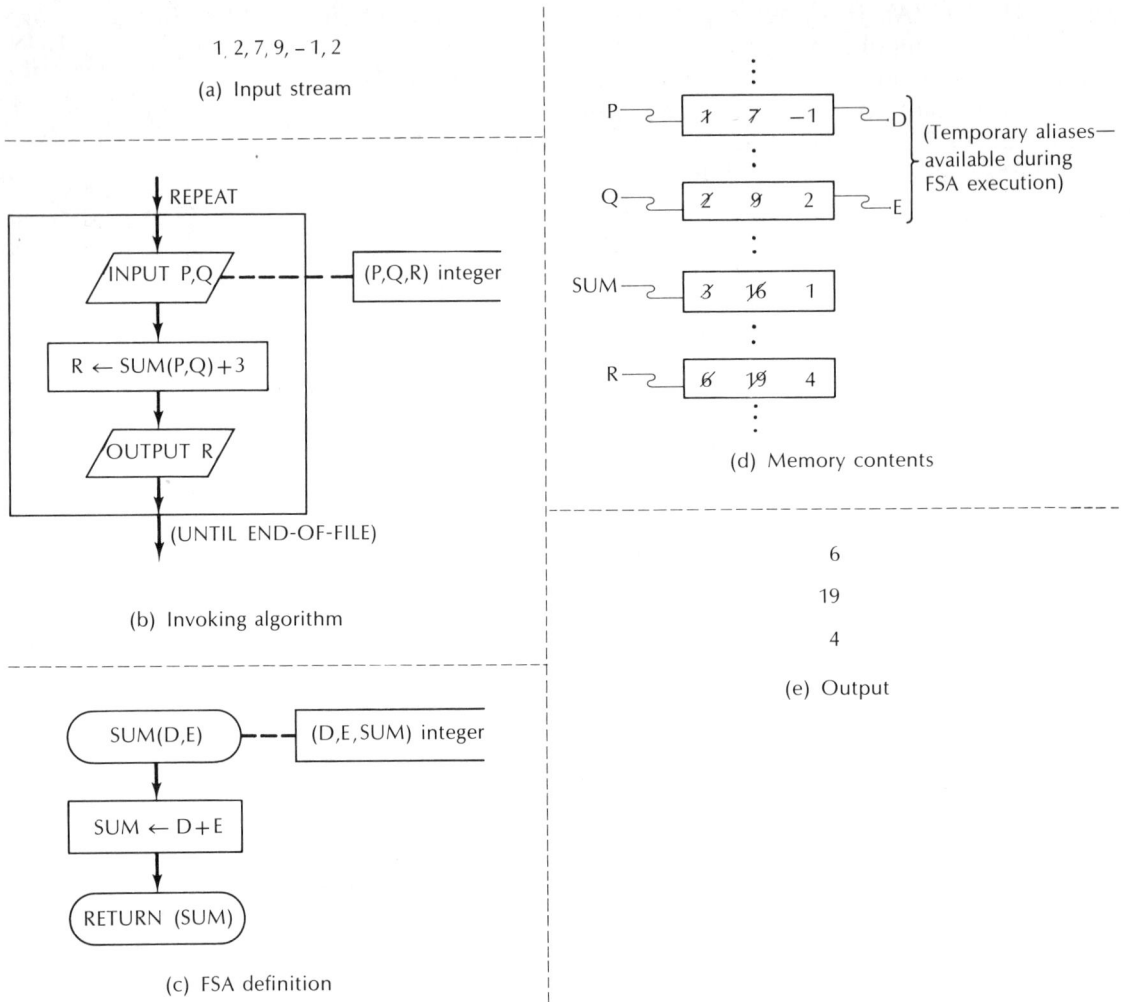

FIG. 8.2-3. Illustration of FSA SUM(D,E) which evaluates D + E.

means for the corresponding result to be returned. It is important to note that only one value is returned. This value can either be assigned to the name of the FSA or appear in the RETURN statement, as illustrated in the examples below, where the value returned in each case is the same.

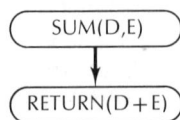

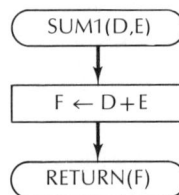

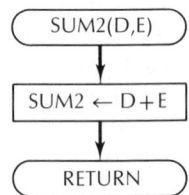

SUM(D,E)

RETURN(D+E)

SUM1(D,E)

F ← D+E

RETURN(F)

SUM2(D,E)

SUM2 ← D+E

RETURN

1, 2, 7, 9

(a) Input stream

1   INPUT P,Q --- (P,Q,R,S,T) integer

2   R ← SUM(P,Q) + 3

OUTPUT R

4   INPUT S,T

5   R ← SUM(S,T) + 6

6   OUTPUT R

(b) Invoking algorithm

SUM(D,E) --- (D,E,F) integer

F ← D + E

RETURN(F)

(c) Definition

P → 1 → D
Q → 2 → E
} (Aliases for invocation in Statement 2)

R → $\cancel{6}$ 22

S → 7 → D
T → 9 → E
} (Aliases for invocation in Statement 4)

⋮

F → $\cancel{3}$ 16

(d) Memory contents

6
22

(e) Output

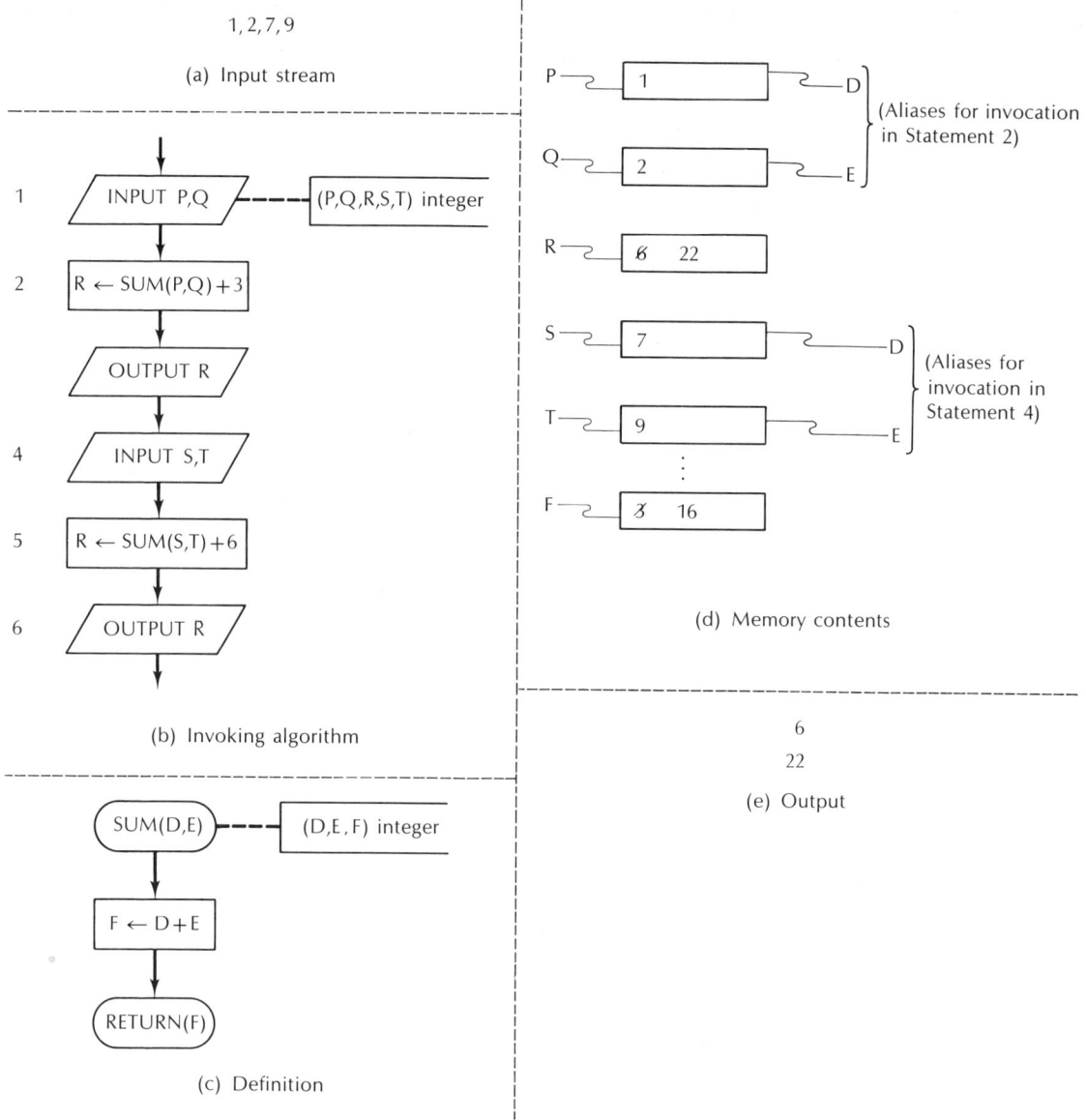

FIG. 8.2-4. Two invocations of FSA SUM.

MORE ON ACTUAL AND FORMAL PARAMETERS 8.3

Having introduced the notion of actual parameters and formal parameters (or dummy variables) by means of examples of FSAs, we can now consider these means of communication in more detail. At the outset, it is important to note that the use of actual and formal parameters is not restricted to FSAs. In Section 8.4 we will see that they are also used in conjunction with general subalgorithms.

187

An actual parameter list may consist of any of the following:

1. variable names;

2. constants;

3. expressions.

On the other hand, a formal parameter list consists only of variable names. These variable names can also refer to arrays, in which case entire arrays of values would be involved.

In order to relate a given list of actual parameters to a corresponding list of formal parameters, it is necessary to satisfy three important rules:

1. The number of parameters in the actual and formal parameter lists must be the same.

2. The types and dimensionality of the corresponding parameters in the actual and formal parameter lists must be identical.

3. The correspondence between formal and actual parameters is established only by their relative position within the list (i.e., the first pair of formal and actual parameters are matched, followed by the second pair, and so on).

Consider the case in which the actual and formal parameters lists are given by

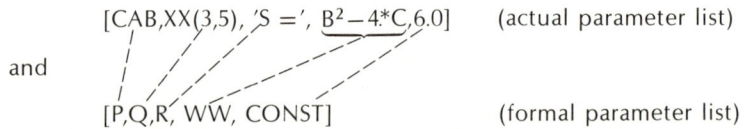

$$[CAB, XX(3,5), 'S =', B^2 - 4*C, 6.0] \quad \text{(actual parameter list)}$$

and

$$[P, Q, R, WW, CONST] \quad \text{(formal parameter list)}$$

respectively. Assume that all variables have been declared as integer variables, except XX and R which have been declared as integer array and character string, respectively. The relationship, then, between the actual and formal parameters would be as follows:

☐ The first formal parameter P in the subalgorithm will take a value equal to the current value of the variable CAB in the invoking algorithm.
Comment: Note that if P and CAB were declared as arrays consisting of the same number of elements (say M), then each time P is used, the value taken by P(I) would be the current value of CAB(I), I=1,2,...,M.

☐ Whenever the second formal parameter is referenced in the subalgorithm, its value will be the current value of XX(3,5) in the invoking algorithm. Both Q and XX(3,5) are simple integer variables.

☐ The character string constant 'S=' would be used everywhere in the subalgorithm where the formal parameter element R is referenced.

☐ The expression $B^2 - 4.*C$ in the invoking algorithm is evaluated first, and the resulting numeric constant is used as the value for WW in the subalgorithm.

□ The value taken by CONST in the subalgorithm is equal to the constant, 6.0.

To conclude our discussion, we present the additional example shown in Figure 8.3-1, which involves an FSA. It should be instructive for the reader to verify that the final output and the intermediate results are as shown in Parts (e) and (d), respectively, of the figure. The reader will then find that the FSA in Figure 8.3-1(c) returns either the sum of the element values of JAY, or their average value, depending upon whether L is 0 or 1, respectively.

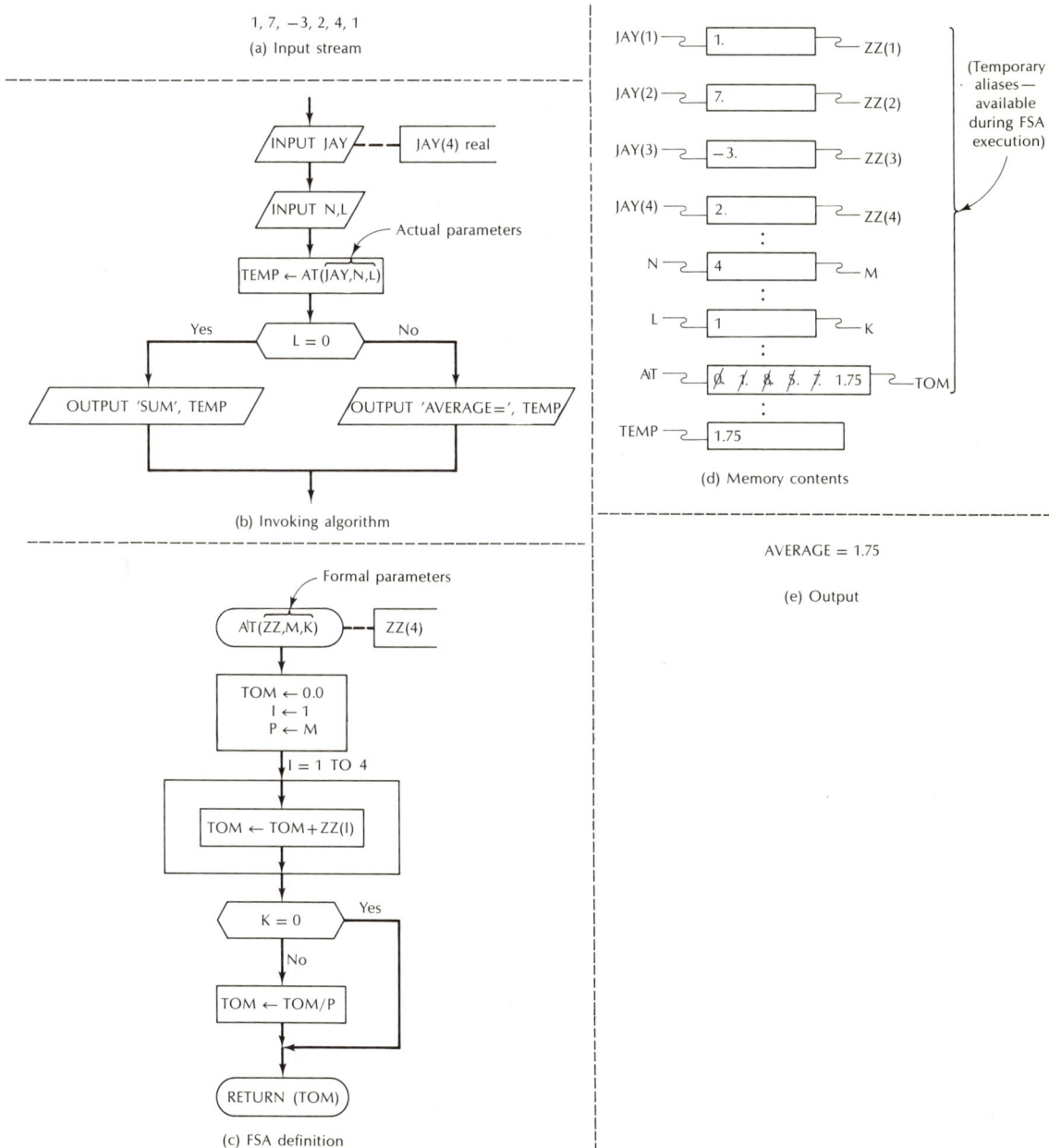

FIG. 8.3-1. On relating actual and formal parameters.

From the preceding discussion, it is apparent that an FSA is invoked *implicitly,* in the sense that it is invoked by the appearance of its name in an expression. In contrast, a **general subalgorithm (GSA)** is invoked *explicitly* via a CALL statement, as follows:

CALL "name of GSA" (parameter list)

It should be noted that the verb used in the invoking statement varies, depending upon the programming language. We will use CALL in connection with the flowchart language.

There are two other basic differences between general and function subalgorithms:

1. A GSA can return a set consisting of *none* or *one or more* values, instead of exactly one value, as is the case with an FSA.

2. While return from an FSA causes control to be returned to the point of invocation, return from a GSA causes control to be returned to the statement that immediately follows the point of invocation, as illustrated below:

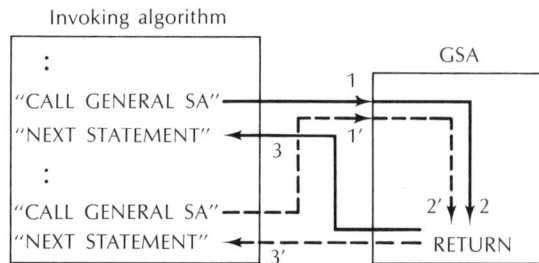

The four examples that follow illustrate the manner in which GSAs are invoked. The first of these examples is shown in Figure 8.4-1, and demonstrates how a GSA can be used to achieve the same action as the FSA SQUARE in Figure 8.2-2. Note that the formal parameters (or dummy variables) A and B are stored in the locations assigned to the actual parameters X and Z, respectively. This assignment is done dynamically during execution in flowchart language, but on return in some others. When the RETURN statement in the invoked GSA is executed, control is transferred to the statement

$$Y \leftarrow Z + 2.0$$

in the invoking algorithm, since this statement immediately follows the point of invocation; that is, the statement

CALL SQUARE (X,Z)

It follows that the current value of B is returned and the output that appears as a consequence of the OUTPUT Y statement is as shown in Figure 8.4-1(e). The corresponding memory contents at each stage of the related processing are shown in Figure 8.4-1(d).

2, 4, 3

(a) Input stream

(b) Invoking algorithm

(c) GSA definition

(Temporary aliases—available during GSA execution)

(d) Memory contents

6

18

11

(e) Output

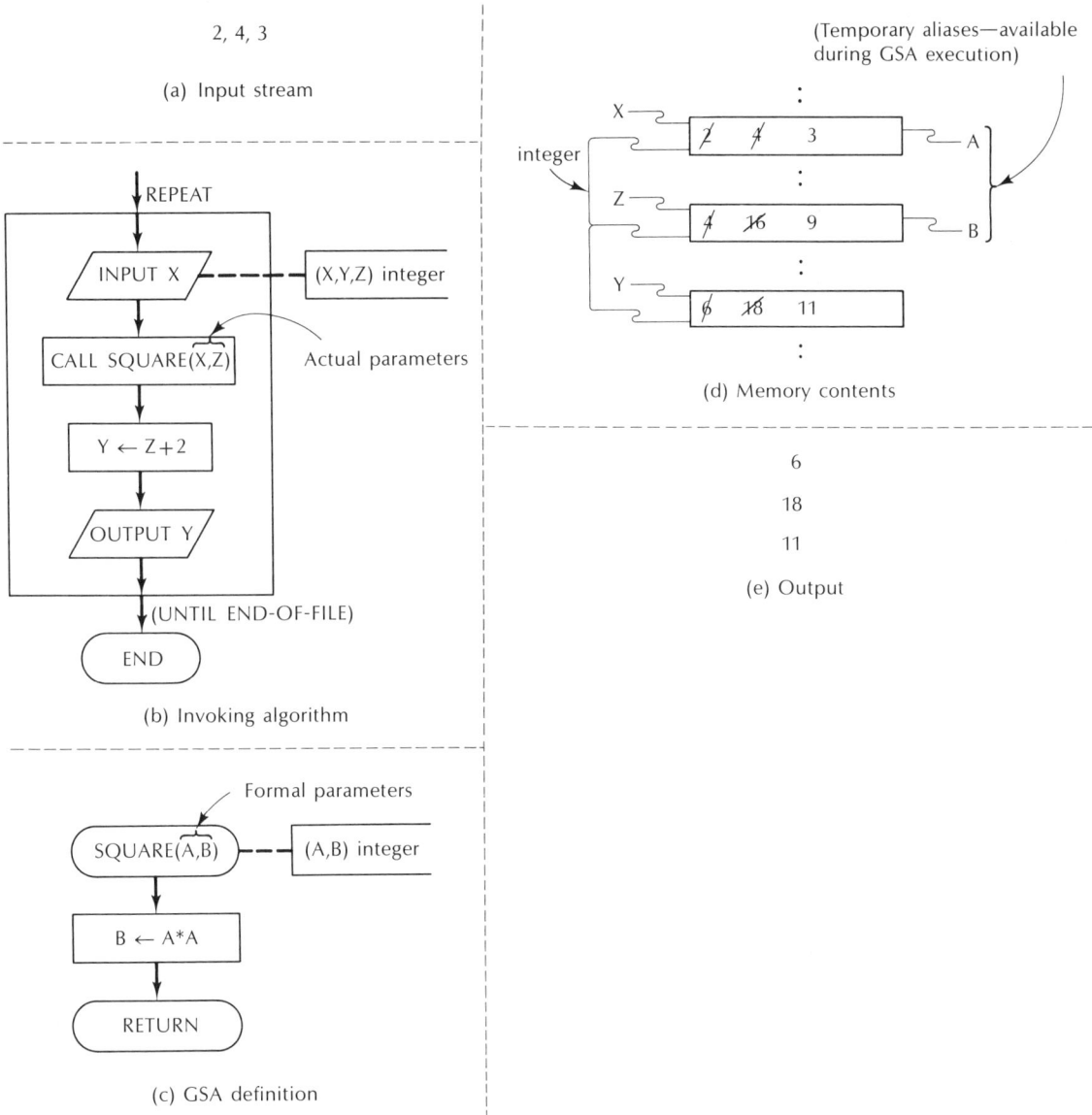

FIG. 8.4-1. GSA called SQUARE(A,B), where A is the input parameter and B is the output parameter.

In the second example, Figure 8.4-2 involves the GSA BOB(D,E,F,G), where the formal parameters (dummy variables) D,E,F, and G are aliases to the memory locations of the actual parameters P,Q,T, and U, respectively. Since D and E correspond to the input variables P and Q, the values returned are those of the variables F and G. The role played by the actual parameters T and U is to accept these two returned values. Execution of the RETURN statement in the invoked GSA causes transfer of control to Symbol 3 of the invoking algorithm. The resulting output and the corresponding memory contents are as shown in Parts (e) and (d), respectively, of Figure 8.4-2.

191

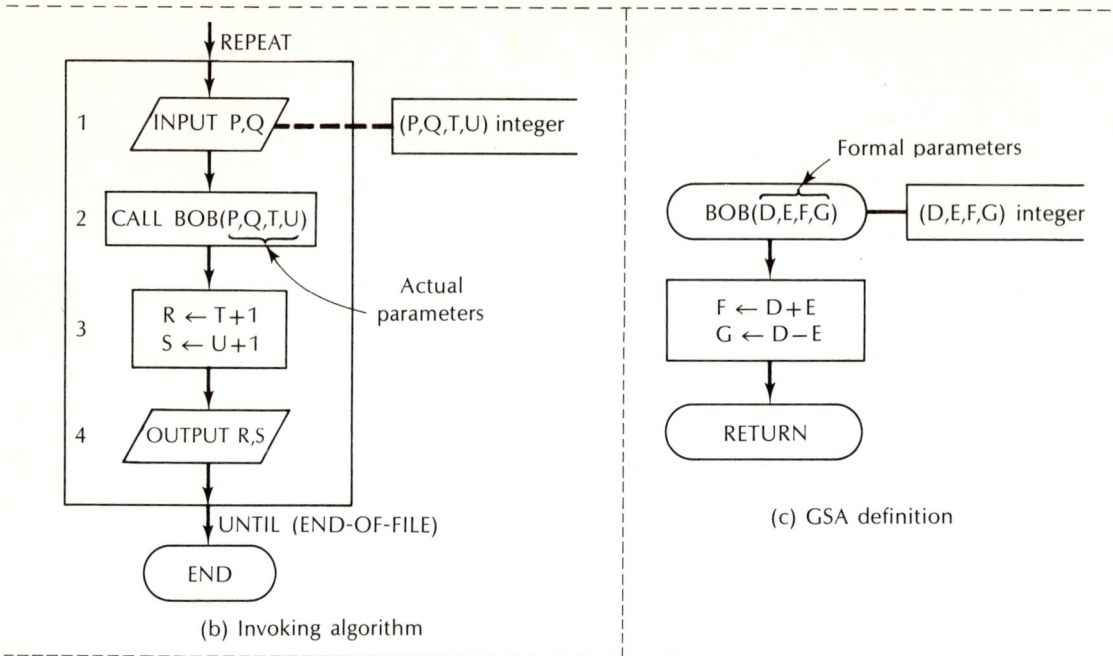

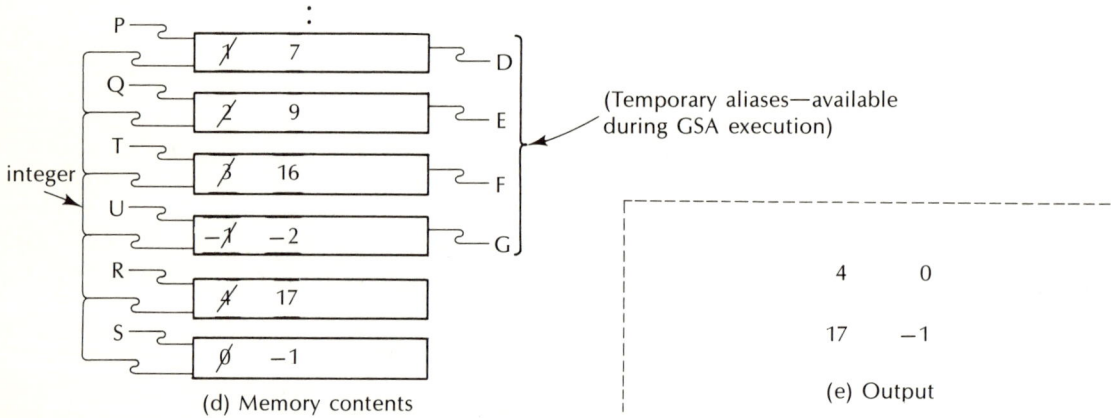

FIG. 8.4-2. GSA called BOB(D,E,F,G), where D and E are input parameters and F and G are output parameters.

The third example concerns the case in which a GSA called SQUARE is invoked at two points in the invoking algorithm with different sets of parameters, as shown in Figure 8.4-3. The formal parameters are A and B, where A is the input parameter and B is the output parameter. Note there is no confusion between the formal and actual parameters, each of which is denoted by B. Only the parameter B associated with the GSA uses the alias name.

192

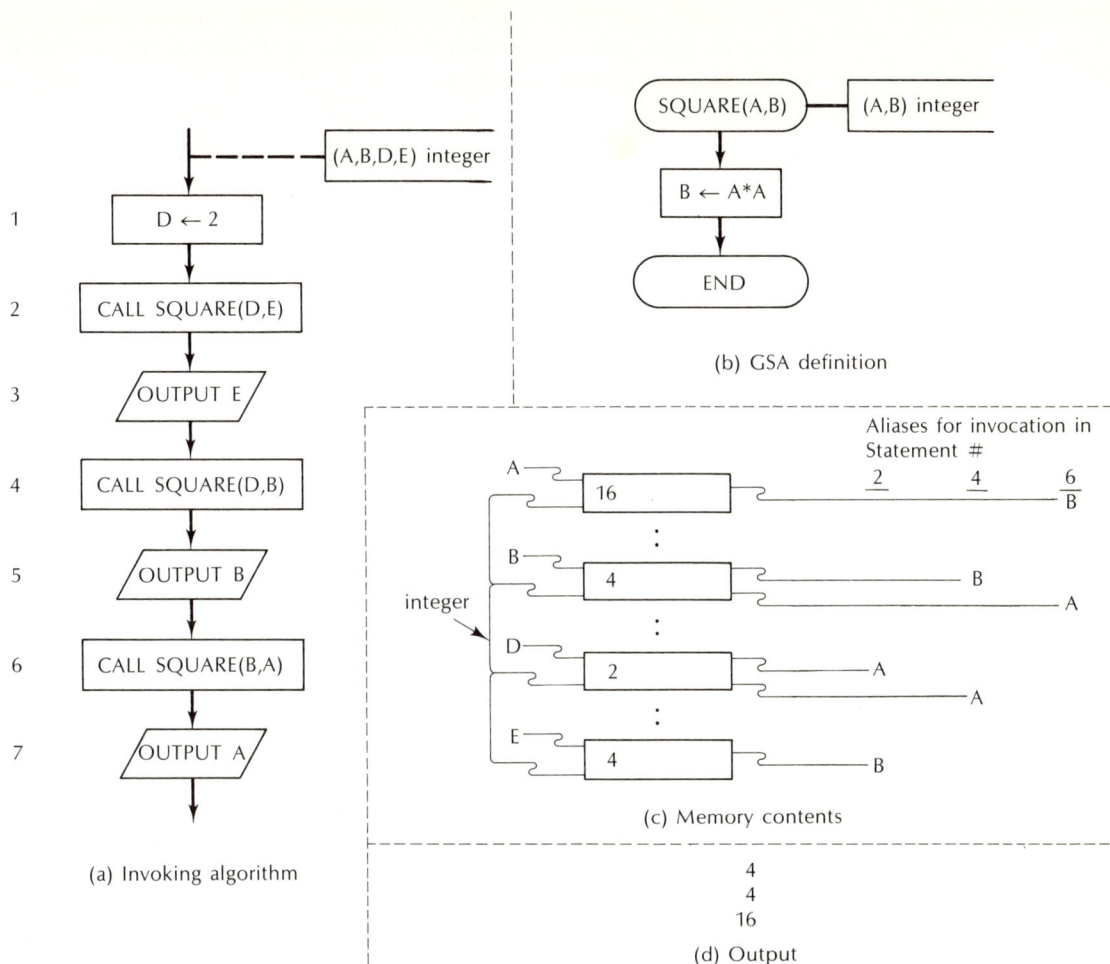

FIG. 8.4-3. Three invocations of a GSA called SQUARE.

To conclude our discussion of GSAs we summarize the following points:

☐ A GSA is more flexible than an FSA in that the former can return a set of values, the set consisting of no values, one value, or more than one value.

☐ Parameters must be supplied in the GSA definition for the needed input values, as well as for the output values that are to be returned by the GSA.

☐ Good programming practice dictates that a parameter used for input should not be changed in value.

☐ An entire statement is required to invoke the GSA. Control is normally returned to the statement following the invoking statement.

193

The fourth and final example requires that we read in an array of names of parts of a motor assembly. These parts are identified with a set of serial numbers, as follows:

Rotor: 1824
Casing: 1825
Winding: 1826
Armature: 1827
Coils: 1828
Slip rings: 1829

We seek a structured algorithm which prints the name of a part when a customer submits its serial number.

A desired solution is as follows:

1. Read names of parts into an array.
2. Read serial number submitted by customer.
3. Determine name of part corresponding to number read in Step 2.
4. Print name obtained in Step 3.

The corresponding flowchart is shown in Figure 8.4-4, from which it is apparent that a GSA called FINDER is used to implement Step 3 in the solution. In this GSA there are three formal parameters; namely, NAME, I, and N. The parameter NAME denotes the array in which the names of the parts are stored, while the parameter I denotes the number of the part which is obtained from the customer. The name of the part that is returned by the GSA is denoted by the parameter N. The variable K is used only in FINDER and is known only there.

Parameters are used in general subalgorithms to transfer results out of the subalgorithm. In such cases, the following rules contribute to good programming practice:

1. Formal parameters can be used either to input information to the subalgorithm or to output information from it. Using them for both these purposes should be avoided.
2. When possible, those formal parameters that are used to input information to the subalgorithm should only be referenced—i.e., their values should not be changed.

To illustrate the preceding rules, consider three versions of a subalgorithm SUM as shown in Figure 8.4-5, where A and B are the input parameters, and the output parameter is C. Here, neither A nor B should be changed, since values in the invoking algorithm are also changed. Errors of this type are often not detected by the programmer. They are referred to as an unexpected **side effect** of sugalgorithm usage. Such errors can be avoided by not changing an input parameter within a subalgorithm.

'ROTOR', 'CASING', WINDING', 'ARMATURE', 'COILS', 'SLIP∧RINGS',
1827, 1824, . . .

(a) Input stream

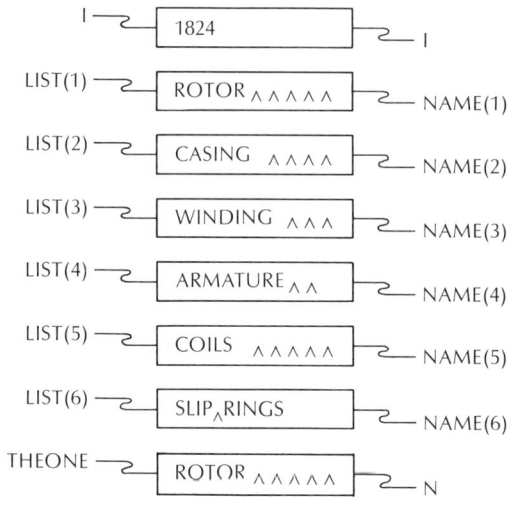

(b) Invoking algorithm

(c) GSA definition

(d) Memory contents

(e) Output

FIG. 8.4-4. GSA called FINDER, where NAME and I are the input parameters
and N is the output parameter.

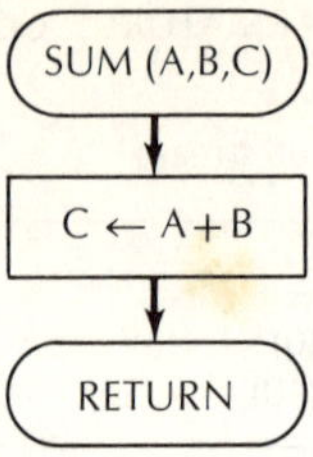

(a) GSA SUM with no undesirable
side effects (A and B are
input parameters, C is an
output parameter).

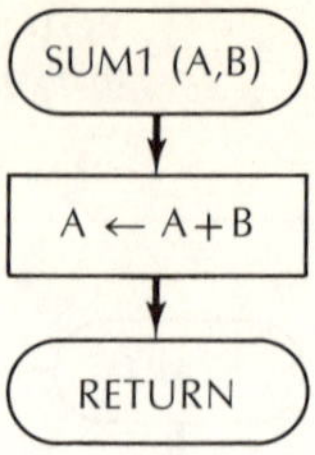

(b) GSA SUM1 with an undesirable
side effect caused by using
A as the input parameter,
and also as the output
parameter.

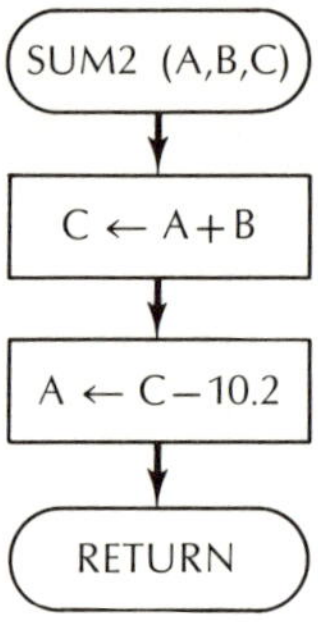

(c) GSA SUM2 with a side effect caused
by the change of an input parameter
within the GSA.

FIG. 8.4-5. Undesirable side effect resulting from changing input parameters.

To conclude, we note that it is possible in some languages to combine the features of function and general subalgorithms. The resulting "combined subalgorithms" can return one value in place of their names and, at the same time, return values via their parameters.

8.5 INVOKING CONSIDERATIONS

In the preceding discussion of subalgorithms, the formal parameters were invoked using an alias scheme. This form of communication of information via parameters is referred to as **call by location.** We also looked upon formal parameters as dummy variables in that they were temporarily assigned the same memory locations as the corresponding actual parameters. Thus, if a formal parameter in a subalgorithm is assigned a new value, then the corresponding actual parameter also takes this new value.

Formal parameters can also be invoked by value. In **calling by value,** the current value of an actual parameter (rather than its location [or address]) is transferred to the subalgorithm. As a consequence, calling by value is essentially a one-way communication—that is, from the invoking algorithm to a subalgorithm.

2.5, 5.0, 1.6

(a) Input stream

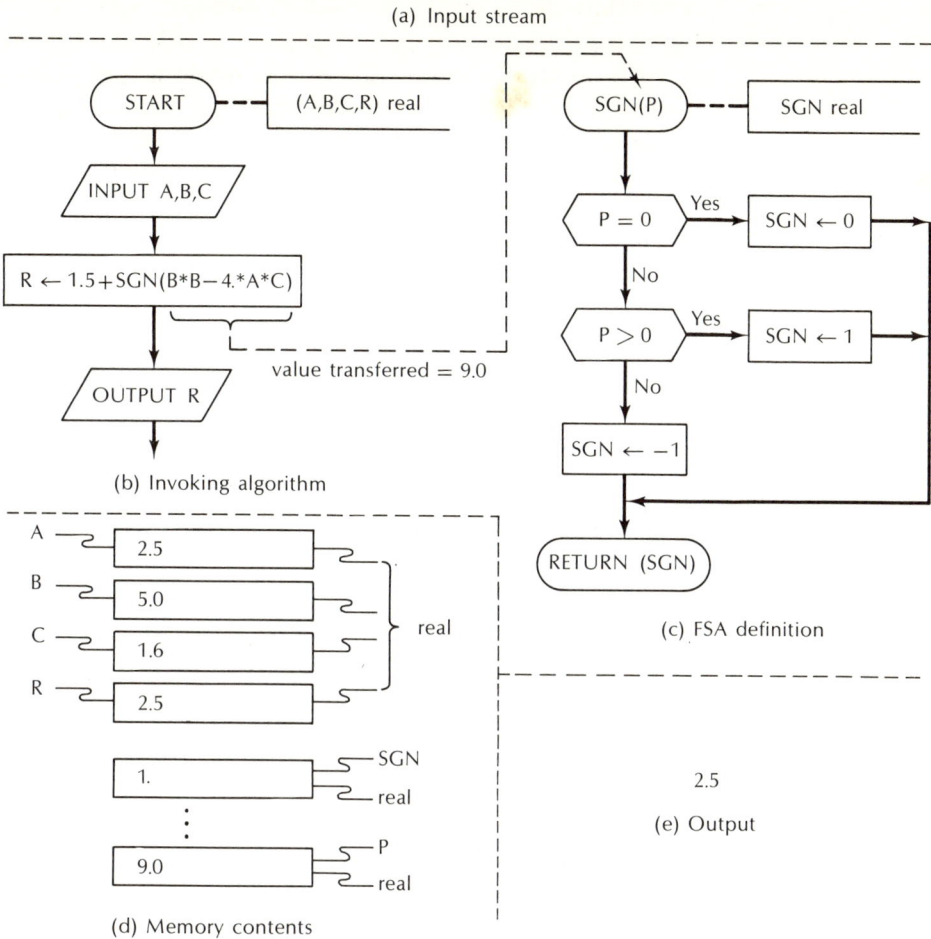

(b) Invoking algorithm

(c) FSA definition

(d) Memory contents

2.5

(e) Output

FIG. 8.5-1. Calling by value.

Calling by value is very useful when an actual parameter is an expression. To illustrate, let us consider the evaluation

$$R \leftarrow 1.5 + SGN(B*B - 4.0*A*C) \qquad (8.5\text{-}1)$$

where R, A, B, and C are real variables and SGN is the name of a FSA. Since the actual parameter associated with SGN is an expression, its *value* is made available to the FSA. The only value that is returned from the FSA to the invoking algorithm is that which results from evaluating the quantity SGN(B*B−4.0*A*C) in Equation (8.5-1). For example, if SGN is the FSA given in Figure 8.5-1(b), then the value transmitted to the FSA from the invoking algorithm is given by

$$B*B - 4.0*A*C = 9.0$$

since A = 2.5, B = 5.0, and C = 1.6, as indicated in Figure 8.5-1(a). The corresponding memory contents and output are shown in Parts (d) and (e), respectively. We should observe that a location was created

197

for P because the actual parameter corresponding to P is an expression, a condition which results in a call by value.

An additional case in which calling a formal parameter by value would be useful is when the corresponding actual parameter is a constant. This follows from the fact that a constant is a special case of an expression. In our flowchart language, all constants and expressions are transferred by value. However, in some languages, the type of reference is an option.

Calling by location is especially useful when arrays appear as actual parameters. To illustrate, consider the case of an array consisting of 30 elements, in which case calling by value would require the following steps:

1. communication of each of the 30 values to the subalgorithm, and
2. the establishment of 30 separate temporary locations in the subalgorithm for storing the array.

On the other hand, calling by location would require no additonal storage in the subalgorithm.

Calling by either location or value is equally convenient for simple variables. Most languages have these methods of communicating values by parameters, and perhaps additional methods.

8.6 LOCAL AND GLOBAL VARIABLES

In addition to formal parameters (or dummy variables), the notion of local and global variables must be understood when implementing subalgorithms. The following discussion will serve only as an introduction to some basic ideas on this subject. Our convention will be to explicitly declare both local and global variables. Global variables may be shared among an invoking algorithm and some/all the subalgorithms it invokes. The use of global variables is generally discouraged in an environment in which good programming practices are followed. This is because global storage can result in variable values being changed by subalgorithms unknown to the program or other subalgorithms. This constitutes another class of undesirable side effects. If we avoid the use of global storage, this type of error is not introduced.

Before presenting examples of local and global variables, let us state the definitions:

1. **Local variables** are variables that are defined as local explicitly or implicitly within an invoked subalgorithm; hence, their memory locations are available only to the subalgorithm.
2. **Global variables** are variables that are explicitly defined as global in the invoking algorithm *as well as* explicitly defined in an invoked subalgorithm. Thus, the same memory location is shared by both.

The first example we will consider is shown in Figure 8.6-1, where JAN(A,B) is a GSA, the formal parameters (or dummy variables) of which are A and B, while C is a local variable—local to JAN. This is because C is declared exclusively in the GSA. It is important to note that although A and B are also declared in JAN, they are formal parameters, and that status takes precedence over their use in either a local or global framework. We note that C in the invoking algorithm is distinct from the C that is available only to the invoked algorithm.

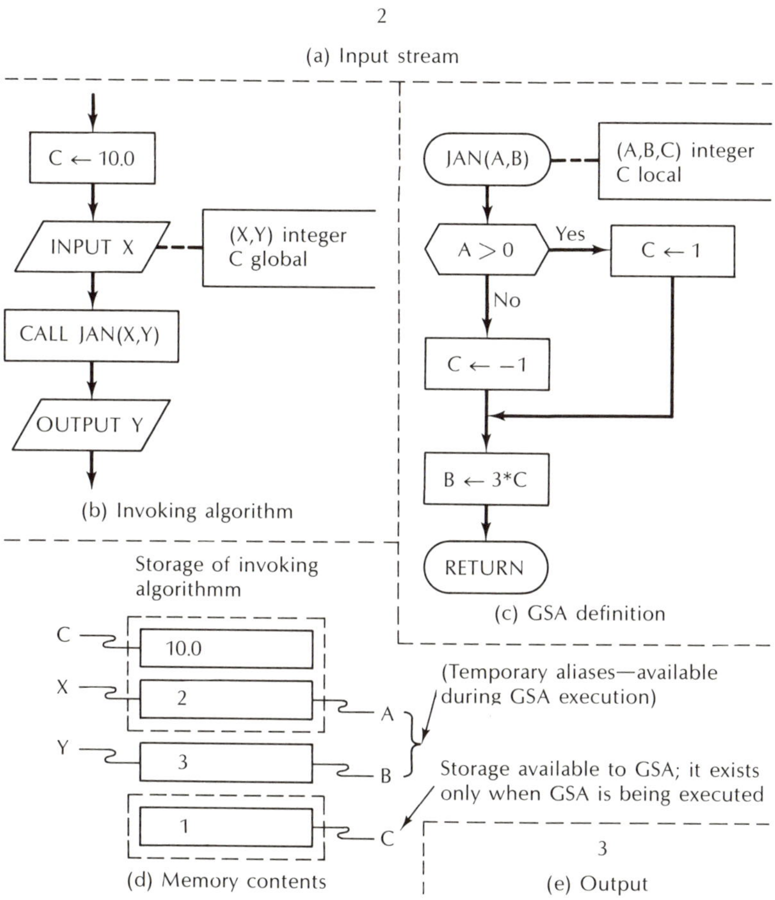

FIG. 8.6-1. Example of a local variable.

It is straightforward to verify that the input stream in Figure 8.6-1(a) results in the memory contents and output shown in Parts (d) and (e), respectively, of Figure 8.6-1.

As a second example, consider the two GSAs, COLE and TOY, shown in Parts (b) and (c), respectively, of Figure 8.6-2. From Symbol 1a it follows that B is a formal parameter to COLE, and the fact that C is also a formal parameter is evident from Symbol 1b. Thus, C, B, and X will share the same location in memory as indicated in Figure 8.6-2(d). The

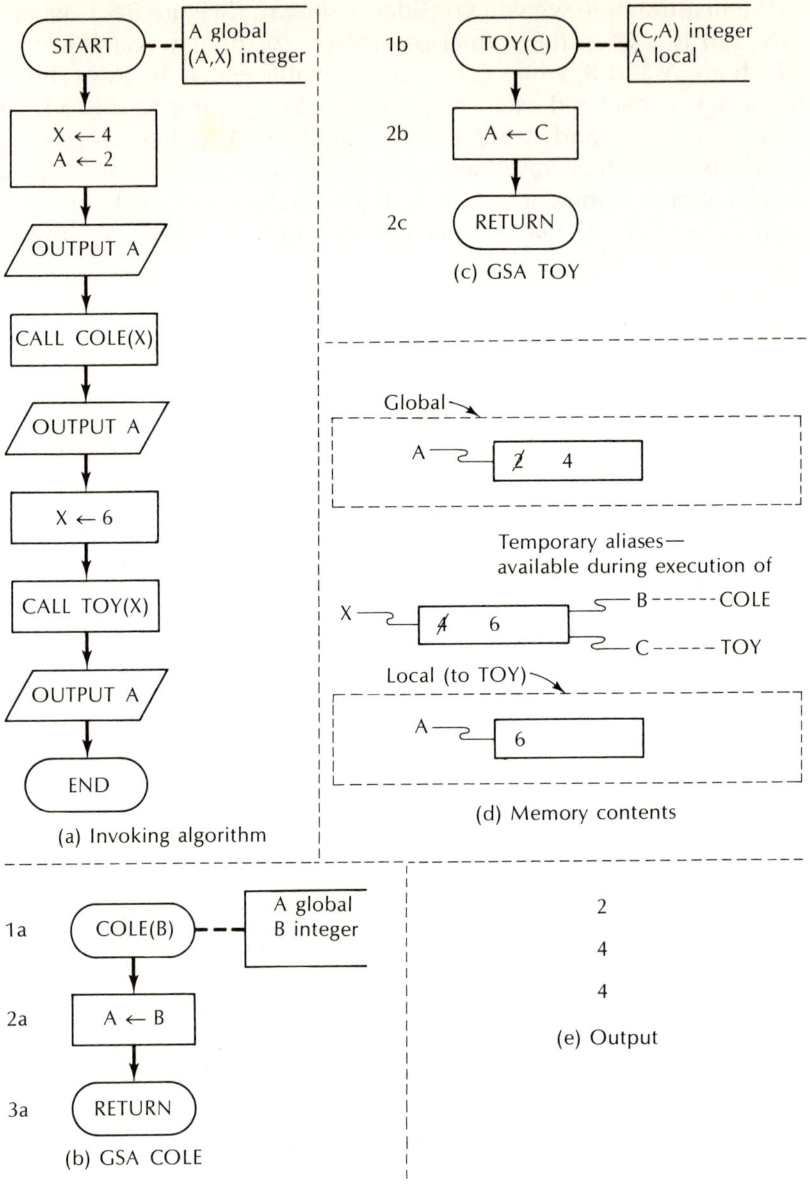

FIG. 8.6-2. Example of local and global variables.

following observations are related to the local and global variables in this example:

☐ The variable A in Figure 8.6-2(c) is a *local* variable—local to TOY, since it is not a formal parameter, and is declared in TOY as a local variable.

☐ The variable A in Parts (a) and (b) of Figure 8.6-2 is a global variable, since it is so declared. It is "global" to the invoking algorithm (or subalgorithm) and the GSA COLE in that any references to A in these routines will reference the same location in global storage.

It follows that the resulting output is as shown in Figure 8.6-2(e).

> *Remark:* It is important to note that the local variable A and the global variable A are *distinct* in that they are stored in separate locations in memory, as shown in Figure 8.6-2(d).

AN ILLUSTRATIVE EXAMPLE 8.7

As a concluding example pertaining to function subalgorithms, let us consider the problem of converting a given roman numeral to its decimal equivalent. The symbols that will be used as roman numerals and their decimal equivalents are shown in Table 8.7-1. All other symbols will be considered invalid.

TABLE 8.7-1. Roman numerals and their decimal equivalents

Roman	I	V	X	L	C	D	M
Decimal	1	5	10	50	100	500	1000

Although this problem, in itself, is rather academic, it illustrates how a certain type of problem in the field of **cryptography** is solved. This field has been the source of many challenging problems, an example of which is the now famous work of Alan Turing related to the theory of computing. Turing was the British scientist who broke the German coding system during World War II. Cryptography is also a field that plays a significant role in providing a means of communicating confidential information; hence, it is of interest to intelligence agencies. In recent years, however, cryptography has also become an issue in modern computer systems; for example, when financial transactions are communicated between computer systems at various banks.

Before presenting a general algorithm that enables conversion of a roman numeral to its decimal equivalent, it is instructive to consider the basic idea of the desired conversion process. This idea is best described by referring to Figure 8.7-1. We seek the decimal equivalent of the roman number XIV. Let S_1, S_2, and S_3 denote the decimal equivalents of X, I, and V respectively. From Table 8.7-1 it follows that $S_1 = 10$, $S_2 = 1$, and $S_3 = 5$. We now define a variable called SUM, the final value of which will be the desired decimal equivalent. Let the initial value of SUM be 0, as indicated in Figure 8.7-1(b). Then SUM is either increased or decreased in accordance with the following rules:

Case 1: $i < 3$. If

$$S_{i+1} > S_i$$

then

$$SUM \leftarrow SUM - S_i \qquad (8.7\text{-}1)$$

else

$$SUM \leftarrow SUM + S_i \qquad (8.7\text{-}2)$$

201

'X', 'I', 'V'

(a) Input stream

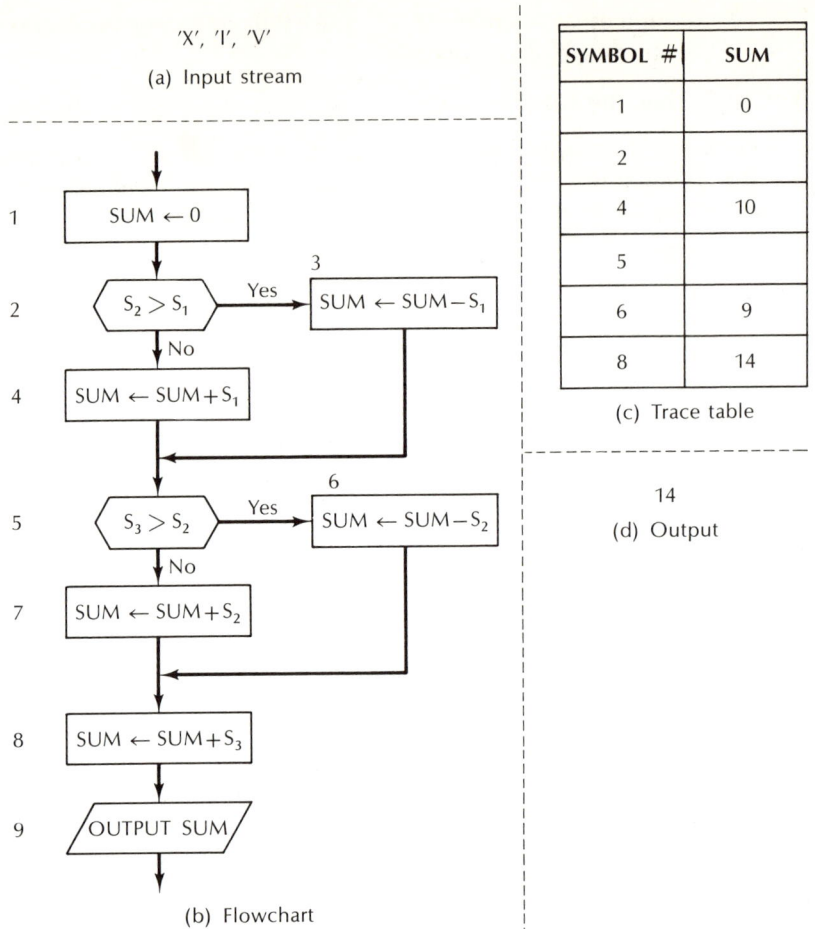

SYMBOL #	SUM
1	0
2	
4	10
5	
6	9
8	14

(c) Trace table

14

(d) Output

(b) Flowchart

(*Note:* It is assumed that the values $S_1 = 10$, $S_2 = 1$, and $S_3 = 5$ have been assigned by a prior process.)

FIG. 8.7-1. Converting a roman numeral to its decimal equivalent.

Case 2: $i = 3$

$$SUM \leftarrow SUM + S_3 \qquad (8.7\text{-}3)$$

From Equations (8.7-1) and (8.7-2) it is apparent that the value of SUM can be increased or decreased in a systematic way by merely comparing successive values of S_i, for $i < 3$. When $i = 3$, SUM is incremented according to Equation (8.7-3). The preceding steps are summarized in Figure 8.7-1(b), and the corresponding trace table is shown in Part (c) of the figure. The final output is thus 14, which is the decimal equivalent of XIV.

We will now present two solutions to the problem of converting a given roman numeral into the more common decimal form which utilizes the arabic numerals 0,1,2, ... ,9. These solutions will involve GSAs and, without loss of generality, it will be assumed that a given roman numeral consists of a maximum of 30 characters.

Solution #1: We consider the case in which the input is a series of individual characters. Each series represents a roman numeral, and successive roman numerals are separated by the blank character—e.g., 'X', 'I', 'V', '$_\wedge$', 'L', 'I', '$_\wedge$'. The output will be the roman numeral input followed by either (a) the corresponding decimal equivalent or (b) an error message. Then, the initial solution is as follows:

0. Repeat Steps 1–4 until you run out of input data.
1. Read roman numeral and store it in an array, NUM.
2. Echo check the roman numeral.
3. Convert the roman numeral.
4. Print the results of the conversion.
5. Terminate.

A refinement of this initial solution would be

0. Repeat Steps 1–4 until you run out of input data.
1. Read roman numeral and store it in an array, NUM.
   1.1 Set counter value of I to 1.
   1.2 Repeat Steps 1.3 through 1.4 until NUM(I) equals '$_\wedge$'; then transfer to Step 2.
   1.3 Read I-th character of roman numeral, (NUM).
   1.4 Increment I by 1.
2. Echo check the roman numeral.
   2.1 Print value of NUM(J) for values of J from 1 through (I − 1).
3. Convert the roman numeral.
   3.1 Let AVALUE denote desired decimal equivalent.
   3.2 Set AVALUE to zero and J to 1 (J will be used as a subscript to access roman numerals). Set I to 1 (I will be used as an error indicator; I = 0 implies that an error has occurred during conversion, while I = 1 implies no error has occurred).
   3.3 Convert J-th roman numeral to its decimal equivalent.
   3.4 If a digit is not valid, set I to zero and go to Step 4.
   3.5 Adjust AVALUE to account for value of current roman numeral as indicated in Figure 8.7-1).
   3.6 If conversion is complete, then go to Step 4; else increment J by 1 and return to Step 3.3.
4. Print results of the conversion.
   4.1a If roman numeral is not valid,
   4.1b Then print error message and return to Step 1.
   4.2 Else print value of converted roman numeral (i.e., its decimal equivalent) and return to Step 1.
5. Print termination message.

203

We now implement Step 3 in the preceding solution via a GSA which has the following two inputs:

1. the roman numeral in the form of single characters up to a maximum of 30 characters, and

2. the number of characters in the roman numeral.

The corresponding outputs are

1. an indicator for a correct conversion, and

2. the desired converted value.

We shall refer to this GSA as TRANS, and associate four formal parameters with it, as indicated below:

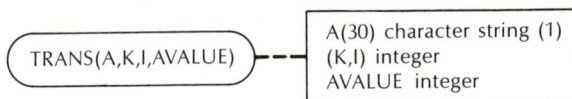

$$\text{TRANS(A,K,I,AVALUE)} \quad - - - \quad \begin{array}{l} \text{A(30) character string (1)} \\ \text{(K,I) integer} \\ \text{AVALUE integer} \end{array}$$

The overall algorithm to convert a given roman numeral (up to 30 characters long) to its decimal equivalent is shown in Figures 8.7-2 and 8.7-3. These figures show the invoking algorithm and the corresponding GSA, respectively. Symbols 1.1 through 1.4 in the invoking algorithm are concerned with reading in sequences of roman numerals, where successive sequences are separated by blanks. The invocation of TRANS in Symbol 3 of Figure 8.7-2(b) communicates the values of NUM, SW, and VALUE by location, as the actual parameters are variable names. The second actual parameter is $I - 1$ which will have its value transferred on a value basis. We leave it to the reader to determine why the value of $I - 1$ is transferred instead of that of I. Because of the form of the parameter lists, the formal parameter array A has the same contents as the array NUM in the invoking algorithm. That is,

$$A(1) = 'X', A(2) = 'I', A(3) = 'V', A(4) = '_{\wedge}'$$

In the GSA there are two variables; namely, LDIG and DIG. The decimal values associated with successive roman numerals are stored in the locations assigned to LDIG and DIG. The portion of the GSA in Figure 8.7-3(a), is very similar to the algorithm represented by Figure 8.7-1(b). We observe that the process of comparing LDIG and DIG is identical to that of comparing successive values of S_i.

In the GSA shown in Figure 8.7-3, we note that there are *two* RETURN statements:

1. The RETURN statement in

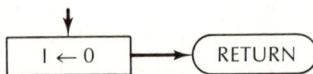

$$\boxed{I \leftarrow 0} \longrightarrow \text{RETURN}$$

is executed if an illegal character is found in a given sequence of roman numerals. Clearly, $I = 0$ causes SW to equal zero

'X', 'I', 'V', '∧'

(a) Input stream

SYMBOL #	I	NUM(I)
1.1	1	
1.3		'X'
1.4	2	
1.3		'I'
1.4	3	
1.3		'V'
1.4	4	
1.3		'∧'
2.1		
3		
4.1a		
4.2		
5		

(c) Trace table

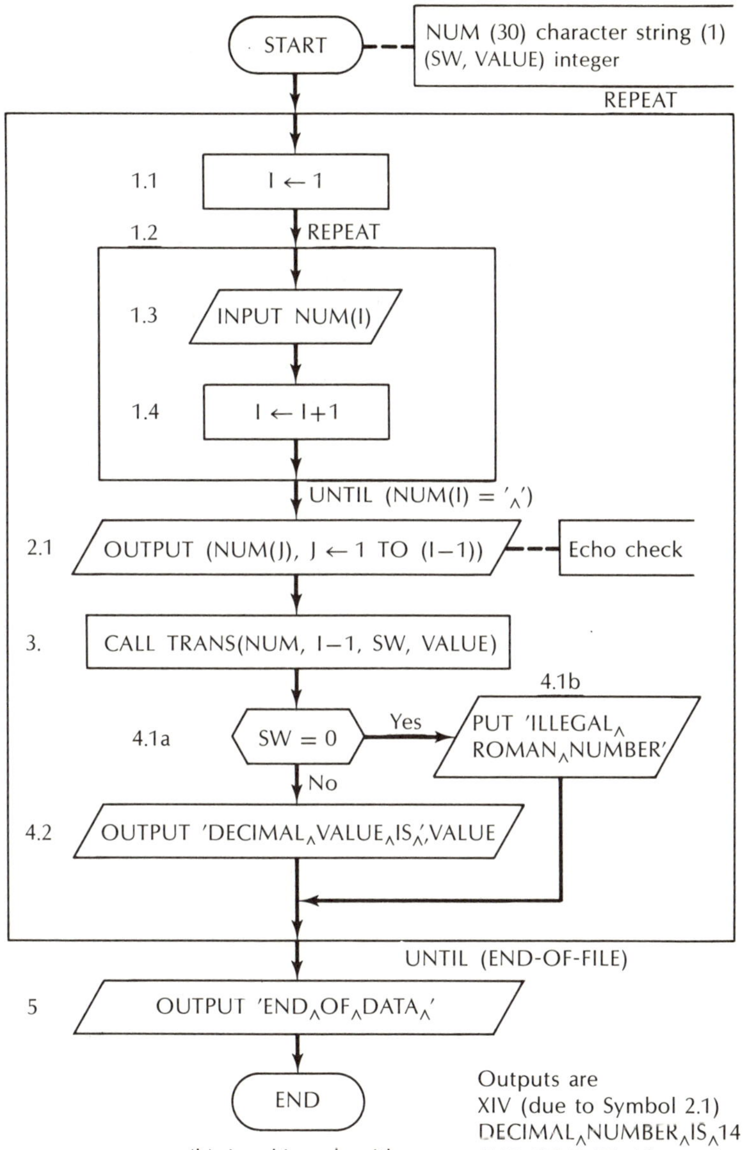

START ──── NUM (30) character string (1)
(SW, VALUE) integer

REPEAT

1.1   I ← 1

1.2   REPEAT

1.3   INPUT NUM(I)

1.4   I ← I+1

UNTIL (NUM(I) = '∧')

2.1   OUTPUT (NUM(J), J ← 1 TO (I−1)) ──── Echo check

3.   CALL TRANS(NUM, I−1, SW, VALUE)

4.1b

4.1a   SW = 0 ──Yes──→ PUT 'ILLEGAL∧ ROMAN∧NUMBER'

No

4.2   OUTPUT 'DECIMAL∧VALUE∧IS∧'VALUE

UNTIL (END-OF-FILE)

5   OUTPUT 'END∧OF∧DATA∧'

END

(b) Invoking algorithm

Outputs are
XIV (due to Symbol 2.1)
DECIMAL∧NUMBER∧IS∧14 (due to Symbol 4.2)
END∧OF∧DATA (due to Symbol 5)

FIG. 8.7-2. Invoking algorithm for roman numeral conversion (Solution #1).

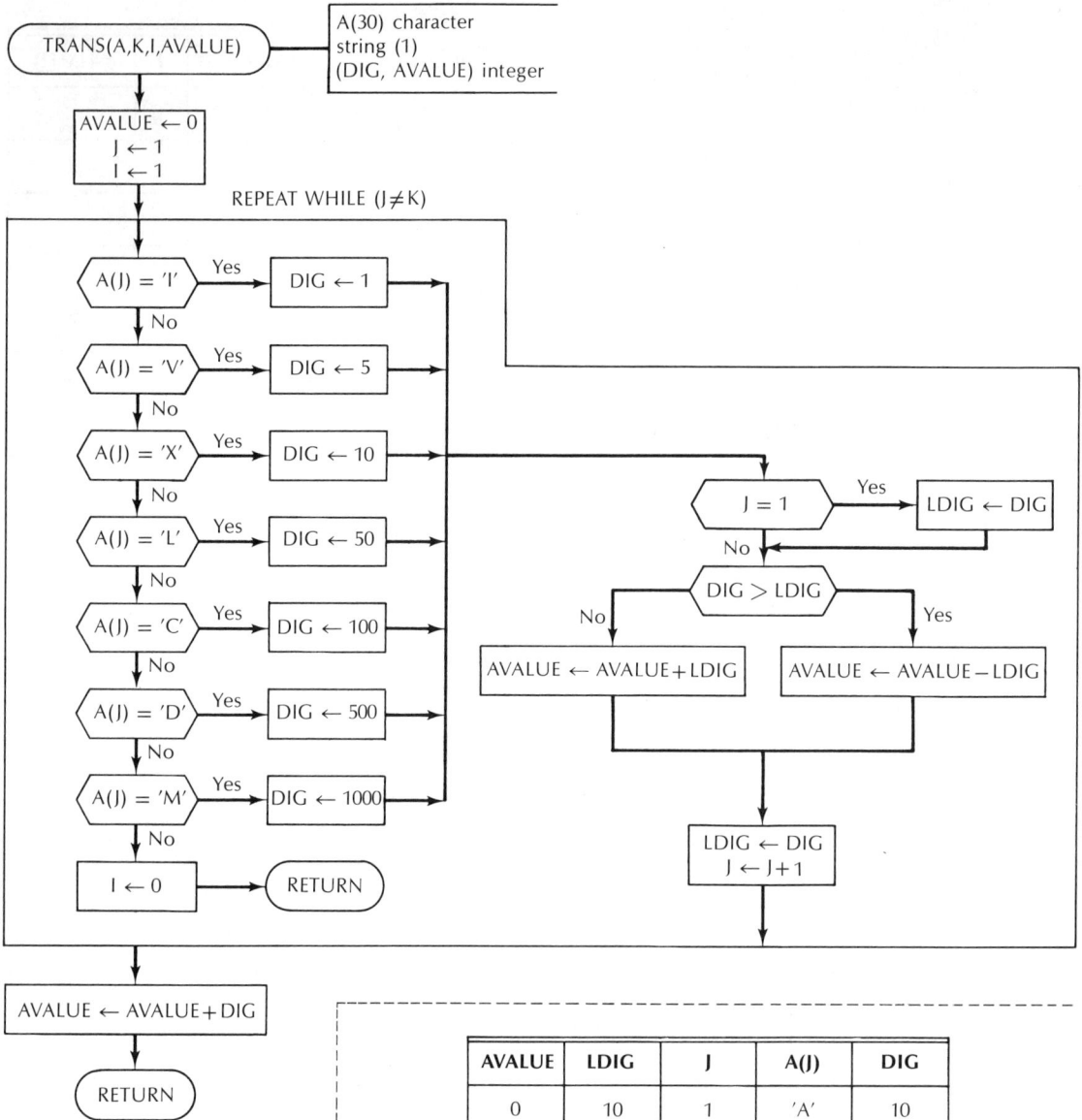

FIG. 8.7-3. GSA for roman numeral conversion (Solution #1).

The flowchart shows:

TRANS(A,K,I,AVALUE) → A(30) character string (1), (DIG, AVALUE) integer

AVALUE ← 0
J ← 1
I ← 1

REPEAT WHILE (J≠K)

A(J) = 'I' Yes → DIG ← 1
No
A(J) = 'V' Yes → DIG ← 5
No
A(J) = 'X' Yes → DIG ← 10
No
A(J) = 'L' Yes → DIG ← 50
No
A(J) = 'C' Yes → DIG ← 100
No
A(J) = 'D' Yes → DIG ← 500
No
A(J) = 'M' Yes → DIG ← 1000
No
I ← 0 → RETURN

J = 1 Yes → LDIG ← DIG
No
DIG > LDIG
No → AVALUE ← AVALUE+LDIG
Yes → AVALUE ← AVALUE−LDIG

LDIG ← DIG
J ← J+1

AVALUE ← AVALUE+DIG

RETURN

(a) GSA TRANS

AVALUE	LDIG	J	A(J)	DIG
0	10	1	'A'	10
10	2	2	'I'	1
9	1	3	'V'	5
14				

(b) Condensed trace table for the input 'X', 'I', 'V', '∧'

206

(see Symbol 4.1a), resulting in a transfer of control to the Symbol 4.1b, and subsequently a new roman numeral is read in.

2. The RETURN statement in

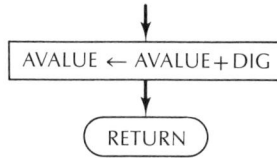

```
        ↓
  ┌──────────────────────────┐
  │ AVALUE ← AVALUE+DIG       │
  └──────────────────────────┘
        ↓
     ( RETURN )
```

is executed when the conversion of a given roman numeral is complete. The value returned is the current value of AVALUE. Thus,

$$SW = 1$$

and

$$VALUE = \text{current value of AVALUE.}$$

Since SW is equal to 1, the OUTPUT statement in Symbol 4.2 is executed, and the output that appears is the value of AVALUE that was returned from the GSA TRANS through the fourth parameter.

For the purpose of illustration, trace tables associated with the specific input 'X', 'I', 'V', '$_\wedge$' are included in Figures 8.7-2 and 8.7-3. The corresponding output (which equals 14) is shown in Figure 8.7-3(b); it is returned to VALUE in Figure 8.7-2(c) at Symbol 3.

Solution #2: We now assume that the input is not presented in the form of individual characters, but rather as a single string of roman numerals, such as 'XIV'. The structured algorithm presented in the former solution must be modified by replacing Steps 1 and 2 with the following sequence of steps:

1. Read a roman numeral, extract single characters from it, and store them in an array of single characters.
    1.1 Read roman numeral into a character string variable capable of storing a varying number of characters up to 30.[4]
    1.2 Repeat Steps 1.3 through 1.4 for the number of characters in the string.
    1.3 Using the SUBSTR operation (see Section 4.4), extract the first character and store it in an array element; echo this character.
    1.4 Modify the original string by removing the first character.

---

[4]Some languages allow strings to vary in length. The length is changed dynamically to the exact number of characters in the constant assigned to (or read into) that variable. The maximum length the string will attain must still be indicated.

Step 1 in the revised algorithm is now implemented using a GSA called APART, the input of which is the given roman numeral string. The corresponding output consists of

(a) the length of the string, and
(b) the roman numeral.

The formal parameters associated with APART are as indicated below:

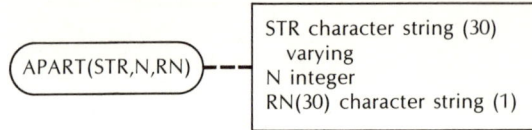

```
APART(STR,N,RN) --- STR character string (30)
                        varying
                    N integer
                    RN(30) character string (1)
```

The GSA is shown in Figure 8.7-5, while Figure 8.7-4 shows the corresponding revised invoking algorithm. The reader should verify that this

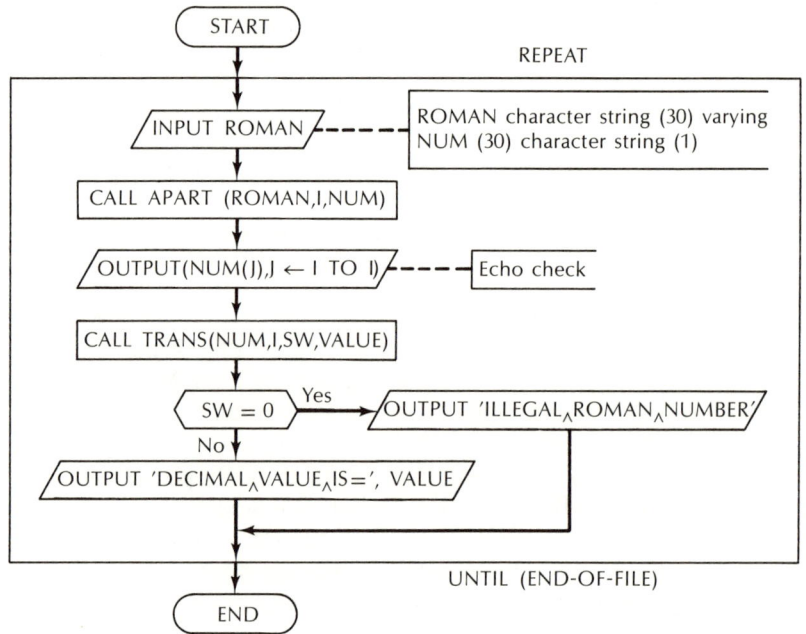

```
                    START

                              REPEAT

          INPUT ROMAN --- ROMAN character string (30) varying
                          NUM (30) character string (1)

        CALL APART (ROMAN,I,NUM)

        OUTPUT(NUM(J),J ← I TO I) --- Echo check

        CALL TRANS(NUM,I,SW,VALUE)

                    SW = 0  Yes  OUTPUT 'ILLEGAL ROMAN NUMBER'
                    No
        OUTPUT 'DECIMAL VALUE IS=', VALUE

                                    UNTIL (END-OF-FILE)

                    END
```

(a) Invoking algorithm

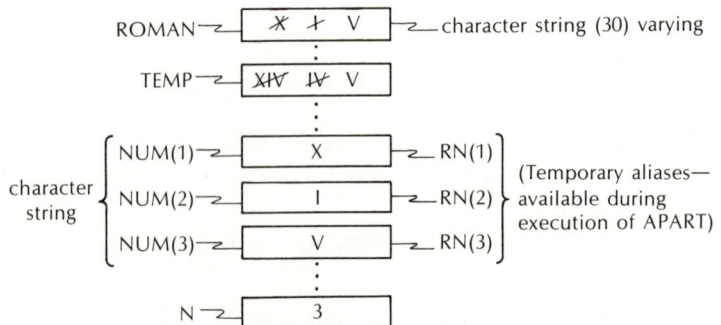

```
ROMAN —   X X V   — character string (30) varying

TEMP —    XIV IV V

         NUM(1) —    X     — RN(1)
character
 string  NUM(2) —    I     — RN(2)   (Temporary aliases—
                                      available during
         NUM(3) —    V     — RN(3)   execution of APART)

         N —    3
```

(b) Memory contents for the input 'XIV'

FIG. 8.7-4. Invoking algorithm for roman numeral conversion (Solution #2).

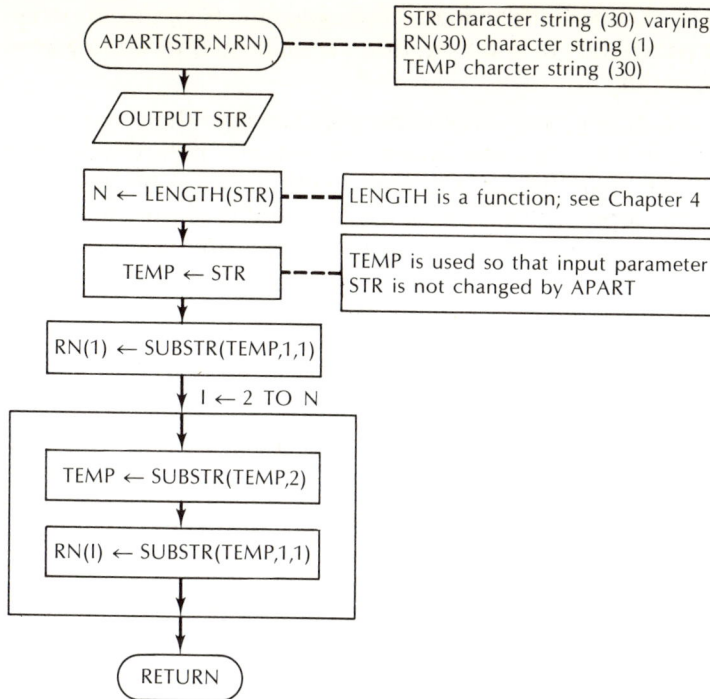

FIG. 8.7-5. One of the GSAs used in the roman numeral conversion (Solution #2).

revised algorithm is indeed correct. It should be observed that the amount of detail necessary is appreciably reduced in this case, as is evident by comparing Figure 8.7-4 with Figure 8.7-2.

SUMMARY 8.8

We introduced the important notion of using subalgorithms in this chapter. Two types were considered; namely, function and general subalgorithms. We discussed communication between invoking algorithms and corresponding invoked subalgorithms using actual and formal parameters. That such parameters are related in a one-to-one manner was illustrated by a variety of examples, with the assumption that the formal parameters were passed by location. In Section 8.5 the concept of communicating through formal parameters by value was also introduced. We pointed out that formal parameters which are constants or expressions are commonly invoked by value. On the other hand, invoking by location is relevant to actual parameters and is useful when arrays are involved in parameter lists. If a formal parameter is a simple variable, invoking it by either name or value is equally convenient.

In Section 8-6 we considered elements of global storage as a means of communication, mainly by means of examples. We noted that corresponding side effects (i.e., unexpected errors) can occur, especially when the main algorithm and the subalgorithm are not devel-

209

oped by the same individual. Other side effects can occur when the input parameters of a subalgorithm are changed in value within the subalgorithm.

In conclusion, we emphasize that subalgorithms are extremely useful in reducing the time, effort, and expense of the development of the overall algorithm for a desired solution. This is because the sub-algorithm approach enables a group of people to work simultaneously toward a solution. Yet one programmer can use all previous related work, since it can be made available in the form of subalgorithms.

problems

8-1.   Consider the following algorithm:

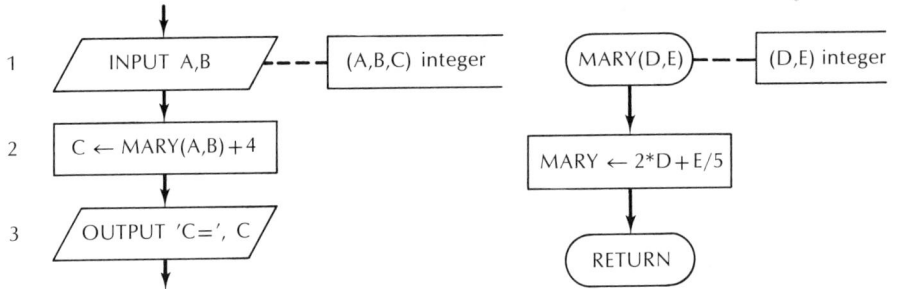

The input stream is 3,10.
(a)   What are the actual parameters?
(b)   What are the formal parameters?
(c)   Is MARY an FSA or GSA?
(d)   What is the value of C when Symbol 3 is executed?

8-2.   (a)   Develop an algorithm that uses an FSA to evaluate the product of the values of two given variables, and returns it to the invoking algorithm.
       (b)   Same problem statement as in (a), except that the algo-rithm should consist of a GSA.

8-3.   The following flowchart sequence is given:

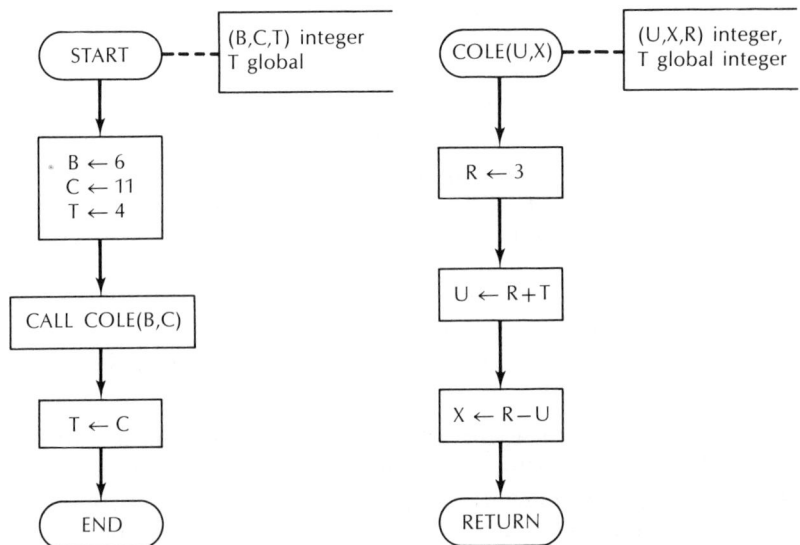

(a) Identify the following:
Local variable(s) _____; Global variable(s) _____;
Formal parameter(s) _____; Actual parameter(s) _____.
(b) Trace through the memory conditions by referring to the variables U, B, X, C, T, and R.

8-4. Given pairs of values for two variables A and B, the FSA QUAD(A,B) is to return an integer value 1, 2, 3, or 4 according to the quadrant in which the point (A,B) lies. Develop an algorithm for determining the quadrant associated with a given pair of values of A and B.

8-5. A loan company lends money at compound interest of R percent per month. That is, each month R percent of what a borrower owes is added to the debt he or she must repay some day. Develop an FSA IRATE(N,R,L) that computes the amount that must be paid after N months for a loan of L dollars at R percent per month interest.

Use this algorithm to find the amount that a borrower pays on a loan of 100 dollars at 1 percent per month interest for a duration of 12 months, 24 months, and 36 months.

8-6. Rework Problem 8-5 by developing IRATE as a GSA.

8-7. Develop an algorithm for the GSA SHUFFLE(X,Y,Z) that takes three numeric variables X, Y, and Z, and places the largest in X, the second largest in Y, and the smallest in Z.

8-8. Develop an algorithm for a GSA to compute the following general compound interest formula:

$$AMT = P(1 + R/M)^{M \cdot N}$$

where

AMT is the amount at the end of N years;
P is the principal;
R is the interest rate; and
M is the number of times per year that interest is paid.

Write an invocation statement with associated variable declarations and explanations.

8-9. The square root of the value of a variable X can be computed iteratively to an accuracy of 0.001 using the FSA SQRT(X) shown in Figure P 8-9.
(a) Develop an appropriate invoking algorithm in flowchart form.
(b) Use the algorithm to find the square root of 18.0, and show the corresponding trace table.

211

FIG. P 8-9

8-10. The factorial of a positive integer is denoted by N!, and is calculated as

$$N \cdot (N-1) \cdot (N-2) \cdot \ldots \cdot 2 \cdot 1$$

For example,

$$6 \text{ factorial} = 6! = 6 \cdot 5 \cdot 4 \cdot 3 \cdot 2 \cdot 1$$

and

$$4 \text{ factorial} = 4! = 4 \cdot 3 \cdot 2 \cdot 1$$

Develop an FSA which computes N!. Use this algorithm to evaluate 5!, and show the corresponding trace table.

8-11. Solve the Dutch National Flag problem for an array of 100 elements, using a GSA. The problem is to separate the elements of an array with the values 'RED', 'WHITE', and 'BLUE', rearranging them so that all the reds come first, the whites next, and finally the blues.

8-12. Redesign the GSA TRANS in Figure 8.7-3 to an FSA form. Recall that only one value can be returned; thus, the error condition must be dealt with within the FSA. The value returned in some way communicates this information to the invoking algorithm.

problems for computer solution

The general learning objectives for this set of problems are

☐ to be able to use a simple subalgorithm;

☐ to gain experience with formatted input/output; and

☐ to become familiar with multidimensional arrays.

These problems are written in a less definitive fashion than those presented previously. The purpose is to gradually introduce the reader to the realities of production programming situations.

CS 8-1. Create a daily production report for the general manager of a firm. Of interest is the production in the three divisions—manufacturing, assembly, and finishing. The company produces eight different products but may not produce all products each day.

You receive punched cards containing the data. Each card is produced as a crew in one of the divisions completes a batch of products; therefore, the cards are not in consistent order. They contain division number (1–3), product number (1–8), quantity produced, and cost of labor for production.

Produce a report of the form

	Manufacturing		Assembly	Finishing	Total	
Product	Units	Labor	...		Units	Labor
Bolts	XXXX	$XXX.XX				
Shoes						
⋮						
Total	XXXXXX	$XXXX.XX	...			

Use a subalgorithm to accomplish the printing of the report.

(COBOL, PL/I [business])

CS 8-2. A very common application program is that of maintaining a credit-card master file. Consider such a program for T. Vanity's Bookstore. The old master file must include the following information:
(a) customer account number;
(b) customer name;
(c) customer address;
(d) customer city, state, zip code;
(e) previous month's balance;
(f) current charges for this month; and
(g) billing date.
The new master file to be produced by your algorithm will include this same information, updated. The transaction file must include
(a) customer account number;
(b) transaction code;
  ☐ a value of 1 indicates a charge sale
  ☐ a value of 2 indicates a payment on account
  ☐ a value of 3 indicates the account has been closed
(c) transaction amount.

(COBOL, PL/I [business])

CS 8-3. Develop a general subalgorithm for updating a payroll file. Test this subalgorithm by enclosing it in a program.

The parameters are the quantity of the item sold and the stock number. Assume that the inventory master file is

213

stored on a magnetic disk. Each master file record contains the following:

(a) name;

(b) SS number;

(c) total sales to date;

(d) base hourly rate; and

(e) commission percentage.

Each time this subalgorithm is invoked, the payroll master file should be searched until the appropriate record is found. After all input data has been processed, the master file should be printed.

(COBOL, PL/I [business])

CS 8-4. Students at the state university can be classified by class and college as follows:

Class code		College code	
Freshman	0	Agriculture	0
Sophomore	1	Arts & Sciences	1
Junior	2	Business	3
Senior	3	Engineering	4

Each student is assigned a class code and a college code. Read the student information (30 students) and print out a cross-tabulation table that indicates the enrollment of students by class and college. Example:

	Agriculture	Arts & Sciences	Business	Engineering	Total
Freshman	1	...			
Sophomore	2	...			
Junior	7	...			
Senior	1	...			
Total	11	...			30

A subalgorithm should be invoked to print this table. Pass the subalgorithm the array (table) of numbers, size of the table, and the lists of column and row labels.

(PL/I [general])

CS 8-5. *Additional learning objective:* to gain experience with a graphics terminal.

Using the graphic subroutines, draw a square located toward the center of the screen. Then construct a dashed line from the upper left corner to the lower right corner of the square. Use a loop to construct this as a series of short vectors. Label the four corners of the rectangle. Example:

(FORTRAN [architecture])

CS 8-6. *Additional learning objective:* to gain experience with nested definite loops.

You are to solve a system of three equations in three unknowns using the Gaussian reduction method. In short, this method is to reduce a system of the form

$$a_{11}x_1 + a_{12}x_2 + a_{13}x_3 = C_1$$
$$a_{21}x_1 + a_{22}x_2 + a_{23}x_3 = C_2$$
$$a_{31}x_1 + a_{32}x_2 + a_{33}x_3 = C_3$$

to

$$x_1 + a'_{12}x_2 + a'_{13}x_3 = C'_1$$
$$x_2 + a_{23}x_3 = C'_2$$
$$x_3 = C'_3$$

This is accomplished by dividing the first equation by a_{11} (assuming $a_{11} \neq 0$) and using the result to eliminate x_1 from the two succeeding equations. Similarly, the second equation is divided by a'_{22} and x_2 is eliminated from the third equation, etc. A back substitution of x_3 results in the solution.

Read the data using formatted input and echo the input using formatted output. The main algorithm is to read the input and print the results. A subalgorithm is to form the reduction, accomplish the back solution, and transmit the results x_1, x_2, and x_3 back to the main algorithm.

(FORTRAN [engineering], PL/I [scientific])

CS 8-7. One of the important advantages of the computer is the speed and flexibility of its graphic output. For this problem you will produce a graphical representation of components of soil composition. Use a simple subalgorithm to generate the graphical part of the output. The labels that will appear on the graph and the graphic symbol (asterisks, in this case) that will compose the bars of the graph should be read as data. The scale will range from 0 to 50. The input will consist of several sets of data, each consisting of the four components found in each sample. These four values will represent

215

the amounts of SAND, LOAM, CLAY, and ROCK, respectively.

Use the following form for your output:

```
              SOIL COMPOSITION INPUT VALUES


    SAND          LOAM            CLAY           ROCK
 ************************************************************

    32            19              45             11
 ************************************************************


                  SOIL COMPOSITION


                                                          ─ 50



                                        *
                                        *
                                        *                 ─ 40
                                        *
                                        *
                                        *
                                        *
                                        *
                                        *
            *                           *                 ─ 30
            *                           *
            *                           *
            *                           *
            *                           *
            *                           *
            *                           *
            *              *            *                 ─ 20
            *              *            *
            *              *            *
            *              *            *
            *              *            *
            *              *            *          *
            *              *            *          *       ─ 10
            *              *            *          *
            *              *            *          *
            *              *            *          *
            *              *            *          *
            *              *            *          *
            *              *            *          *
            *              *            *          *       ─ 0
          SAND           LOAM         CLAY        ROCK
```

FIG. CS 8-7

(FORTRAN [engineering, general])

CS 8-8. Game playing by computer has become increasingly popular. Besides its entertainment value, it has a very serious purpose of providing insight into the way people solve problems. One subproblem from this area comes from the game of chess. A knight in chess can make L-shaped moves on a board with eight columns and eight rows. It can move two spaces forward, backward, and then an extra square to either side; or, it can move two spaces sideways and then one up or down. See the accompanying diagram for all legal moves from a position.

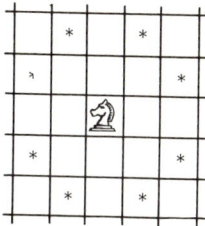

*Possible move, record by (i,j),
where i = row number
and j = column number.

Number the columns one through eight and likewise the rows. Given a position by row and column number, have a subalgorithm return in an array the eight or fewer moves (final square) the knight may make. Remember, the board has edges.

Let the main program call the subalgorithm for a series of positions. Let the process terminate when there are no more positions in the input.

(FORTRAN [general], PL/I [general])

CS 8-9. *Additional learning objectives:* to use a built-in subroutine RANDU and to become acquainted with simulation techniques.

To help finance your education, you and several friends have just purchased a pizza parlor. You have been asked to figure out how many waitresses/waiters you should hire. Obviously, if you have too few, then customers will have to wait a long time to be served and you will lose business. On the other hand, too many and you will lose money paying them. Thus, you decide to simulate the process. You speculate that every minute the odds are 50/50 that a new customer will come in, and that it takes four minutes to serve each customer. Use one, two, and three waitresses/waiters to see how many it would be best to have. Print out a table using formatted I/O showing total time customers spend waiting and the amount of time waitresses/waiters spend waiting.

Use a simple subalgorithm and the built-in subroutine for generating random numbers to simulate arrivals.

Your main program will consist of a loop, reflecting what happens each minute. *Hint:* It will be helpful if you keep track of
(a)   how many customers are waiting;
(b)   how many waitresses/waiters are waiting;

217

(c) how many waitresses/waiters have just started serving;

(d) how many waitresses/waiters have been with customer one minute;

(e) how many waitresses/waiters have been with customer two minutes;

(f) total time customers have spent waiting; and

(g) total time waitresses/waiters have spent waiting.

Note: This is a difficult problem.

(FORTRAN [general, scientific], PL/I [computer science])

CS 8-10. In many scientific computing problems, matrices are manipulated. This problem requires a formatted input of data for two matrices A and B. These matrices are to be passed to a subalgorithm MATMU1 that will form the matrix product of the two matrices and return this product to the calling routine. The resultant matrix is to be printed out using formatted output.

Recall that matrix multiplication is calculated as

$$
\begin{bmatrix}
c_{11} & c_{12} & \cdots & c_{1N} \\
c_{21} & c_{22} & \cdots & c_{2N} \\
\vdots & \vdots & c_{ij} & \vdots \\
c_{N1} & & \cdots & c_{NN}
\end{bmatrix}
=
\begin{bmatrix}
a_{11} & \cdots & a_{1N} \\
a_{21} & \cdots & a_{2N} \\
\vdots & & \vdots \\
a_{N1} & \cdots & a_{NN}
\end{bmatrix}
\cdot
\begin{bmatrix}
b_{11} & \cdots & b_{1N} \\
b_{21} & \cdots & b_{2N} \\
\vdots & & \vdots \\
b_{N1} & \cdots & b_{NN}
\end{bmatrix}
$$

where

$$c_{ij} = a_{i1} \cdot b_{1j} + a_{12} \cdot b_{2j} + \cdots + a_{iN} \cdot b_{Nj}$$

(FORTRAN [engineering, scientific], PL/I [scientific])

CS 8-11. The problem is to create an algorithm that will solve a system of two linear equations in two unknowns.

Arrange for the input of the coefficients of the following two equations:

$$A_{1,1}X + A_{1,2}Y = A_{1,3}$$
$$A_{2,1}X + A_{2,2}Y = A_{2,3}$$

Also read in the physical definitions of X and Y as string data limited to 12 characters to be used to label your output.

Recall the determinant method for the solution of simultaneous equations. Cramer's rule says if the determinant

$$\begin{vmatrix} A_{1,1} & A_{1,2} \\ A_{2,1} & A_{2,2} \end{vmatrix} = A_{1,1}A_{2,2} - A_{2,1}A_{1,2} \neq 0$$

then

$$X = \frac{\begin{vmatrix} A_{1,3} & A_{1,2} \\ A_{2,3} & A_{2,2} \end{vmatrix}}{\begin{vmatrix} A_{1,1} & A_{1,2} \\ A_{2,1} & A_{2,2} \end{vmatrix}}$$

and

$$Y = \frac{\begin{vmatrix} A_{1,1} & A_{1,3} \\ A_{2,1} & A_{2,3} \end{vmatrix}}{\begin{vmatrix} A_{1,1} & A_{1,2} \\ A_{2,1} & A_{2,2} \end{vmatrix}}$$

For two simultaneous equations, it is not necessary to use a subalgorithm to calculate the various determinants; however, you are required to do so here.

Write a subalgorithm which, when given, will calculate the determinant of A.

Can you generalize the solution to (n × n) matrices?

(FORTRAN [scientific])

CS 8-12. *Additional learning objective:* to use built-in functions.

Write and test four FUNCTIONS called SINE, COSINE, ARCTAN, and EXP, each of which takes a real argument, X, and returns the value of sine(x), cosine(x), arctan(x), and e^x, respectively. The values are computed by adding up enough terms of an infinite series expansion, so that the change caused by adding in the next term is less than the accuracy desired of 1.0E−6. Use a separate subalgorithm to compute each function.

Next, compare the values computed by your functions to the values returned by the built-in functions.

The expansions to be used are as follows:

$$SINE(X) = \frac{X^3}{3!} + \frac{X^5}{5!} - \frac{X^7}{7!} + \cdots$$

$$EXP(X) = 1 + X + \frac{1}{2!}X^2 + \frac{1}{3!}X^3 + \frac{1}{4!}X^4 + \cdots$$

(FORTRAN [engineering, scientific], PL/I [scientific])

CS 8-13. You are to build a table of values of e^x in a vicinity around a given value of x. For each given value of x, print the following table formed as shown in a well-labeled format.

e^{x-1}	$e^{x-.8}$
$e^{x-.6}$	$e^{x-.4}$
$e^{x-.2}$	e^x
$e^{x+.2}$	$e^{x+.4}$
$e^{x+.6}$	$e^{x+.8}$
e^{x+1}	

The value of e^x is to be calculated in a subalgorithm using the Taylor series expansion:

$$e^x = 1 + x + \frac{x^2}{2!} + \frac{x^3}{3!} + \cdots$$

219

The subalgorithm is to add terms to the expansion until the term to be added is less than 0.00001.

The program should process as many values of x as appear in the input stream.

(FORTRAN [scientific])

CS 8-14. The Variety Manufacturing Company produces a multitude of different products, but never more than five simultaneously. Two departments, assembling and finishing, are involved in the manufacturing process. The manager needs a daily report of all the costs incurred for each product grouped by department. Provide this daily cost report showing daily costs by product and department, plus the total cost by department. The input data will consist of the date, the names of all products produced that day, and a series of records containing the following information:

Product code	Cols. 1–3
Department code	Cols. 4–6
Cost (in dollars and cents)	Cols. 10–20

An implied decimal is to occur between Columns 18 and 19. A given product and department may have more than one daily entry.

Create a report similar to the given sample.

DAILY COST REPORT

22 JULY 1979

	Assembling	Finishing
"Name of Product 1"		
"Name of Product 2"		
⋮ ⋮		
	"Total"	"Total"

(COBOL, PL/I [business])

CS 8-15. *Additional learning objective:* to understand the concept of table lookup.

The tax table that follows can be used for computing the tax due for any person with income between 4250 and 4750 and no more than five exemptions. For this problem you will be given the name, income, and exemption information for a number of people.

Write a program that will read and echo check the tax table, read tax-payer information, select from the tax table the correct tax due for each person, and then print all tax-payer information plus the tax due. Prepare the output so that it looks neat.

TAX TABLE

INCOME	EXEMPTIONS				
	1	2	3	4	5
4250–4300	540	410	306	188	96
4301–4350	545	415	316	198	106
4351–4400	561	424	322	204	110
4401–4450	569	431	330	221	117
4451–4500	577	439	338	229	124
4501–4550	586	443	346	236	131
4551–4600	594	454	354	244	139
4601–4650	602	462	362	257	146
4651–4700	610	470	370	259	154
4701–4750	618	477	378	266	161

Use a simple subalgorithm to read the tax table.

(PL/I [business])

CS 8-16. *Additional learning objective:* to gain experience with table lookup.

The table that follows shows the number of tickets remaining in the various sections and at the various prices for an upcoming rock concert. For this problem you will be queried by distributors as to whether a specific number of tickets are available within a section at a price.

Write a program that will use a simple subalgorithm to read and echo check the tickets-available table. Then read a request that will include the number of tickets requested, section, and price. Respond appropriately to the request and prepare the output so that it looks neat. Part of the problem is to update the table.

TICKET PRICE	SECTIONS					
	1	2	3	4	5	6
$5.00	200	250	150	100	125	175
$4.00	300	100	0	50	250	300
$3.00	200	125	150	175	150	150

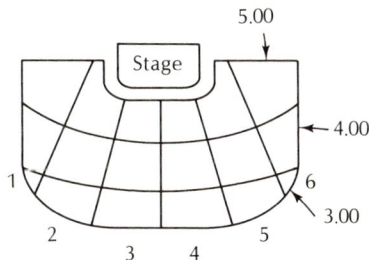

(PL/I [general, business])

CS 8-17. *Additional learning objective:* to gain experience with table lookup.

You are working for a company that has decided to print information sheets from their new computer, so that payroll checks can be more easily typed. Each employee has a salary-range code and number of dependents as two values in their master payroll record. Using these, employees can access the tax table that follows for their tax rates.

Input a record for each employee that also includes employee's number and gross salary. Access the tax table for a tax rate, calculate net pay, and print a report that includes employee ID, gross salary, net salary, taxes, salary-range code, and number of dependents.

Before you begin this process read, store, and print the tax table with a subalgorithm.

Number of Dependents

	0	1	2	3	4	5 or more
1	0.15	0.1	0.09	0.07	0.05	0.02
2	0.16	0.12	0.11	0.09	0.06	0.03
3	0.17	0.13	0.12	0.09	0.07	0.05
4	0.19	0.18	0.16	0.12	0.10	0.08

Use formatted output to print the tax table and the output report.

(PL/I [business])

CS 8-18. You are the programmer for Sam's Yacht Sales, Inc. Develop a program to calculate the tax for the following states:

Missouri	4 percent tax rate
Kansas	3 percent tax rate
Nebraska	$3\frac{1}{2}$ percent tax rate
Oklahoma	$2\frac{1}{2}$ percent tax rate

Accept records from one of these four states, and use a subprogram TAX that requires three parameters. The input parameters are the state and amount of sale. The corresponding tax is the desired output. Use a data structure to store the totals of taxes assessed for the states, and the grand total dollars of sales. Format your output in an easy-to-read report fashion.

(PL/I [business])

CS 8-19. *Additional learning objective:* to become acquainted with Polish postfix notation.

A common way to write arithmetic expressions for easy compilation is Polish postfix notation. This notation eliminates the need for parentheses to denote subexpression evaluation and enables a single left-to-right compilation scan.

Write a subalgorithm that will take an expression in infix notation and translate it into Polish postfix notation. To simplify this assignment, eliminate the parenthesized sub-expression and consider only the binary operators +, −, *, / from your algorithm. Examples:

Infix	Polish postfix
A+B	AB+
A+B*C	ABC*+
A*B+C	AB*C+

In Polish postfix notation the compiler may scan left to right. As it encounters an operator, it takes the two immediately preceding operands (e.g., BC) and acts upon them producing a result (operand). The general algorithm is as follows:

(a) Read the expression character by character. As an operand is encountered, output it.

(b) As each operator is encountered, save it until the next operator is encountered. If the next operator is of precedence less than or equal to the previous operator, then output the previous operator repeat if possible. If the next operator is of higher precedence, then save it also (save these in an array).

(c) When the end of expression is encountered, output in reverse order of storage all the operators in the array.

Note that in translating complicated expressions we usually tag each element of the expression with a precedence number. Do this using another array. One suggestion for precedence numbers is

Operand	0
+,−	1
*,/	2

Use a subalgorithm for the tagging process.

(PL/I [computer science])

CS 8-20. *Additional learning objective:* to become acquainted with some aspects of number systems.

All data stored inside of the computer is stored in binary form. When the contents of memory are displayed for a programmer, it may be printed in binary, octal, or the hexadecimal number system. It would be useful, therefore, to have a program that can read in an integer and convert it to these number bases.

Your program must be able to read in an integer and convert it to binary, octal, and hexadecimal. It should print the original decimal number and all of the calculated con-

223

versions appropriately labeled. The conversion routine must be written as a subalgorithm. The subalgorithm should do *no* input/output operations. Its function is to accept the number, the desired base, and return the converted number in the appropriate base. The main program will do the input/output operations with appropriate headings over the output data.

The binary number system consists of the digits 0 and 1. Some examples are

$$0_{10} = 000_2 \qquad 3_{10} = 011_2$$
$$1_{10} = 001_2 \qquad 4_{10} = 100_2$$
$$2_{10} = 010_2 \qquad 5_{10} = 101_2$$

The octal number system consists of the digits 0, 1, 2, 3, 4, 5, 6, and 7. Some examples are

$$0_{10} = 0_8 \qquad 8_{10} = 10_8$$
$$1_{10} = 1_8 \qquad 9_{10} = 11_8$$
$$2_{10} = 2_8 \qquad 10_{10} = 12_8$$

The hexadecimal number system consists of the digits 0, 1, 2, 3, 4, 5, 6, 7, 8, 9, A, B, C, D, E, and F. Some examples are

$$0_{10} = 0_{16} \qquad 12_{10} = C_{16}$$
$$1_{10} = 1_{16} \qquad 13_{10} = D_{16}$$
$$2_{10} = 2_{16} \qquad 14_{10} = E_{16}$$
$$10_{10} = A_{16} \qquad 15_{10} = F_{16}$$
$$11_{10} = B_{16} \qquad 16_{10} = 10_{16}$$

A general algorithm for converting a decimal number to a different base is

(a) digit = MOD (number, base)
(b) char. digit = CHAR (digit)
(c) new base number = char. digit × new base number
(d) number = number/base
(e) if number = $\emptyset$, then new base number contains the converted base number; else repeat Steps (a) through (e).[5]

The calculations performed on Step (d) must be done in integer mode; i.e., there can be no decimal (fractional) places.

Your program should be able to make the conversion for any number of integers.

(PL/I [computer science])

CS 8-21. One of the most interesting structures for information storage in computer science is that of a binary tree. Conceptually, it is somewhat similar to our so-called family tree. A binary tree is a nonlinear structure that capitalizes on a

---

[5]MOD and CHAR are built-in PL/I functions for modulus determination and developing the character representation of a number.

branching relationship present in the data. Each "record" of information in the structure is commonly referred to as a "node". Our tree consists of one entry pointer that identifies the first node (top node) of the tree. This node and *all* others will have 0, 1, or 2 pointers to subsequent nodes. The node fields include left pointer, information, right pointer (*left* and *right* are commonly chosen as pointer names since there will be two pointers, at most).

Consider the following tree:

The left and right pointers in your problem will contain the subscript of the node being pointed to. Use arrays to store the pointers (subscripts) and the data.

Standard operations performed on binary trees include SEARCHING, INSERTING, TRAVERSING, and DELETING. In this problem you are to develop an algorithm that inputs a code (0 indicating search or 1 indicating insert) and the information field consisting of an integer from 1 to 20. If the code is zero, the algorithm is to search the tree to determine if the information field exists in the tree. If the code is one, the information field is to be inserted into the tree at the appropriate position.

Before you terminate your program, print the array used to store your tree.

(PL/I [computer science])

CS 8-22. *Additional learning objectives:* to experience the use of a structure (a nonhomogeneous collection of data) and to become acquainted with the concept of table lookup.

The atomic weight and atomic number of the first seven elements from the periodic table (alphabetically ordered) are as follows:

Name	Symbol	Atomic number	Atomic weight
Actinium	Ac	89	227
Aluminum	Al	13	26.98
Americium	Am	95	243
Antimony	Sb	51	212.76
Argon	A	18	39.944
Arsenic	As	33	74.91
Astatine	At	85	2.0

225

The input to your program is the symbol for an element, and the output is the name of the element, its atomic weight and atomic number.

The main program should read the table and then successively read an input, call a subalgorithm to accomplish the table lookup, and finally print the output.

The subalgorithm should search the list it is given for the search argument and return the position the argument was found to occupy in the list.

(PL/I [general])

CS 8-23. You work for a large publishing firm. At times you handle manuscripts in which the verb tense must be changed. In the one you are currently editing, each occurrence of *be* must be replaced by *been* and *is* must be replaced by *was*. Code a subalgorithm which, when given the wrong tense verb, the correct tense verb, and the text to be searched, will accomplish this task and return the corrected text.

Read in the manuscript, one record at a time, until you are out of data. As each record is read, call your subalgorithm once for each pair *be-been* and *is-was* and then print out its return value. Read and print the data using formats.

(PL/I [general])

CS 8-24. Your concern for ecology has led to the purchase of a new bike. The gear ratios on your bike are as follows:

Gear	Ratio
1	$3.75 : 1 = 3.75$
2	$3.02 : 1 = 3.02$
3	$2.20 : 1 = 2.20$
4	$1.36 : 1 = 1.36$
5	$1 : 1 = 1.00$

The speed of your bike is given by the formula

Speed (miles per hour) = $.02 \cdot$ rpm/gear ratio

For example, your speed at 3750 rpm in low gear would be

$$.02 \cdot 3750 \cdot \frac{1}{3.75} = 20 \text{ mph}$$

Write a program that reads in records giving rpm's at points when you might shift. For each shift point, have your program print out the top speed in each gear. Use a subalgorithm to calculate the top speeds and an array to store the gear ratios.

(FORTRAN [general], PL/I [general])

CS 8-25. Assume you are involved in the testing of fabrics for various uses. Write a program that will read in the data, do an echo check, calculate the measures specified, and report an acceptance decision.

The data to be considered are the following:
(a) readings of breaking strength on five specimens for both length (warp) and width (filling), measured in pounds; and
(b) readings of reflectance for three specimens after 0, 1, 5, 10, 20, and 40 launderings.

Acceptable breaking strength (ABS(2)) and acceptable percent reflectance loss readings (ARR(5)) at 1, 5, 10, 20, and 40 should be read by the program.

The decisions to be made are as follows:
(a) If average breaking strength is greater than or equal to ABS for both warp and woof, then check reflectance; else terminate after reporting the average measure of breaking strength.
(b) If all average reflectance loss readings are greater than or equal to ARR(I), then accept fabric and report averages. If one average reflectance loss is greater than or equal to APR(I), then print measures indicating some person must decide acceptability. Report measures. If no average reflectance loss is greater than or equal to ARR(I), then reject fabric.

Print all average reflectances and average reflectance loss values.

A subalgorithm should be designed to accept an n × m array and calculate averages of all n rows. For instance,

AVG: PROC (A,M,N,AV)

where A is the N × M array and AV is a list of M locations in length for return of the averages.

(PL/I [general, home economics])

CS 8-26. *Additional learning objective:* to experience the use of recursion.

The Fibonacci sequence of numbers

$$0,1,1,2,3,5,8,13,21,\ldots$$

can be generated by the use of the formulas

$$f_0 = 0, f_1 = 1 \quad \text{and} \quad f_i = f_{i-1} + f_{i-2}$$

Consider two solutions to this problem and code each as a subalgorithm. Solve this problem using an iterated solution and using a recursive solution.

227

Print out the first 20 Fibonacci numbers in an easy-to-read and informative format using each of the solutions.

By examination of the accounting records, estimate which algorithm takes more computer time.

(PL/I [computer science, scientific])

CS 8-27. *Additional learning objectives:* to be able to use variable length strings and become familiar with dynamic storage allocation.

Create a magic square. A magic square is an n × n array in which each of the integers 1, 2, . . . , n² occurs exactly once, and all column sums, row sums, and diagonal sums are equal. For any odd n, the following algorithm will create a magic square:

1. Write down the digits in consecutive order, beginning with 1 in the middle of the top row.
2. Move up one square and to the right one square to put the next integer (k + 1) in, unless one of the following four special cases apply:
   (a) If the move takes you above the top row in the j-th column, move to the bottom of the j-th row and place the integer there.
   (b) If a move takes you outside to the right of the square in the j-th row, place the integer in the j-th row to the extreme left.
   (c) If a move takes you to a square already filled, place (k + 1) immediately below k.
   (d) If you move out of the square at the upper right-hand corner (off of the diagonal), place (k + 1) immediately below k.

Example for n = 3:

$$8 \; 1 \; 6$$
$$3 \; 5 \; 7$$
$$4 \; 9 \; 2$$

Code a subalgorithm that, when given n, will create an n × n magic square and print it before return to the main routine. The main routine is to read in a title of variable length along with a value for n. The title is to be centered on a printer line. Repeat this process for three values of n including 3.

(PL/I [computer science, scientific])

CS 8-28. You are to build a table of values for binomial coefficients in the neighborhood of n and m when the binomial coefficient

$$b_{n,m} = \frac{n!}{m!(n-m)!} \qquad n \geq m$$

$$= \text{undefined} \qquad n < m$$

is desired.

Calculate and print (in the form shown below) a well-labeled table of the following binomial coefficients in the neighborhood of $b_{n,m}$:

$$b_{n-1,m} \qquad b_{n,m-1}$$
$$b_{n,m} \qquad b_{n,m+1}$$
$$b_{n+1,m}$$

Use a two-dimensional array to store these values before you print them.

Create a subalgorithm, say FACT (n), that will calculate

$$n! = n(n-1)(n-2)\ldots(2)(1)$$

where $0! = 1$. Then, in the main program, successively calculate values of the binomial coefficients

$$b_{n,m} = \text{FACT}(n)/(\text{FACT}(m) - \text{FACT}(n-m))$$

(PL/I [scientific])

229

9

Some Commonly Used Algorithms

KEYWORDS

ascending sequence
binary search
descending search
iterative computations
pop
push
search
search argument
search function
serial search
stack processing
table lookup

In this chapter we will present several important algorithms involving techniques that are commonly used in computing. These algorithms employ data structures and language constructs that have been introduced in the previous chapters; hence, the material will also serve as review.

The first class of algorithms to be discussed comprises **search algorithms,** which are introduced in the context of **table-lookup** problems. The problem of looking up an entry in a table, list, or some other structure occurs frequently in practice. Finding distances between places in a mileage table, finding a recipe in a book, and determining the logarithm of a number using a logarithm table are examples. Figure 9.0-1 illustrates the use of a computer for freight routing, which involves the table-lookup process of finding distances between places. In order to solve such table-lookup problems, we need to use search strategies. There are several such strategies available; however, we shall restrict our attention to two—**serial search** and **binary search.** The serial search strategy can be used with any data structure, regardless of the order in which information is stored. On the other hand, the binary search strategy requires the information to be stored in a sorted order, but is usually more efficient than the serial search strategy.

FIG. 9.0-1.   A computer system being used by a freight company to schedule trucks bound for the company's other freight terminals throughout the eastern United States. (Courtesy of Honeywell, Inc.)

Iterative algorithms constitute the second class of algorithms that will be considered. It is well known that iterative computations occur often in computing applications. Such computations require information from the i-th iteration to be used for computing the results of the $(i + 1)$th iteration. We will illustrate this class of algorithms by means of an example concerning the development of an amortization table for a loan acquisition.

We will conclude this chapter with a discussion of computer processing problems that involve lists. The use of lists in computing is very common. For example, stacks are used in the simulation of nonperishable inventory problems and in the translation of programming languages to a form that is understood by computers. Queues are used in the simulation of perishable inventory problems. For purposes of illustration, we will program some of the basic operations on a stack that is stored in a data structure other than a one-dimensional array.

TABLE-LOOKUP ALGORITHM USING SERIAL SEARCH 9.1

The process of looking up an entry in a table arises in a variety of disciplines. In physics and chemistry there are tables of atomic weights, while in the area of internal revenue there are tables of income tax, and so on. Numerous situations in which such table-lookup operations are involved come easily to mind.

To illustrate, we consider a table which lists the distances (in miles) to various cities from some place X, as shown in Table 9.1-1. We make

TABLE 9.1-1. A table of distances

f(i,1) List argument	f(i,2) Function
ALBUQUERQUE	700
CHICAGO	682
ATLANTA	1001
DETROIT	921
⋮	⋮

the following observations about this table:

1. $f(i,2)$ is called the **function value** corresponding to the **list argument** $f(i,1)$.

2. The table is in the form of a two-dimensional array. For example

$$f(2,1) = \text{'CHICAGO'}$$

and

$$f(2,2) = \text{'682'}$$

which are stored as character strings.

233

We refer to the specific list argument being sought in a given table as the **search argument.**

The basic idea of the serial search strategy is to start with the first list argument and determine whether it is the one being sought. If not, we proceed to the second list argument, and so on, as illustrated in Figure 9.1-1. The input to a serial search algorithm consists of the table to be

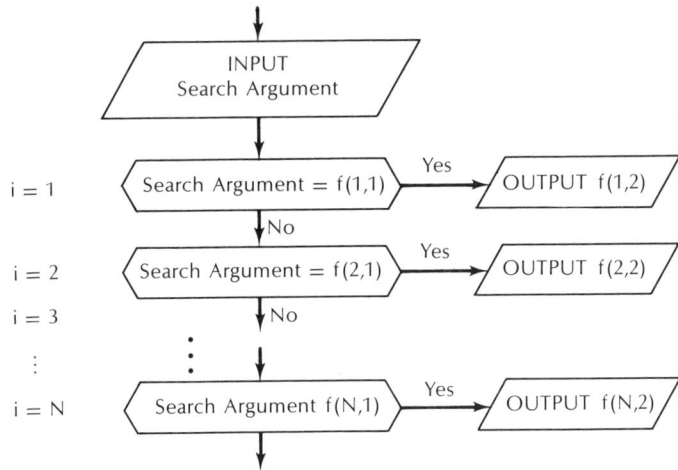

Note: $f(i,1)$ denotes the i-th list argument

FIG. 9.1-1. Basic idea of serial search.

searched, size of the table, and the search argument. The result is the function value that corresponds to the search argument, if it is found, or a message stating that the argument was not found.

We are now in a position to develop a general algorithm that performs table-lookup operations. The following initial solution is proposed:

1. Input number of entries in the table.
2. Input table into a two-dimensional array.
3. Repeat Steps 4 and 5 until arguments are exhausted from input stream.
4. Search table for search argument.
5. If search argument is found, then output function value; else output 'NOT$_\wedge$FOUND$_\wedge$IN$_\wedge$TABLE'.

A refinement of Steps 4 and 5 in the initial solution leads to the following structured algorithm:

1. Input number of entries in the table.
2. Input table into a two-dimensional array.
3. Repeat Steps 4 and 5 until search arguments are exhausted from input stream.
4. Search table for search argument.
   4.1 Read search argument.

4.2 Repeat 4.3 for each list argument in table, starting with the first; increment by 1 for each repetition.

4.3 If search argument equals list argument, then output list function values corresponding to list argument.

5. If search argument is not found, then output 'NOT$_\wedge$FOUND$_\wedge$IN$_\wedge$TABLE'.

The flowchart corresponding to the preceding algorithm is shown in Figure 9.1-2. The fact that the variable I is incremented by 1 each time

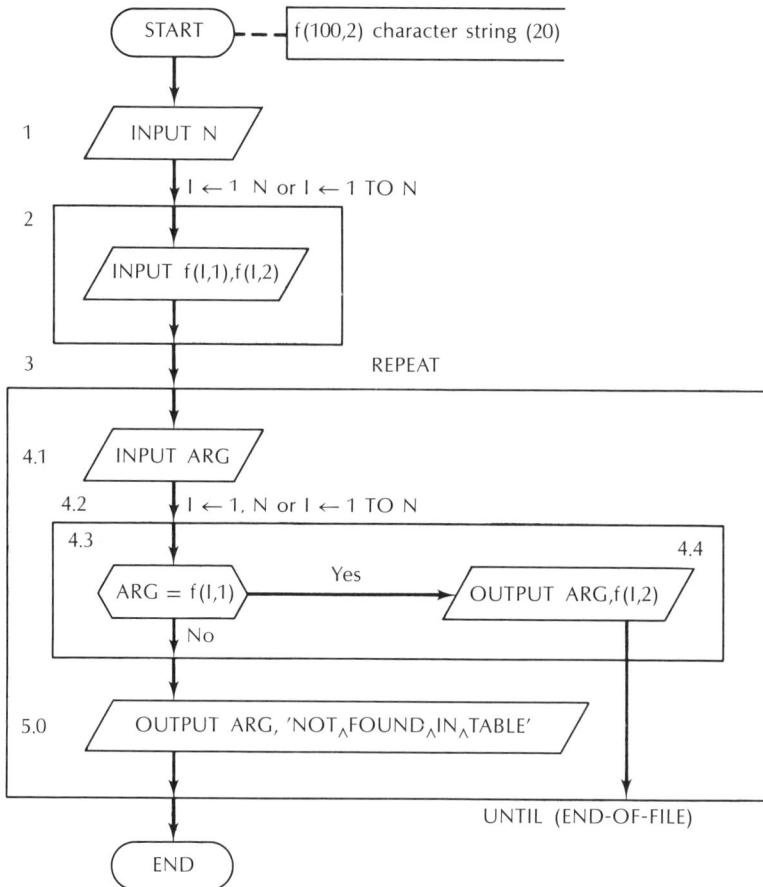

FIG. 9.1-2. Table-lookup algorithm.

implies that the search is carried out in a serial manner. Since this type of search compares the search argument with each list argument serially until a match is found, it is inefficient in applications involving long tables. This shortcoming can be overcome by resorting to a binary search technique, some aspects of which are discussed in the next section.

To illustrate the serial search technique (see Figure 9.1-2), we consider the following table with four entries that are obtained from Table 9.1-1.

List argument	Function
ALBUQUERQUE	700
CHICAGO	682
ATLANTA	1001
DETROIT	921

If the search argument is ATLANTA, then the corresponding trace table is as shown in Table 9.1-2.

9.2 TABLE-LOOKUP ALGORITHM USING BINARY SEARCH

The binary search technique enables us to carry out a table-lookup operation in an efficient manner for large tables. The only requirement for performing this type of search is that the related list argument values (numeric or character string constants) form an **ascending** or **descending sequence.**[1]

The basic idea underlying the binary search process is that the table to be searched is divided into two subtables. We then determine which subtable contains the search argument. That subtable is again divided into two subtables. Again, we determine which of these subtables contains the search argument and divide that subtable into two more subtables. This process is continued until we either locate the list argument that matches the given search argument or exhaust the list.

For the purpose of illustration, let us now locate the name KEN in the following list of names:

ANN, BEN, CAROL, DAVID, GARY, KEN, RITA, SAM

The binary search process leads to sublists of names, as summarized in the form of a tree diagram (see Figure 9.2-1). We observe that the final sublist consists of only one name—KEN.

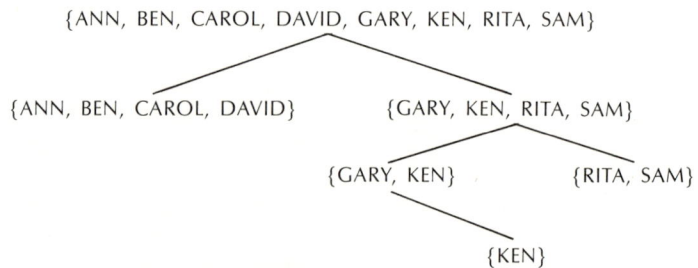

```
              {ANN, BEN, CAROL, DAVID, GARY, KEN, RITA, SAM}
                   /                              \
{ANN, BEN, CAROL, DAVID}              {GARY, KEN, RITA, SAM}
                                        /                \
                            {GARY, KEN}              {RITA, SAM}
                                    \
                                  {KEN}
```

FIG. 9.2-1. An illustration of the notion of binary search.

---

[1] In an *ascending sequence,* the first list argument is the smallest value, or comes first in a collating sequence when compared with every other list argument in the table; the next smallest is the second list argument, and so on.

In a *descending sequence,* the first list argument is the largest, followed by the second largest, and so on.

TABLE 9.1-2. Trace table for table-lookup algorithm

Symbol #	N	I	F(1,1)	F(1,2)	F(2,1)	F(2,2)	F(3,1)	F(3,2)	F(4,1)	F(4,2)	ARG	Comments
1	4	1	'ALBUQUERQUE'	'700'								
2		2			'CHICAGO'	'682'						
		3					'ATLANTA'	'1001'				
		4							'DETROIT'	'921'	'ATLANTA'	
4.1		1										
4.2												
4.3		2										arg ≠ 'ALBUQUERQUE'
4.2												
4.3		3										arg ≠ 'CHICAGO'
4.2												arg = 'ATLANTA'
4.3												
4.4												
4.1												

Output is 1001

237

SOME COMMONLY
USED ALGORITHMS

We now consider the development of a GSA that enables a table lookup on a character string table via the binary search technique. The inputs to the GSA are

(a) the size or length of the table to be searched, denoted by the formal parameter N;

(b) the table to be searched, denoted by the formal parameter F; and

(c) the search argument, denoted by the formal parameter ARG.

The output of the GSA is an indicator of the location in the table where the search argument was found. This indicator is denoted by the formal parameter PTR. In the event that the search argument is not found, a zero is returned as the value of PTR. Thus, if the desired GSA is called BINSCH, then the first statement of its definition is given by

$$BINSCH(F,N,ARG,PTR)$$

From the preceding discussion of the basic principle underlying the binary search procedure, we can construct a structured algorithm for the desired GSA as follows:

1. Determine whether ARG is in the range of the table.
   1.1 Determine whether the value of ARG is less than the first list argument or greater than the last list argument.
   1.2 If less than the first argument or greater than the last argument, then assign 0 to PTR and return.
2. Delimit the subtable to be processed with two integer values indicating the upper and lower bounds.
   2.1 Create pointers TP and BP for the top and bottom of the table respectively. That is,

$$TP \leftarrow 0$$

and

$$BP \leftarrow N + 1$$

(The reader should note that the table is between the pointers.)

3. Repeat Steps 4 and 5 until ARG is found, or table is exhausted.
   3.1 Repeat Steps 4 and 5 if $TP + 1 \neq BP$.
   3.2 If $BP = TP$, then assign 0 to PTR.
   3.3 Return.
4. Find element in the middle of the table and determine if the corresponding list argument has a value equal to that of ARG.
   4.1 Create a pointer PTR for middle of table; that is

$$PTR \leftarrow (TP + BP)/2$$

   4.2 Determine whether ARG has the same value as the list argument.
   4.3 If it does, then return to invoking algorithm.

238

5. Select the subtable in which to continue search.
    5.1 If ARG < list argument, then continue to search in upper subtable.
        5.1.1 If ARG < F(PTR,1), then BP ← PTR.
    5.2 If ARG > list argument, then continue to search in lower subtable.
        5.2.1 If ARG > F(PTR,1), then TP ← PTR.

The flowchart corresponding to the above structured algorithm is shown in Fig. 9.2-2.

Next, we develop an invoking algorithm for the GSA BINSCH, the input to which consists of

(a) N, the length of the table;
(b) TABLE, the table to be searched; and
(c) SCHARG, the search argument.

The output of this invoking algorithm is either the function value corresponding to SCHARG, or a message to the effect that SCHARG

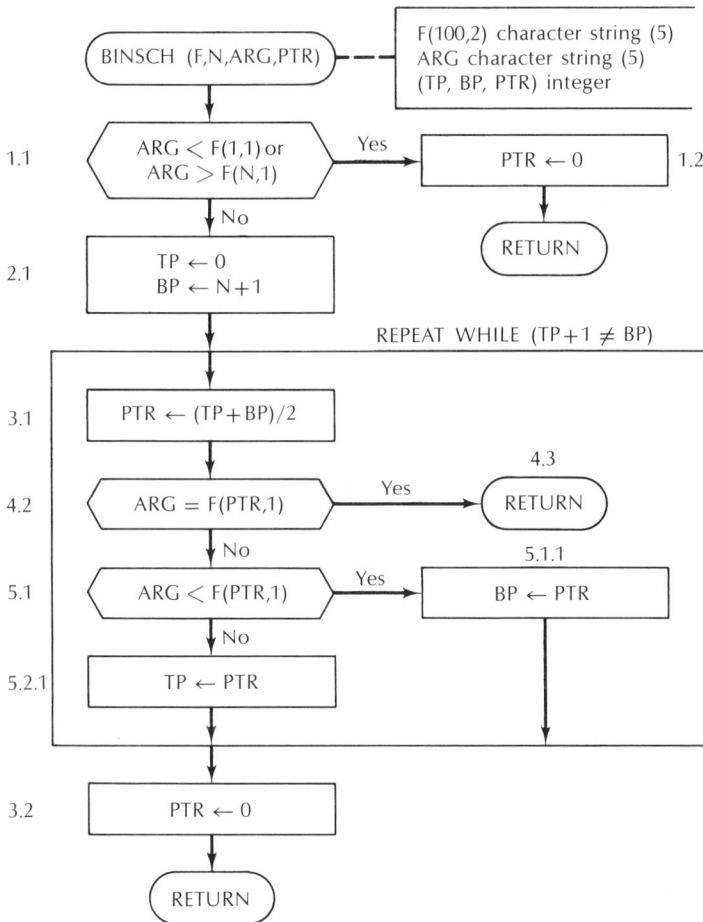

FIG. 9.2-2. An ascending binary search GSA for searching a character string table.

239

was not found. An appropriate structured algorithm is as follows:

1. Read the table.
   1.1 Read N.
   1.2 Read the N table values.
2. Repeat Step 3 until the input is exhausted.
3. Perform a search.
   3.1 Read SCHARG.
   3.2 Perform a binary search on table.
   3.3 Process result of search.
       3.3.1 If search successful, then output function corresponding to SCHARG.
       3.3.2 If search unsuccessful, then output message.

A flowchart representation of this algorithm is given in Figure 9.2-3.

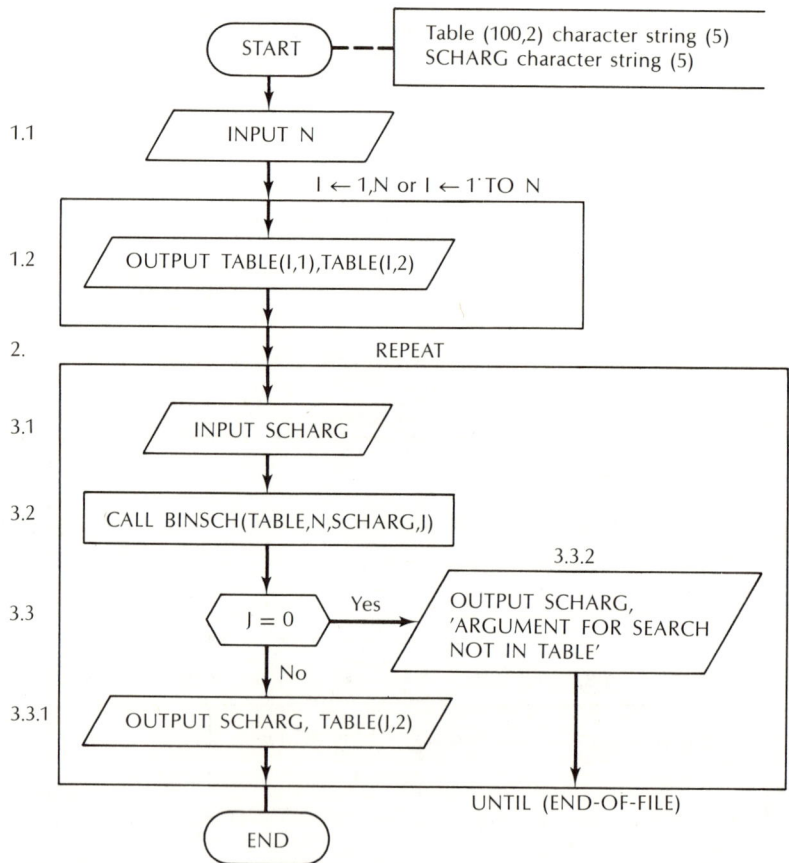

FIG. 9.2-3. An invoking algorithm for the GSA BINSCH in Figure 9.2-2.

9.3 A TABLE-LOOKUP EXAMPLE

To illustrate the manner in which the binary search algorithm for table lookup is used, let us consider the following table of the test scores of

Table(I,1)	Table(I,2)
ANN	93
BEN	28
CAROL	77
DAVID	65
GARY	47
KEN	88

It is observed that the list arguments ANN, BEN, etc. are in alphabetic order; hence, they form an ascending sequence—a condition that is required if we are to use the binary search algorithm. The value of N in the invoking algorithm (see Figure 9.2-2) equals 6, since there are 6 list arguments in the table.

Let us assume that the given search argument is 'GARY'. That is, SCHARG = 'GARY'. The statement

$$CALL_\wedge BINSCH(TABLE,N,SCHARG,J)$$

causes the GSA BINSCH in Figure 9.2-2 to be invoked. Hence, F, N, ARG, and PTR in this GSA are formal parameters that use a call by location. The value of the formal parameter PTR thus equals the value of the actual parameter J in the invoking algorithm, which is arbitrary. This implies that the value of PTR is also arbitrary when Symbol 2.1 in Figure 9.2-2 is executed. However, when Symbol 3.1 is executed, we obtain a specific value of PTR; namely,

$$PTR = (0 + 7)/2$$
$$= 7/2 = 3$$

since PTR has been declared as an **integer variable,** thereby causing only the *integer portion* of $7/2 = 3.5$ to be retained.

It is now straightforward to verify that the trace table that results when Symbols 1.1 through 5.2 in the GSA BINSCH (see Figure 9.2-2), are executed with ARG = 'GARY' is as shown below:

Symbol #	TP	BP	PTR	Comments
1.1	arbitrary	arbitrary	arbitrary	
2.1	0	7		
3.1			3	
4.2				ARG $\neq$ 'CAROL'
5.1				ARG $\not<$ 'DAVID'
5.2.1	3			
3.1			5	
4.2				ARG = 'GARY'
4.3				RETURN

241

From this trace table it is apparent that when the RETURN statement in Symbol 4.3 is executed, the value of PTR equals 5, which implies that J has the value 5 in the invoking algorithm in Figure 9.2-3. The next statement that is executed in Figure 9.2-3 is

OUTPUT SCHARG, TABLE(J,2)

which results in the output

GARY 47

One search is now complete, and another value for SCHARG can be read and processed as we have just described.

9.4 AMORTIZATION TABLE CALCULATION

Using information from one repetition of a loop in the next repetition is a common computational process. The amortization of a loan involves this type of computation. An amortization table is one that provides information pertaining to the balance on a loan on a period-by-period basis. It is commonly encountered in business and finance. For example, the amortization table for a loan of $4000 at an interest rate of 9 percent and monthly payments of $50 per month is shown in Table 9.4-1. Details of how this table is developed will be discussed later.

TABLE 9.4-1. An amortization table

Due date	Beginning balance	Interest payment	Principal payment	Ending balance
1/1/76	4000.00	30.00	20.00	3980.00
2/1/76	3980.00	29.87	20.13	3961.87
3/1/76	3961.87	29.71	20.29	3941.58
⋮	⋮	⋮	⋮	⋮

For purposes of discussion, we make the following assumptions:

1. The loans are repaid on a monthly basis.
2. A beginning principal balance is given.
3. The annual interest rate and the monthly payment are fixed.

Let us consider the case when the quantities that are to appear in the amortization table are as follows:

1. Due date—the date when a payment is due.
2. Beginning balance—the principal balance at the beginning of a month.
3. Interest payments—the portion of a month's payment that is used to cover the interest charge.
4. Principal payment—the portion of a month's payment that is applied toward the principal.

5. Ending balance—the principal balance following the remittance of a payment.

Given these assumptions and the desired outputs, we arrive at the following structured algorithm that can be used repeatedly to produce amortization tables:

1. Input the given information for one table that includes the principal, interest rate, payment amount, payment due date.
2. Initialize variables and repeat Steps 3–6 if there is input.
    2.1 Determine monthly interest rate and set variable to indicate that this is the first loan period.
    2.2 Print headings for table.
3. Repeat Steps 4–6 until the table is complete.
4. Calculate monthly payment and test for validity. If monthly payment is less than or equal to interest, then monthly payment is invalid.
    4.1 Calculate interest for month, amount of payment applied to principal, ending balance on the principal.
    4.2 If this is first loan period, test to see whether payment is greater than interest; if it is not, then provide error message and start another calculation.
    4.3 If ending balance on principal is less than zero, then this is the last month; adjust last payment accordingly.
5. Output a line of the table and see if table is complete.
    5.1 Output a line.
    5.2 If ending principal balance is equal to zero, then return to Step 1.
6. Change date and beginning balance.
    6.1 If month is December, then change month to January and increment year by 1.
    6.2 If month is not December, then increment month by 1.
    6.3 Set the principal balance to the ending principal balance of last period; go to Step 3.1.

The flowchart corresponding to the preceding structured algorithm is given in Figure 9.4-1, where the variables Prn, Rate, Monr, Payt, Mo, Day, Yr, Pamt, and Endbal are defined as follows:

Prn: beginning balance
Rate: interest rate
Monr: monthly interest rate
Payt: monthly payment
Mo: month
Day: day
Yr: year
Pamt: principal payment
Endbal: ending balance

In the solution, we note that Steps 3–6 (which produce the table) have been incorporated to form a GSA.

243

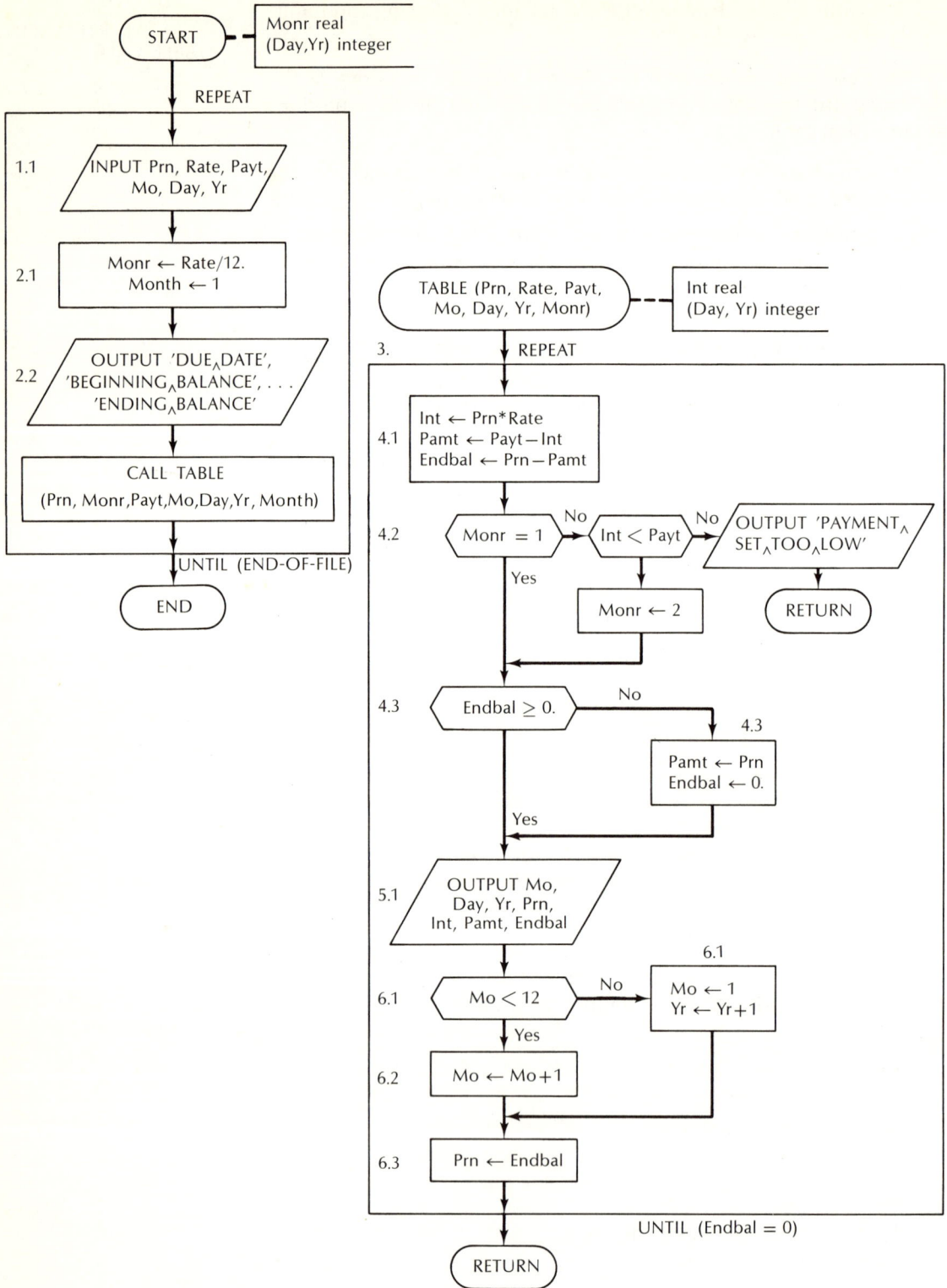

FIG. 9.4-1. Algorithm for generating amortization tables.

To demonstrate the manner in which the above amortization algorithm is used, we consider a specific case, as follows:

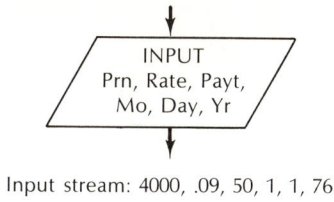

INPUT
Prn, Rate, Payt,
Mo, Day, Yr

Input stream: 4000, .09, 50, 1, 1, 76

From the INPUT statement and input stream, we obtain

$$\text{Prn} = 4000.0 \qquad \text{Rate} = .09 \qquad \text{Payt} = 50.0$$
$$\text{Mo} = 1 \qquad \text{Day} = 1 \qquad \text{Yr} = 76$$

Using these values, we obtain the following results, which are summarized in tabular form:

	Prn	Int	Pamt	Endbal
Due date	Beginning balance	Interest payment	Principal payment	Ending balance
1/1/76	4000.00	4000 × .09/12 = 30.00	50 − 30 = 20	4000 − 20 = 3980.00
2/1/76	3980.00	3980 × .09/12 = 29.87	50 − 29.87 = 20.13	3980.00 − 20.13 = 3961.97
. . .	. . .	. . .	. . .	. . .

We note that the preceding information corresponds to the first two lines of the amortization table (Table 9.4-1). Similarly, as many lines as necessary can be computed via the algorithm in Figure 9.4-1 to generate the whole amortization table.

STACK PROCESSING 9.5

In Section 6.1, we introduced the notion of linear lists, citing four types: (a) one-dimensional arrays, (b) stacks, (c) deques, and (d) queues. In this section we discuss the problem of **stack processing**, which involves adding or removing an element from a stack—i.e., **pushing** or **popping** a stack.

We recall that a stack is analogous to a plate stacker in a cafeteria, as shown in the following illustration. If an item is added at one end, it must be removed from the same end. Hence, stacks are sometimes referred to as LIFO (Last In, First Out) lists, and are commonly used to

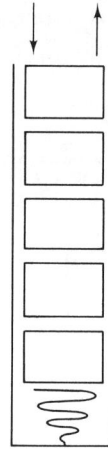

store items of nonperishable inventory.[2] An attempt to remove an item from a stack when it is empty results in a **stack underflow;** an attempt to add an item when a stack is full results in a **stack overflow.**

We now consider the development of a stack processing algorithm that can be used to push or pop items of a stack and print out the final stack configuration. We will use a one-dimensional array to store the stack, assuming that stacks with pointers are unavailable to us. We will also assume that the input stream will consist of a sequence of push and pop commands, where each push command is immediately followed by the item to be pushed.

We arrive at the following initial solution for the desired stack processing algorithm:

1. Initialize variables for use in the process.
2. Read a command into a variable called ACTION.
3. If end-of-file is found, then dump (i.e., print) stack contents, one item at a time, and then stop.
4. If ACTION has the value 'POP', then remove an item from the top of the stack.
5. If ACTION has the value 'PUSH', then add the corresponding item from input stream to the stack.

This solution can be refined to obtain the following structured algorithm:

1. Initialize variables for use in the process.
   1.1 Set pointer (PT) to zero (i.e., indicate that the stack is empty).
   1.2 Let SW be a label variable and set it to 1; SW = 1 implies that it is not necessary to dump the stack.

---

[2]Conversely, items of perishable inventory are stored in the form of *queues* which allow items to be entered at one end and removed at the other end. Queues are therefore called FIFO (First In, First Out) lists.

2. Read a command into a variable called ACTION.
   2.1 Read input stream item into the location assigned to a variable called ACTION.

3. If end-of-file, dump stack contents one item at a time, and then stop.
   3.0 If end-of-file, then do Steps 3.1–3.5.
   3.1 Print a message that stack items will be printed.
   3.2 Set SW to 2 (this causes a repetition of Steps 3.3 and 3.4 until all stack items have been printed).
   3.3 Write number of the stack element to be printed and the value of that element.
   3.4 Pop an element by invoking GSA POP.
   3.5 Transfer control to the statement labeled SW.

4. If ACTION has the value 'POP', then remove an item from the top of the stack.
   4.1 If ACTION has value 'POP', invoke GSA POP.
   4.2 Transfer to statement labeled SW.

5. If ACTION has value 'PUSH', then add the corresponding item from input stream to the stack.
   5.1 If ACTION has value 'PUSH',
   5.2 then read a value from input stream.
   5.3 Push value read in Step 5.2 and add it to stack by invoking GSA PUSH.
   5.4 Transfer to statement labeled SW.

Note: The label variable SW causes transfer of control to various points in the algorithm as summarized below.

- ☐ If SW = 1, then transfer to algorithm inconnector 1 to read another value for ACTION.

- ☐ If SW = 2, then transfer to algorithm inconnector 2 to pop an item off the stack.

- ☐ If SW = 4, then transfer to algorithm inconnector 4 to report a stack underflow.

- ☐ If SW = 5, then transfer to algorithm inconnector 5 to report a stack overflow.

From the preceding algorithm, it is apparent that we propose the use of two subalgorithms—one for pushing the stack, and the other for popping it. These are general subalgorithms, and are called PUSH and POP, respectively, as shown in Figure 9.5-2. For the corresponding invoking algorithm, see Figure 9.5-1.

STRUCTURED ALGORITHM CONSIDERATIONS

Examination of the flowchart in Figure 9.5-1 shows that label variables have been used effectively to cause unconditional branching to multiple points. Resorting to unconditional branching, however, leads to less structured algorithms than can be achieved. This is the case with the algorithm represented in Figure 9.5-1, since the overall flowchart consists of three separate components.

247

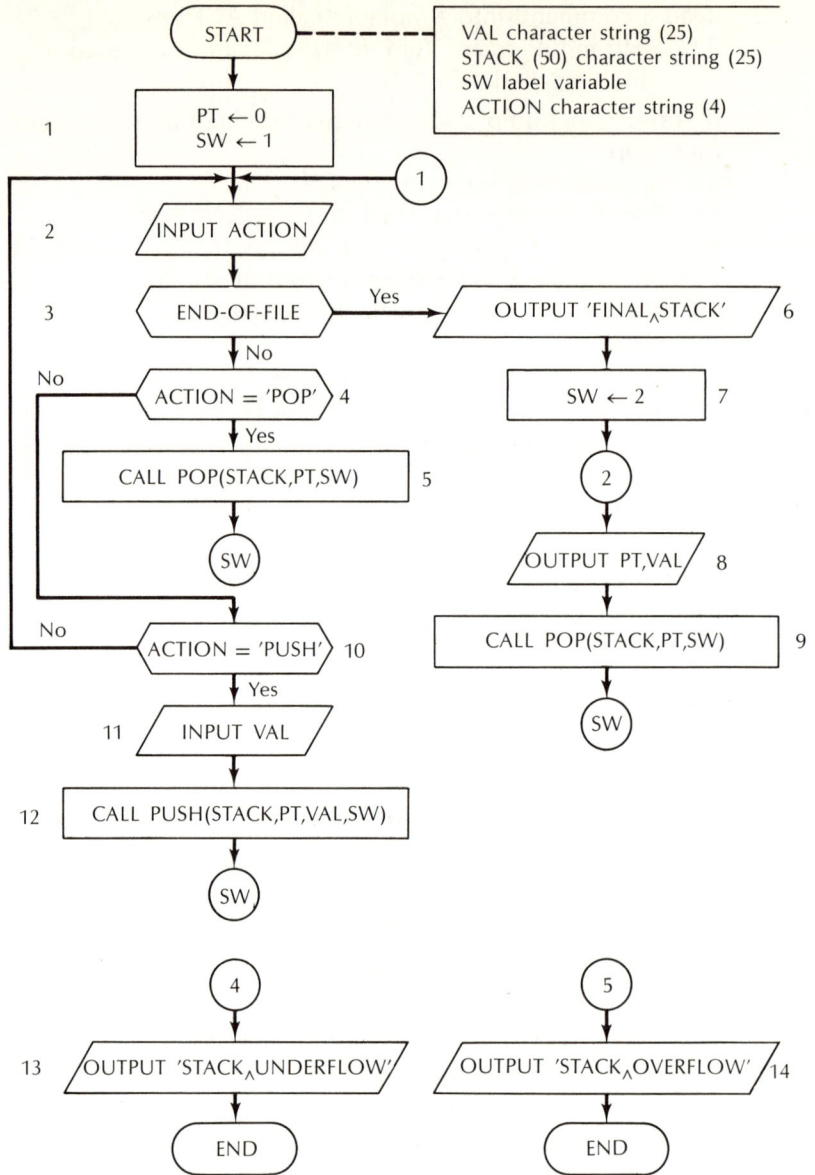

FIG. 9.5-1. Invoking algorithm for stack processing.

We can construct an algorithm that is more structured for the preceding stack processing problem. The motivation for doing so is to emphasize that an algorithm that is more structured is relatively more compact, and easier to follow, create, debug, and prove correct.

1. Initialize PT to a value of 0 and SW to a value of 1.

2. Repeat Step 3 until input is exhausted.

3. Process an action request.
   3.1 Input a value for action.

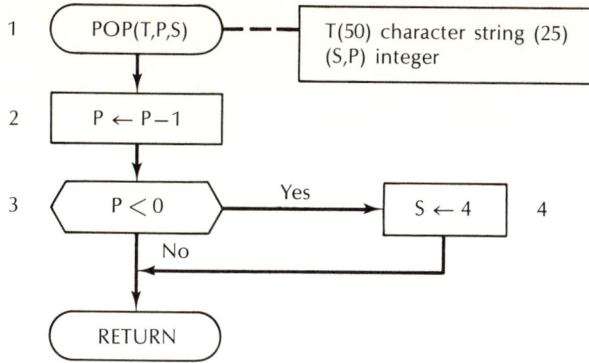

(a) GSA POP

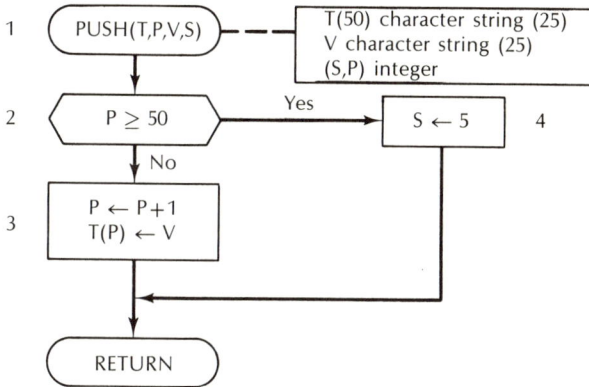

(b) GSA PUSH

FIG. 9.5-2. Subalgorithms used for stack processing.

    3.2 If action is to pop a value from stack, then remove element from stack, if possible.
        3.2.1 If action is pop, then
        3.2.2 move pointer down one in stack,
        3.2.3 and if pointer moved past end of stack, output error message and indicate an error.
    3.3 If action is to push a value on to stack, then read value and put it on stack.
        3.3.1 If action is push, then
        3.3.2 read an input value,
        3.3.3 and if stack is full, output error message and stop.
        3.3.4 Else move pointer to next element in stack and add value read there.
    3.4 If action is not to pop or to push, print error message.
  4. Output final stack contents.
    4.1 Output message.
    4.2 Repeat Steps 4.3 and 4.4 until stack is empty.
    4.3 Output pointer and value of stack in that position.

249

4.4   Pop the stack to reveal next element.
    4.4.1   Move pointer down one in stack.
  4.4.2   If pointer moves past end of stack, output error message and stop.

5.   End

The resulting invoking algorithm is shown in Figure 9.5-3. Step 3.2 is implemented as GSA POP in Figure 9.5-4(a); while GSA PUSH in Part (b) of the figure implements Steps 3.3.3–3.3.4 and Steps 4.4.1–4.4.2.

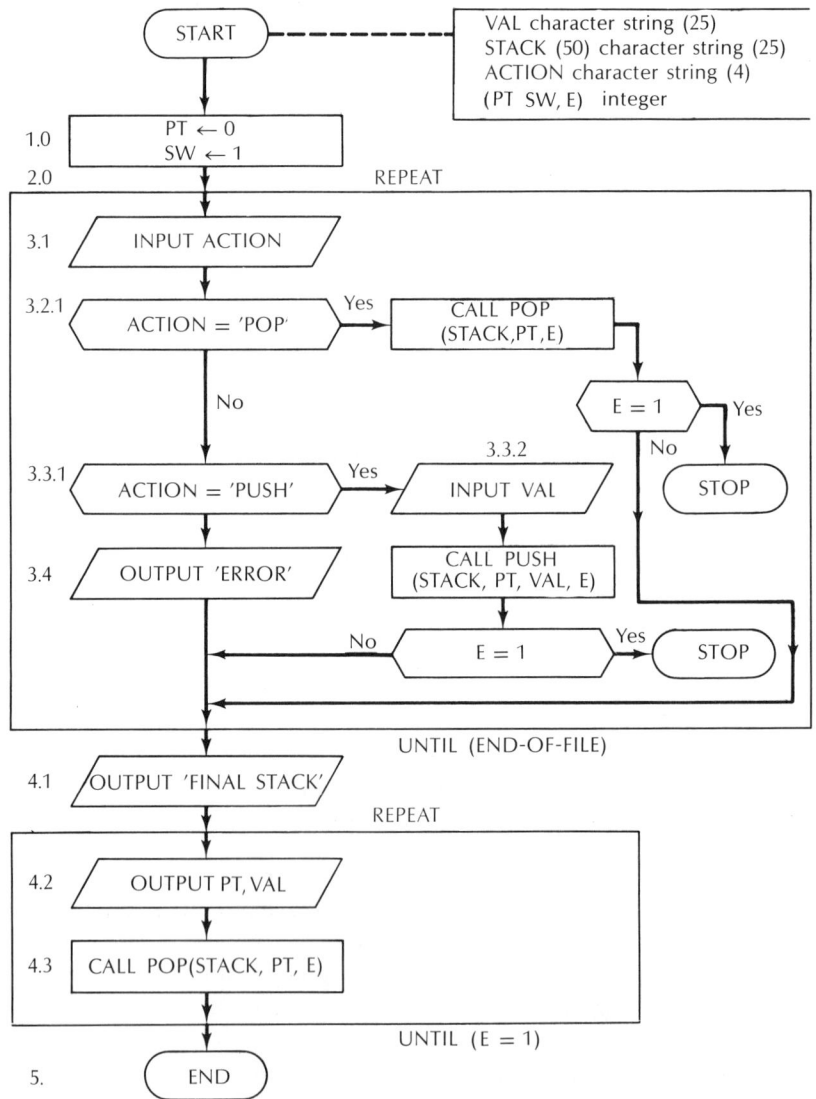

FIG. 9.5-3.   A structured invoking algorithm for stack processing.

POP(T,P,E) ----- T (50) character string (25)
(E,P) integer

E ← 0

3.2.2   P ← P−1

'3.2.3   P ≤ 0   — Yes →   OUTPUT 'STACK∧UNDERFLOW'

E ← 1

RETURN

(a) GSA POP

PUSH (T, P, V, E) ----- T (50) character string (25)
V character string (25)
(E,P) integer

E ← 0

(4.4.1) 3.3.3   P ≥ 50   — Yes →   OUTPUT 'STACK∧OVERFLOW'

(4.4.2) 3.3.4   P ← P+1
T(P) ← V

E ← 1

RETURN

(b) GSA PUSH

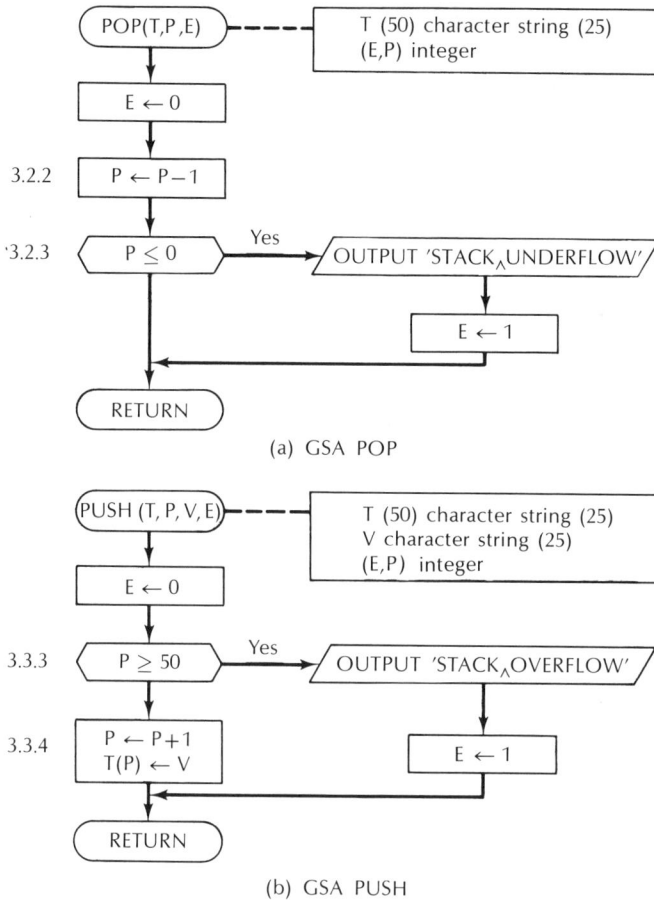

FIG. 9.5-4. Subalgorithms used with invoking algorithm for stack processing in Figure 9.5-3.

A STACK PROCESSING EXAMPLE 9.6

To demonstrate the manner in which the first stack processing algorithm developed in Section 9.5 is used, let us consider a specific input stream, as follows:

'PUSH', 'STOP', 'PUSH', 'THAT', 'POP',
'PUSH', 'THIS', 'PUSH', 'BUSINESS'

It is straightforward to verify that the trace tables associated with the invoking algorithm and the subalgorithms are as summarized in Figure 9.6-1.

We note that the table in Figure 9.6-1(a) is concerned with processing the stack. On the other hand, the table in Part (b) of the figure is concerned strictly with dumping the stack, during which time the value of SW equals 2, while that of PT is initially equal to 3, but is

251

decremented by 1 each time the subalgorithm POP is called.[3] Thus Symbol 8 in Figure 9.5-1 is executed three times, thereby resulting in the following output:

$$3 \quad \text{BUSINESS}$$
$$2 \quad \text{THIS}$$
$$2 \quad \text{STOP}$$

Finally P = 0 in subalgorithm POP, which causes

$$S = SW = 4$$

Hence, a transfer of control to the inconnector 4 occurs and the statement executed is that in Symbol 13 of Figure 9.5-1. This results in the message

STACK$_\wedge$UNDERFLOW

Invoking algorithm					GSA POP				GSA PUSH			
Symbol #	SW	PT	ACTION	VAL	Symbol #	P	S	T(P)	Symbol #	P	V	T(P)
1	1	0										
2			PUSH									
11				STOP								
12									1	0	STOP	
									3			STOP
2			PUSH									
11				THAT								
12									1	1	THAT	
									3	2		THAT
2			POP									
5					1	1						
2			PUSH									
11				THIS								
12									1	1	THIS	
									3	2		THIS
2			PUSH									
11				BUSINESS								
12									1	2	BUSINESS	
									3	3		BUSINESS

(a)

FIG. 9.6-1. Trace table for stack processing example.

[3]Note that the value of PT in the invoking algorithm and P in the subalgorithm POP are equal since P is a formal parameter.

Invoking algorithm					GSA POP				GSA PUSH			
Symbol #	SW	PT	ACTION	VAL	Symbol #	P,	S	T(P)	Symbol #	P	V	T(P)
2						P						
3												
6												'
7	2											
8												
9					1							
7												
8												
9					1		1					
7												
8												
9					1		0					
					4		4					

(b)

FIG. 9.6-1. Continued.

which indicates that the stack is currently empty. We summarize the various configurations of the stack by means of the following "stack progress," where the symbol → denotes the stack pointer:

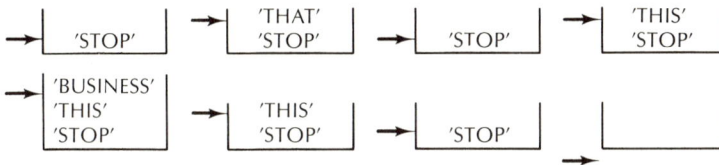

In this chapter, we introduced several important algorithms involving techniques that are commonly used in computing. First, search algorithms were introduced using the table-lookup problem. Both serial and binary search strategies were considered. While the serial search approach is adequate for searching short tables, it is inefficient when large tables are encountered. In such cases, the binary search approach is effective, although it requires that the list arguments form an ascending or descending sequence. Algorithms for arranging list arguments in an ascending or descending sequence are available (see Chapter 10).

Next, an algorithm that enables the preparation of loan acquisition amortization tables was developed in order to provide an example of

253

the class of iterative algorithms. We concluded the chapter with a discussion of the processing of lists. A specific list (stack) was considered, and an algorithm that processes it was developed and illustrated by an example.

9-1. A list of integers A(I) is to be sorted, the largest number being placed in the first position, followed by the next largest, and so on. Develop an algorithm for sorting the given list.

9-2. A set of 100 test scores is available on punched cards. Assume that the first eight columns of each card contain the name of a student, and the next four columns contain the corresponding score. Develop an algorithm that enables the number of scores between 70 and 80, inclusive, to be counted.

9-3. Suppose the number of test scores in Problem 9-2 is *not known*. Thus, as a means of detecting the last test score, assume that an additional card is added at the end of the deck with a 1 punched in Column 13 and all other columns left blank. Modify the algorithm in Problem 9-2 so that it will be able to find the last card; hence, determine that all the scores have been read.

9-4. Work Problem 9-3 for the case in which the additional card does not appear at the end of the data.

9-5. Modify the amortization algorithm so that Steps 3, 4, and 5 of the structured algorithm are each a subalgorithm.

9-6. Develop a trace table for the flowchart given in Figure 9.5-3 using the data from Section 9.6.

The learning objectives for this set of problems and its counterpart in Chapter 10 are as follows:

☐ to benefit from a more comprehensive problem-solving experience;

☐ to be able to use multiple subalgorithms—one for each major step in the structured algorithm solution;

☐ to integrate much of the material covered in the first nine characters; and

☐ to gain experience with complete sets of documentation.

Students who have worked through the previous chapters in this text with good comprehension have now reached the level at which they should be encouraged to describe their own programming projects. Those students whose projects are approved should be allowed to work them in lieu of some of the problems that follow.

This set of problems (and those at the end of Chapter 10) are not defined fully. The motivation for this approach is that, in a real-world environment, problem definitions are almost always incomplete.

CS 9-1. *Additional learning objective:* to become familiar with the binary search method.

Abbe Tritheme developed a code consisting of the three digits 1-2-3. The code is:

A	111	J	211	S	311
B	112	K	212	T	312
C	113	L	213	U	313
D	121	M	221	V	321
E	122	N	222	W	322
F	123	O	223	X	323
G	131	P	231	Y	331
H	132	Q	232	Z	332
I	133	R	233	.	333

Write a program using the binary search method for table lookup to decode messages in the Tritheme alphabet or to encode messages into it. The messages to be decoded will be input as sets of three-digit integer numbers. A code of 333 terminates the message.

The period will be used as a filler between messages as needed; blanks, however, will be ignored. Messages to be encoded will appear as single character strings. No maximum length may be assumed. *Note:* The binary search algorithm is discussed in Chapter 9, Section 9.2.

(PL/I [general, scientific])

CS 9-2. *Additional learning objective:* to gain experience in using strings.

Create a program that plots a graph on one coordinate axis system of a number of functions of a single variable. The independent variable (X) extends vertically and the dependent variable (Y) extends horizontally. Scale the values so as to keep the range of the dependent variable as large as possible, but no larger than a width specified by the user. Label the graph with value notations. Assume the subprogram is called GRAPH with one parameter A. Let A be an array with rank 1 or 2. If A is a vector, then it is to be plotted against its own indices and the number of points is ρA (the rank of A). If A is of rank 2, then the columns are taken as equally spaced abscissa points. The first row is the set of abscissa labels; the remaining rows represent functions that are plotted with different symbols. If several points have the same value, then the lowest row number is plotted.

(APL)

CS 9-3. Develop a program that computes depreciation schedules using the sum-of-digits method of depreciation. For example, suppose that you have an asset (i.e., an item of equivalent value) of $7000 that depreciates over a three-year period. After three years the item can be sold as scrap with a value of $1000. The accompanying table shows how the calculation is performed by the sum-of-digits method.

The given data:

Original cost $7000
Scrap $1000
Total amount of
  depreciation $6000
Useful life   3 years

Year	Depreciation rate	Depreciation
1	$\dfrac{3}{1+2+3} = \dfrac{3}{6}$	$6000 \times \dfrac{3}{6} = 3000$
2	$\dfrac{2}{1+2+3} = \dfrac{2}{6}$	$6000 \times \dfrac{2}{6} = 2000$
3	$\dfrac{1}{1+2+3} = \dfrac{1}{6}$	$6000 \times \dfrac{1}{6} = 1000$

Notice the depreciation rate changes from year to year, its numerator being the number of years left in the useful life of the item to be depreciated. If an item is to be depreciated for n years, then the depreciation rates for each year of the depreciation period would be arrived at as follows:

$$\frac{n}{n(n+1)/2}, \frac{(n-1)}{n(n+1)/2}, \frac{(n-2)}{n(n+1)/2}, \cdots, \frac{1}{n(n+1)/2}$$

Input description:

Column	Contents
1–6	Asset identification
7–15	Original cost of the asset
16–23	Scrap value
24–25	Useful life in years

(COBOL, PL/I [business, general])

CS 9-4. *Additional learning objective:* to use the interactive graphics facility.

Interactively ask a user to input five X,Y coordinates. From this information, draw a pentagon (five-sided figure). Label all five points in a clockwise fashion, starting with the left and top-most point:

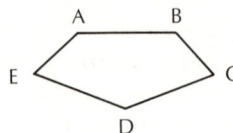

Next, draw lines from vertex A to the midpoints of sides BC, CD, and DE. Similarly, from vertex B draw lines to the midpoints of lines CD, DE, and EA:

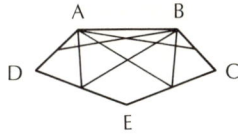

This program should be written so as to allow you to scale and/or rotate the figure, re-create the original figure, generate a new one, and terminate when desired.

(FORTRAN [architecture])

CS 9-5.   Develop a subalgorithm that will find the roots of an equation using the method of bisection. This method requires picking an interval $[x_1, x_2]$ in which one of the roots, say x_0, lies. The method is based upon the fact that if a function crosses the x-axis once in an interval bounded by x_1 and x_2, the signs of $f(x_1)$ and $f(x_2)$ will be different.

Thus, to find x_0

(a)   bisect the interval $[x_1, x_2]$;

(b)   calculate $f(x_0')$, where $x_0' = \dfrac{x_2 - x_1}{2}$;

(c)   replace x_1 or x_2 by x_0' (if x_1 is the same sign as x_0' replace it, etc.);

(d)   consider the new subinterval and repeat the process until f(midpoint) is sufficiently close to zero.

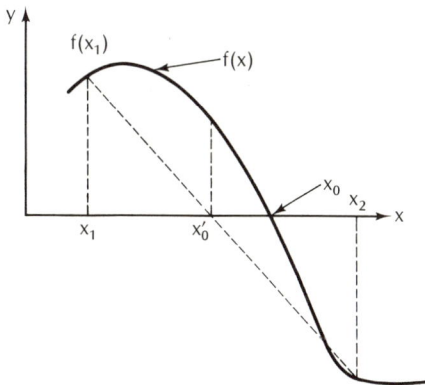

Enclose this subalgorithm in a program containing the function definitions for a few functions. This program should repeatedly call the subalgorithm to find the roots of the equations represented by the functions.

257

CS 9-6.   A popular visual form in which to present data is a bar graph, a sample of which is as follows:

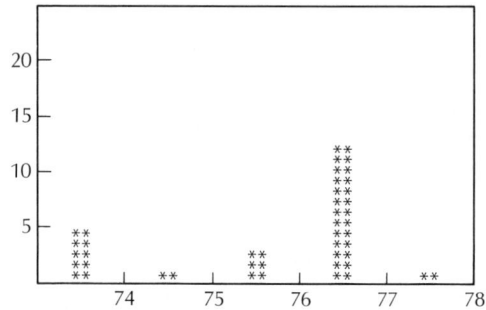

```
    ┌─────────────────────────────────────────────┐
    │                                              │
 20─┤                                              │
    │                                              │
 15─┤                                              │
    │                                   **         │
    │                                   **         │
 10─┤                                   **         │
    │                                   **         │
    │                                   **         │
  5─┤   **                              **         │
    │   **                   **         **         │
    │   **                   **         **         │
    │   **         │      **  **      │ **       │ │
    └───┴──────────┴─────────┴──────────┴──────────┘
       74         75        76        77        78
```

HORIZONTAL AXIS—MONTH
VERTICAL AXIS—TONS

You are to create an algorithm that will read in the labels for the horizontal and vertical axes. It will then read in the five values for the years 1974–1978. Based upon these values, it will scale the values of the data so that a bar graph can be presented in no more than 30 vertical lines. For instance, suppose the data were

74	17.75
75	26.5
76	33.0
77	38.0
78	27.8

Then, the high value is 36. This forces each line to be worth 1.2 units.

Write a subalgorithm that will determine the scales of the data it is given and return a list containing the number of lines of asterisks to print for each year.

The main routine is then to use an array to store the bar graph as it is created. Axis lines, axis division marks, and notations are optional output.

(FORTRAN [general], PL/I [general])

CS 9-7.   Produce a labeled scattergram for a series of point pairs. Read in any number of positive point pairs, find the maximum ordinate and abscissa values, and scale the values into a graph that is 60 spaces wide and 60 spaces high. Mark the graph axes every 10 spaces with a value. Label each axis as to meaning. Print a title for the graph. Use several subalgorithms for the process.

(FORTRAN [general], PL/I [general])

CS 9-8.   Develop a program to do the individual matching for a computer dating service. The dating service uses a 20-

question form, the applicant indicating the degree of agreement with each question on a scale of one to five. Thus, each data record will contain the name of the applicant, sex ('M' or 'F'), and responses to the questions. The last data record will contain the string 'NO MORE APPLICANTS' in place of a name.

Match each person with the two most compatible people of the opposite sex. As a measure of compatibility, use the cosine of the angle between their two response vectors. The larger the cosine, the more compatible the couple.

Use a subalgorithm to compute the cosine of the angle between two given vectors. First convert the given vectors into unit vectors in the same direction. The unit vector in the same direction as $(x_1, x_2, \ldots, x_n)$ is $(u_1, u_2, \ldots, u_n)$, where

$$u_k = x_k/r \quad \text{and} \quad r = \sqrt{x_1^2 + x_2^2 + \cdots + x_n^2}$$

Next, use the two unit vectors $(u_1, u_2, \ldots, u_n)$ and $(v_1, v_2, \ldots, v_n)$ to compute the cosine of the angle

$$u_1 v_1 + u_2 v_2 + \cdots + u_n v_n$$

Your program should output a list of the applicants along with the names of their two best prospective dates.

(FORTRAN [general], PL/I [general])

CS 9-9. *Additional learning objective:* to become acquainted with a random number generator and the idea of Monte Carlo integration.

Integrate the function $f(x) = x^2 + 2x + 1$ between $x = 1$ and $x = 2$ using a Monte Carlo integration scheme. In this method an approximate area is obtained by determining the number of randomly generated points that fall in the area to be approximated. Hence,

$$A = \int_a^b f(x)dx \simeq \left(\frac{p}{n}\right) \cdot (b - a) \cdot \text{limit}$$

where p is the number of points in the shaded region, and limit is the maximum value (or greater than maximum value) that the function takes on in the interval. The total number of points generated randomly is given by n. The process is to generate a pair of uniform random numbers (x,y), $a \leq x \leq b$, $0 \leq y \leq \text{limit}$. The point (x,y) is within the rectangle shown in Figure CS 9-9.

Determine the number of points in the shaded area. If $y \leq f(x)$, then x falls in the shaded area; otherwise it does not. The random number generator should produce a uniform distribution, so that there is an equal probability of the point falling anywhere in rectangle. Integrate several functions using point values of 100 and 200.

259

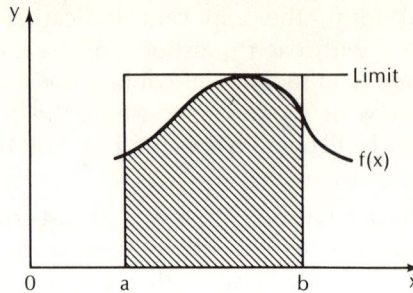

FIG. CS 9-9

(APL, FORTRAN [scientific]), PL/I [scientific])

CS 9-10.   Develop a simplified airline reservation system. Your system includes just 10 flights, and there are only 10 seats per flight. For each set you are to keep the name of the passenger and a service class. There are four classes of service. Any seat can be any class, depending on the price of the ticket.

The system should accept just four requests:

Request	Input	Output
Space available	1	Number of seats available
Add passenger	2 Passenger name, class of service (1-4)	Passenger name and class of service added or the message 'No space available'
Remove passenger	3 Passenger name	Passenger 'passenger name,' removed or 'name not found'
Flight roster	4	A list of passengers' class of service and number of available seats

Create a subalgorithm to accomplish each function. Store the flight information in a structure in global storage, but pass the passenger name and class of service as parameters to your four subalgorithms. Let all printing indicated above be done within an appropriate subalgorithm; any other printing is to be done in the main routine.

(PL/I [business])

CS 9-11.   *Additional learning objective:* to be able to create an interactive computer programming solution.

Develop a program that plays tic-tac-toe with a human opponent. You are not required to make the program learn as it plays. Let the person make the first move and the program respond by checking for a winner. If a winner is detected, print out a message and the board configuration. If no winner occurs, the program plays, checks for winner, and prints out the current game board. Be sure to check for tie games.

One way to check for a winner is to keep an auxiliary board that is initially all zeros. When Player 1 plays, a 1 is inserted into play position; when Player 2 plays, a −1 is inserted into play position. If any row, column, or diagonal sums to 3 or −3, Player 1 or Player 2 wins, respectively. Check for invalid moves by the human player.

(PL/I [computer science])

CS 9-12. Develop a program that will simulate the action of a slot machine—a "one-armed bandit." Have the gambler bet in even-dollar amounts and press RETURN to pull the handle. The possibilities for window contents are lemons, oranges, cherries, and bars. The probability of getting a lemon should be .4; of getting a cherry, .3; of getting an orange, .2; and of a bar, .1. The payoff schedule is as follows:

Combination			Payoff
lemon	lemon	lemon	20 : 1
cherry	anything	anything	5 : 1
cherry	cherry	anything	10 : 1
lemon	anything	anything	1 : 1
anything	orange	anything	5 : 1

Print the contents of the windows on each spin and report the "winnings" each spin. Keep an accumulated win/loss in dollars for the gambler and report the accumulated total each spin.

(PL/I [computer science])

CS 9-13. Cryptography, the science of coding and decoding messages, can be assisted through the use of the computer. Create a program that can accept an English-language statement and encipher it or accept an enciphered statement and decipher it and present the message.

Encipher and decipher your messages using a two-decimal digit representation for the symbols as follows:

Comma	Period	Asterisk	A–Z	0–9
01	02	03	04–30	31–40

When in the enciphering mode, echo original messages and the resultant coded message. When in the deciphering mode, echo and then print a clear message. Devise a way your program can be told to encipher or decipher from the input stream.

(PL/I [general])

261

CS 9-14.   Many people believe that contagious diseases spread faster if they start on one of the coasts rather than in the middle of the country. We will hypothesize a disease-spread mechanism and test it on the computer to see if the results support the presumption that disease spreads faster from the coasts.

We will assume that diseases spread by physical contact, and that a person's contact range does not extend beyond a small region around the person's home. Basically, we are assuming that the percentage of people in a particular region who have the disease is affected (becoming either higher or lower) by the percentages in neighboring regions. More specifically, consider the country divided up into square regions by laying a 40 by 100 grid over a map of the United States. (Assume the United States is rectangular, filling the grid completely.) With the exception of border regions, each region has eight immediate neighbors: north, northeast, east, southeast, south, southwest, west, northwest. Consider that diseases spread by an averaging mechanism: if the percentage of people in a particular region who get the disease is R% today, and neighboring region percentages are N%, NE%, E%, SE%, S%, SW%, S%, NW%, then the percentage in the region tomorrow will be

$$R\% = (R\% + N\% + NE\% + E\% + SE\% + S\% + \\ SW\% + W\% + NW\%)/9$$

For border regions, leave missing neighbors out of the average. As a starting condition assume that 100% of the people in a particular region have the disease on day 1, and that no one else has contracted the disease until the next day, when the disease starts to spread. The original region is a special case and always stays at 100%. Let the simulation run for 30 days.

Use a two-dimensional array to store the percentages. Initialize the starting region to 100% and all others to 0%, using a subalgorithm to compute the next day's percentages—the most complex part of program. In trying to update percentages for a region with an east-west coordinate stored in EW and with a north-south coordinate stored in SN, you will need to use two nested loops—one to go from SN − 1 to SN + 1 and one to go from EW − 1 up to EW + 1. At each of the nine places, you must test for negative values (indicating a part outside the United States). Use two arrays for percentages, one for today and another for tomorrow.

At the end of the 30 days, use another subalgorithm to print the results in the form of a rectangular display of symbols, one for each region. Use the symbols as described on the following page:

Symbol	Meaning
blank	0–10%
1	10–20%
2	20–30%
3	30–40%
4	40–50%
5	50–60%
6	60–70%
7	70–80%
8	80–90%
9	90–100%
*	Original region

(PL/I [scientific])

CS 9-15. *Additional learning objective:* to understand the concept of numerical integration using the computer.

One method for finding the area under a curve is to use the numerical integration technique called the **trapezoidal rule.** Given the curve S below, we can estimate the area A under it between x_0 and x_n as shown below:

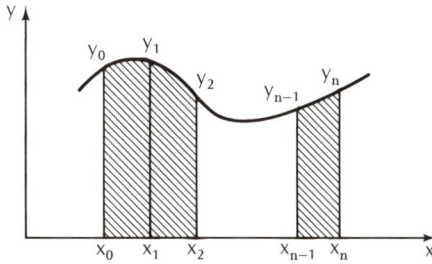

$$A = \frac{y_1 + y_0}{2}(x_1 - x_0) + \frac{y_2 + y_1}{2}(x_2 - x_1)$$

$$+ \cdots \frac{y_n - y_{n-1}}{2}(x_n - x_n - 1)$$

You are to develop a subalgorithm for accomplishing this integration and enclose it in a program that uses it to find the areas under several curves.

(APL, FORTRAN [scientific], PL/I [scientific])

CS 9-16. *Additional learning objective:* to use a recursive subalgorithm.

The problem is to solve a system of four equations in four unknowns of the form

$$[A][X] = [Y] \quad \text{then} \quad [X] = [A]^{-1}[Y]$$

The coefficient matrix [A] and the righthand sides [Y] are given. Solve for [X]. To accomplish this use a number of

263

subalgorithms. One should find the determinant, as we know

$$[A]^{-1} = \frac{1}{|A|} \begin{bmatrix} A_{11} & A_{21} & A_{31} & A_{41} \\ A_{12} & A_{22} & A_{32} & A_{42} \\ A_{13} & A_{23} & A_{33} & A_{43} \\ A_{14} & A_{24} & A_{34} & A_{44} \end{bmatrix}$$

where $|A|$ is the determinant of $[A]$, and

$$|A_{ij}| = (-1)^{i+j}|M_{ij}|$$

M_{ij} is a cofactor of A; i.e., the matrix remaining after Row i and Column j are crossed out in A.

Use a recursive algorithm to compute the A_{ij}. Remember,

$$|A| = a_{i1}A_{i1} + a_{i2}A_{i2} + a_{i3}A_{i3} + a_{i4}A_{i4}$$

(PL/I [scientific])

CS 9-17. When a large number of students take an examination, we can transform the distribution of grades to a normalized distribution. To accomplish this process, first calculate the average grade

$$\text{AVERAGE} = \frac{1}{N}\sum_{I=1}^{N}\text{Grade}_I$$

where N is the number of students and Grade_I is the grade of student number I. Then calculate the standard deviation, which gives a measure of dispersion of the grades around the average value. Standard deviation is calculated by

$$\sigma = \sum_{I=1}^{N}\left(\frac{(\text{Grade}_I - \text{AVERAGE})^2}{(N-1)}\right)^{1/2}$$

From this, the normalized grade of the I-th student can be calculated by

$$\text{GradeN}_I = \frac{\text{Grade}_I - \text{AVERAGE}}{\sigma}$$

Create a program with three subalgorithms to read in a list of indeterminant length of student grades and output average, σ, and the normalized grade of each student.

(FORTRAN [general], PL/I [general])

CS 9-18. *Additional learning objective:* to become familiar with sorting techniques.

Read the text and select one of the internal sort methods. Encapsulate a sorting algorithm in a subalgorithm. Write a program to use the subalgorithm.

Your program must be able to read in a multiple number

of sets of data. These sets consist of a mixture of numbers and letters (see the data given below). The program should then sort these data items into ascending sequence. You must print the symbols when they are read and print them again after they are sorted. Each printing should be appropriately labeled.

You may use a sorting algorithm of your own choosing. Identify the technique in your documentation.

First set of data:

ES15X53	SYLI327	2N2222	16768Ø
2SD548LC	1H2	TVSØA7Ø	921-2748
ZEN2Ø2	815Ø36	2SB117K	IN914
NT3C1ØØ	EQBØ1-11Z	532ØØ23H	2N5296A
GØ5-Ø5Ø	2N3Ø55	57A1ØØ	IN4ØØ1

Second set of data:

MDT	ALCO1ØØØ	ML-4ØØØ	U33C
MM	F9	RS2	GP19
BL-2	SW15ØØ	FP-45	E-9
DD-4Ø	SW-1	U3ØB	PA1

(COBOL, PL/I [scientific])

10

Computer and Data Processing Systems

The fact that a computer can do something man can do does not mean we will employ the computer instead of man.
(H. A. Simon, 1965)

KEYWORDS

batch processing
blocking of records
channel
conversational
data processing
data representation
data security
data verification
direct-access file
field
file backup
file organization
file security
generation backup
interpreter
item
logical record
management information
  system (MIS)
multiprogramming
offline
online
paging
physical record
record
response time
real-time computing
sequential file organization
spool
sorting
time-sharing
turnaround time
virtual storage

The purpose of this chapter is to introduce some concepts of currently available software and hardware systems. We will discuss two types: (a) data processing software systems and (b) computer hardware systems.

By the term **system,** we imply an interacting and interdependent group of entities, which, when taken together, form a whole. **Data processing** is concerned with the storage, processing, and retrieval of large volumes of data associated with businesses and institutions. The commonly used term for this area is "business data processing." The systems used are usually called **data processing systems.** Basic concepts and related terminology of such systems are introduced in Section 10.1. A few algorithms that facilitate the storage and retrieval of information in data processing systems are discussed in Section 10.2. In addition, certain management information system concepts are introduced in Section 10.3.

Section 10.4 concerns some computer system terminology; a sampling of available computer system hardware is introduced in this section, mainly by means of annotated photographs.

10.1 BASIC CONCEPTS AND TERMINOLOGY

Data processing problems typically involve large volumes of data that are subjected to relatively few calculations; for example, payroll processing. In such cases individual units of information called **items** or **fields** make up the data, such as the name, address, and earnings of an individual employee. Items are arranged in the form of contiguous sequences to form **records.** Each record is organized in a well-defined manner, and is often fixed in terms of the order, sizes, and types of the items included (see Figure 10.1-1, illustrating a sample payroll record).

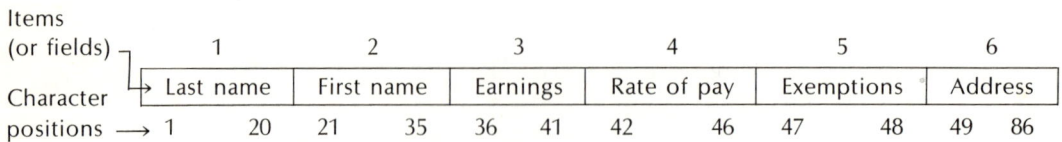

Items (or fields)	1	2	3	4	5	6
Character positions	Last name	First name	Earnings	Rate of pay	Exemptions	Address
	1 20	21 35	36 41	42 46	47 48	49 86

FIG. 10.1-1. Sample payroll record.

The record in Figure 10.1-1 consists of logically related items; it is called a **logical record.** On the other hand, a **physical record** is described in terms of physical or efficiency considerations of the available storage media. For example, an 80-column punched card is a physical record of 80 characters. Several physical records may be required to store one logical record; conversely, several logical records may fit into one physical record.

In order to increase the efficiency of input/output operations, several logical records are often stored together **(blocked)** in one physical record. For example, following each physical record on a magnetic tape, some space is left to provide the tape drive—a device that enables us to read and write on a tape—sufficient slack to stop/start itself. Such gaps are referred to as **interrecord gaps** (IRGs) and typically occupy 0.65 inches (1.65 centimeters) per gap. If information is stored on tape at 800 characters per inch, then 0.1 inch of tape is required to store an 80-character logical record; however, a total of 0.75 inches (1.9 centimeters) of tape is required as a consequence of the related IRG. Similarly, a total of 1.65 inches (4.19 centimeters) of tape would be required to store 10 contiguous logical records (see Figure 10.1-2).

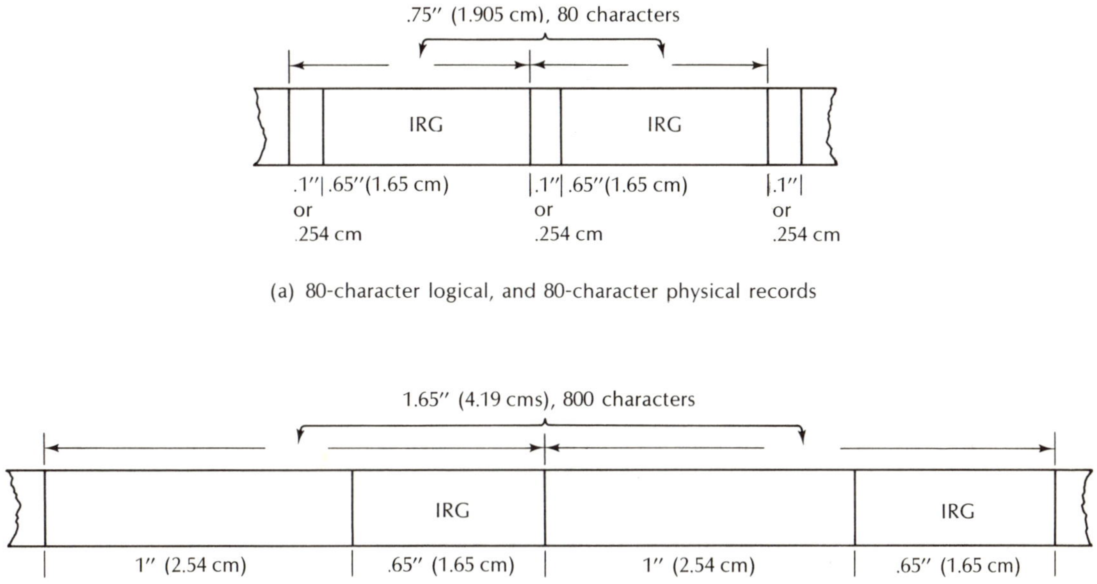

.75" (1.905 cm), 80 characters

	IRG		IRG	

.1" |.65"(1.65 cm) |.1"| .65"(1.65 cm) |.1"|
or   or   or
.254 cm   .254 cm   .254 cm

(a) 80-character logical, and 80-character physical records

1.65" (4.19 cms), 800 characters

	IRG		IRG

1" (2.54 cm)   .65" (1.65 cm)   1" (2.54 cm)   .65" (1.65 cm)

(b) 80-character logical, and 800-character physical records

FIG. 10.1-2. Blocking logical records into physical records on a magnetic tape. Tape density equals 800 characters per inch (or 315 characters per centimeter).

Logical records are aggregated into **logical files,** which are usually just called files. Physical devices are used to store logical files. A logical file may reside on one or more physical files, or may share a physical file with other logical files.

In the past, permanent storage of information on punched cards was common practice. It now has been superceded by the use of auxiliary memory such as magnetic tape and magnetic disk (see Figure 10.1-3). Such storage media, for instance, are used in record creation systems, an example of which is shown in Figure 10.1-4. It is not unusual for a standard 2400-foot (731.52-meter) magnetic tape to hold 15–20 million characters of information, while a common disk pack may hold 133

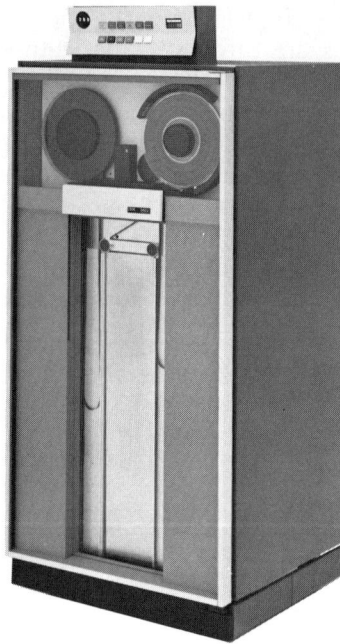

(a) Magnetic tape drive unit with tape reels mounted on it

(b) Disk drive unit (at far right), and disk pack (in woman's left hand)

FIG. 10.1-3. Two storage media and read/write devices. (Courtesy of International Business Machines Corporation.)

million characters. It should be noted that auxiliary memories access information much more slowly than the main memory of a computer. On the other hand, auxiliary memories are substantially cheaper than main memories.

270

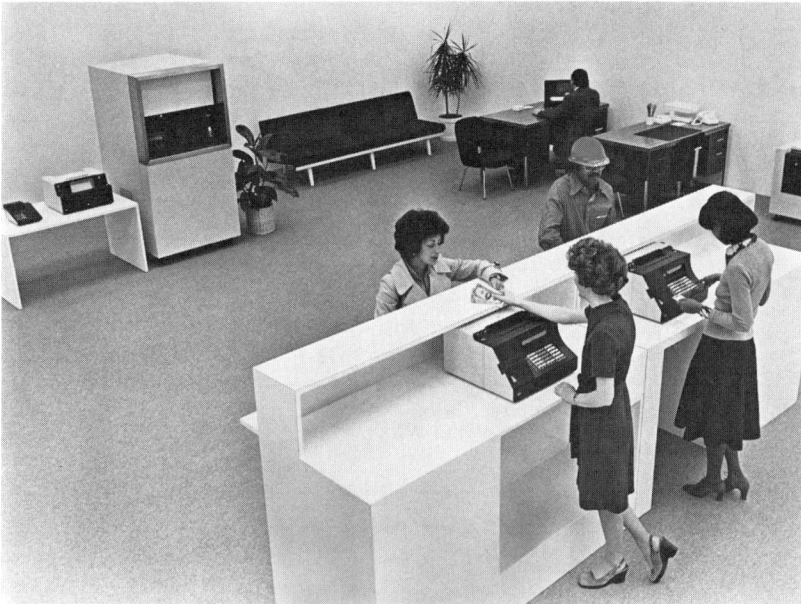

FIG. 10.1-4. A record creation system with storage on magnetic tape or magnetic diskette. (Courtesy of International Business Machines Corporation.)

Data to be stored in the machine are checked for accuracy by a process called **data verification.** This process can be carried out by mechanical means for exact verification, or by use of a program for the detection of gross errors. The mechanical approach consists of key stroking the input data a second time, using a key-stroke verifier device. Each time an error is detected, the input record is changed. Using punched cards, each column in error is notched, so that it can be easily identified for correction. A completely correct card is notched on the right-hand end (see Figure 10.1-5). In addition, program error detection could involve checks to see if the pertinent data are meaningful; for example, the number of hours worked by an individual cannot be negative. Other forms of checking can be used, such as hash totals for a sequence of bank transactions.

Following the creation of a file, its integrity must be guaranteed during the time it is used. To this end, changes to the file must be verified as being correct prior to entering them. Since most files are considered important, a restricted number of individuals and usually only one program should be allowed access rights to change the file. In addition, all changes to critical files should be logged for future reference. This phase of a data processing operation is called **data security;** it refers to the process of maintaining the integrity of information. In addition to physically protecting the files by restricting access to them, data security should also provide a means of recreating lost data.

Physical protection of files may involve special personnel who protect and handle files (librarians), vaults, and other site protection

271

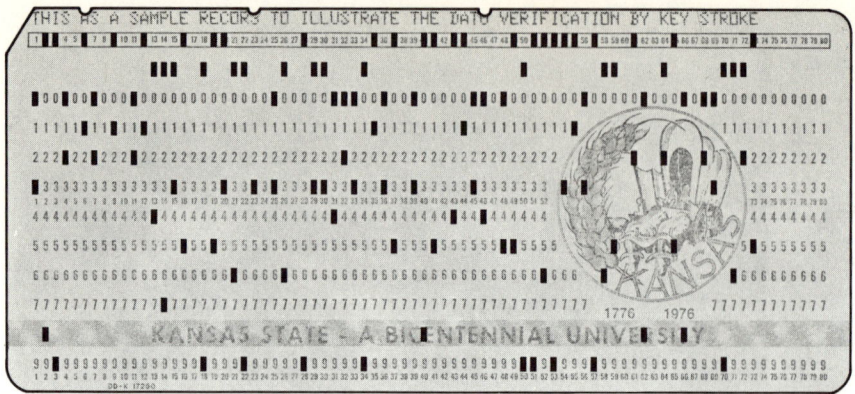

(a) Card columns 6, 23, and 46 in error

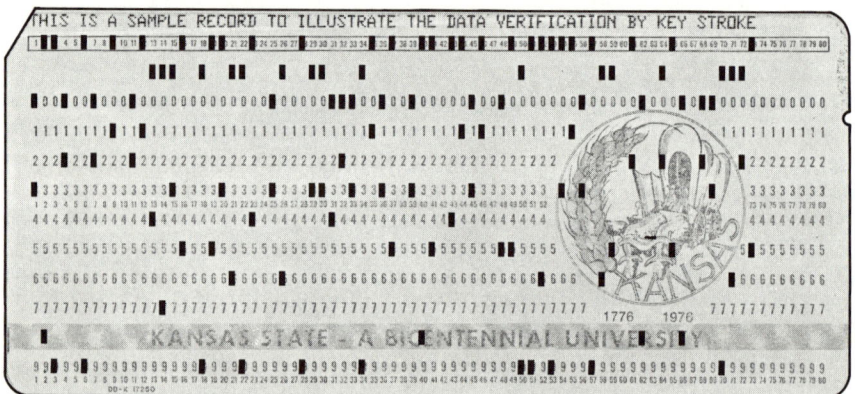

(b) Completely verified card

FIG. 10.1-5. Verification of punched cards.

features. Restricting computer access to files implies allowing access to them only through one common access program that checks the authenticity of the user. Other means, such as passwords as special identification, are also in common use. Any programmer wishing access to a file has to provide the correct identification before access is granted. Such methods do assist in combating fraud and computer crime, but they have met with only partial success.

Next, we consider the notion of **file backup,** which implies that there is a method for recovering files that are somehow lost. Lost files may be the result of vandalism, operator or programmer error(s), or natural disasters, such as fires. For a backup scheme to be totally effective, it is essential that materials effecting backup not be stored in close proximity to the original files. A simple method of creating a backup for files that change infrequently is to produce one or two copies of them at the time of each change. These copies can then be stored in a remote area. A second backup scheme, which eliminates the necessity of the copy process, is referred to as the **generation backup system.** It can be used when a completely new file is to be created each time changes are made, as is the case with magnetic tape files. Here, the first step is to create a new file that consists of the old file plus the changes, via an update program. The old file and the changes can then be saved as backup. Such files are called **father files,** in that they are one generation of change away from the current files. Similarly, the term **grandfather file** implies a file two generations of changes away from the current file.

The schemes we have discussed for providing file security are the most rudimentary. In addition, there are several others of greater sophistication. Because of space limitations, these will not be considered here.

To conclude our discussion of basic concepts and definitions of data processing systems, we will turn to some aspects of **file organization.** There are basically two schemes available for this purpose, although several combinations of these schemes are also used. The two forms that we will consider are (a) sequential files and (b) direct-access files. **Sequential files** are those in which records are physically stored contiguously and can only be read, written, or accessed in sequence; that is, to access a record so stored, we first must access all records preceding it. Files stored on magnetic tapes, for example, are always sequential in nature. On the other hand, **direct-access files** must be stored on devices that are capable of accessing a record directly—without first accessing records that precede it in the file. Examples of such storage devices include the main storage of a computer, magnetic drums, and magnetic disks. These storage devices may also be used to store sequential files.

INFORMATION STORAGE AND RETRIEVAL 10.2

Information storage and retrieval is an area of data processing that concerns (a) the process of making information available, and (b) the problems of organizing, storing, and modifying information. For example, it may be necessary to determine the number of items of a certain product that a company has on inventory on a daily basis. This task can be carried out by developing an appropriate information retrieval algorithm. Examples of basic information retrieval algorithms include the serial and binary search techniques, which were discussed in Chapter 9. We recall that in order to use the binary search technique, the list arguments must be ordered. Thus, if binary search were employed in business data processing, it would be necessary that the records of a file be ordered by one or more of the data items. The process that enables us to achieve the desired ordering is called **sorting.**

Many of the excellent sorting programs that are readily available take the form of problem-oriented languages. They require the user to specify characteristics of the file to be sorted, the hardware to be used, along with the data item(s) on which the records are to be ordered. The sort language then selects a combination of sorting techniques to

273

carry out the desired sort in an efficient manner. It is worthwhile, however, to be familiar with the basic types of sorts available, and to understand of the type of logic that is used in the related algorithms.

Files can be ordered in **ascending order** (i.e., smallest sort item first) or **descending order** (i.e., smallest item last) by sorts that can be classified into two categories: (a) internal sorts and (b) external sorts. **Internal sorts** are algorithms in which the file to be sorted is stored in the main memory of the computer. Sort algorithms in which the file to be sorted is stored in an auxiliary memory, such as tape or disk, are known as **external sorts.** Internal sorts are much faster to execute than are external sorts, but the former are limited by the amount of main memory available to store the file and the sort algorithm. External sorts, on the other hand, can be used to sort files of practically any size. Now, let us consider some specific versions of internal and external sorts.

CLASSIFICATION AND MERGE SORT

This is an external sort that uses the auxiliary storage as a series of temporary storage locations. For instance, if five tape drives are available, then one could be used initially to hold the given input file. The remaining four could be allocated to store records as they are encountered on the input tape in an alphabetical sort as follows:

Tape drive number	Records with sort items starting with
1	A through D
2	E through K
3	L through P
4	Q through Z

Each of the above tapes would, in turn, become the unsorted input tape until each is either completely sorted or is small enough to fit into the main memory. It could then be sorted using an internal sort. The sorted subfiles on tape are then merged back together into one large file. Several strategies exist for this type of sort, but all are based on the idea of reasonably breaking the file into subfiles (often partially sorted), and then recombining these subfiles into a whole sorted file.

A variation of the classification and merge sort is referred to as a **merge sort.** It is relatively more efficient, in that fewer secondary storage files are required. A discussion of the merge sort is beyond the scope of this book. Two other sources for the interested reader are "Sorting" by William A. Martin in *Computing Surveys*, 1971, and *Computer Sorting* by I. Flores, Prentice-Hall, 1969.

REPLACEMENT AND EXCHANGE SORTS

These are two types of commonly used internal sorts. **Replacement sorts** may require space in memory to store either one of two files; thus, they are referred to as one-list replacement sorts or two-list replacement sorts, respectively.

For purposes of illustration, we will now consider a GSA that enables us to sort a file consisting of a one-dimensional array of positive integers into an ascending order. It employs a two-list replacement sort. The corresponding flowchart is shown in Figure 10.2-1, from which it is apparent that the GSA is called TWOREP, involving the following parameters:

(a) LIST, a one-dimensional array containing up to 100 integers;

(b) N, the number of items in the current list to be sorted; and

(c) LSORTED, a one-dimensional array containing up to 100 integers and holding the sorted list.

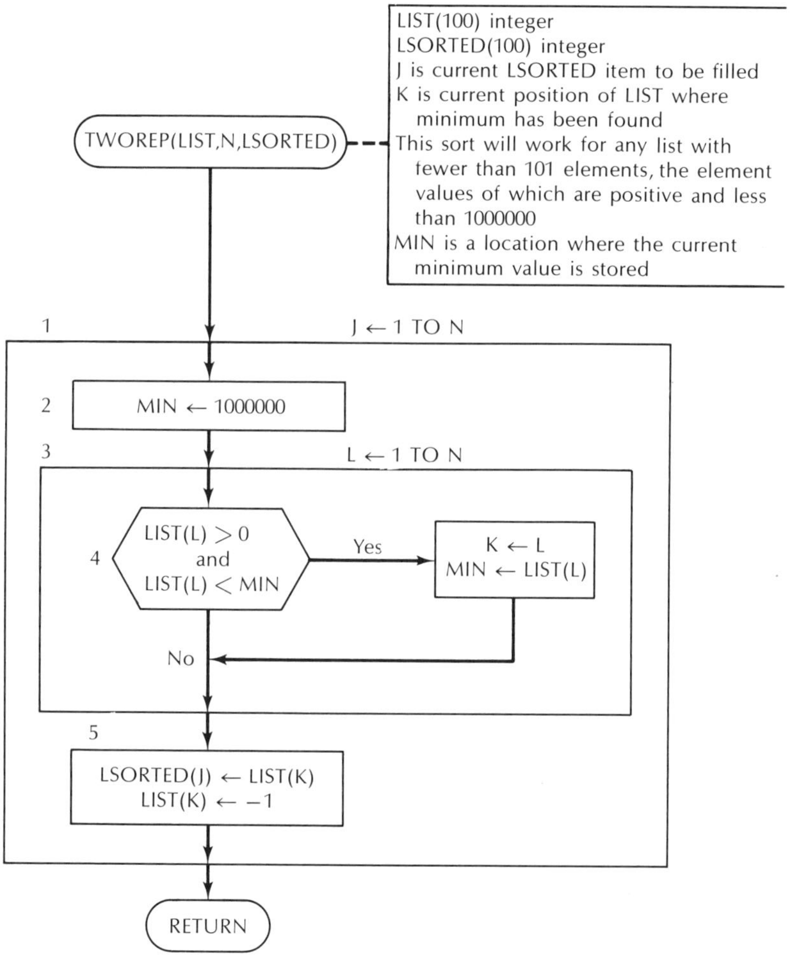

FIG. 10.2-1. Two-list replacement sort algorithm.

A structured algorithm representation of GSA TWOREP is as follows:

1. Repeat Steps 2 through 5 for each element to be filed into the sorted list, LSORTED, by means of a counter J.

2. Set minimum value to a value greater than or equal to any value to be encountered into the list to be sorted.

3. Repeat Step 4 for each element in the unsorted list, LIST.

4. Check to see whether the element is positive and determine whether it is a new minimum.

275

5. Place an element with the minimum value into LSORTED(J); increase J by 1; remove element from LIST.

The one-list replacement algorithm is similar to the two-list replacement sort. It also results in the rearrangement of the elements within the given list in an ascending order. The corresponding flowchart is shown in Figure 10.2-2, from which it follows that the name of the pertinent GSA is ONEREP. This GSA has the parameters LIST and N, the definitions of which are the same as those of the parameters LIST and N in the GSA TWOREP. A structured algorithm for the GSA ONEREP is as follows:

1. Repeat Steps 2 through 5 for each element.
2. Initialize minimum element encountered to first element of the yet unsorted list or sublist.
3. Repeat Step 4 for each unsorted element in the list.
4. Find the smallest element in the remaining unsorted list or sublist.
   4.1 If element considered is less than minimum, then store location of minimum element in a variable K, and store list value in minimum.

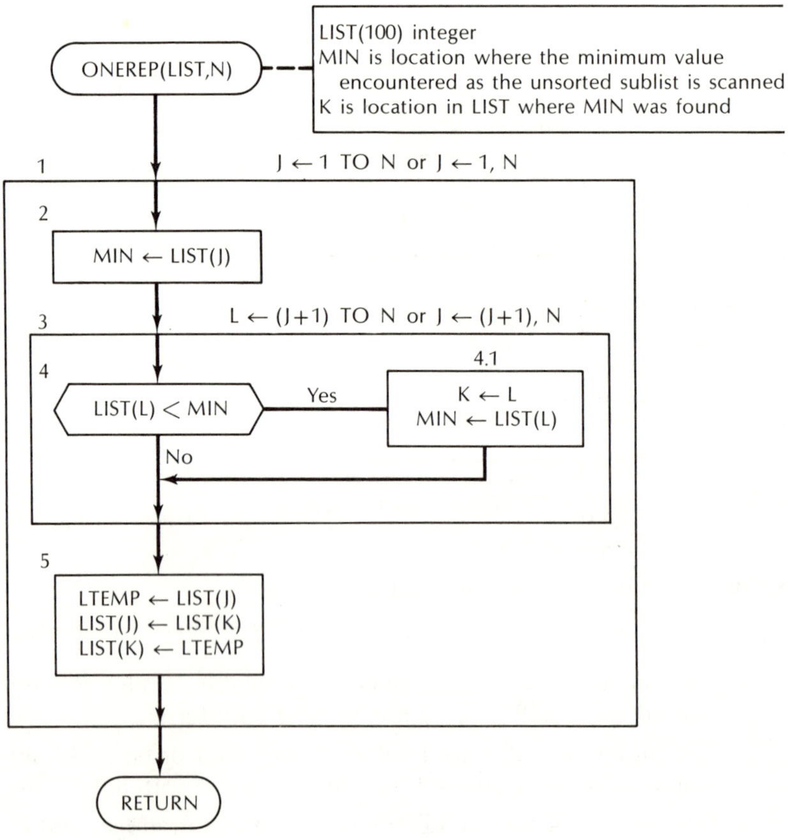

FIG. 10.2-2. One-list replacement sort GSA.

5. Interchange locations of minimum element found in Step 4 with list element to be put next into sorted order.

We now consider an **exchange sort algorithm,** called the **sinking sort.** The basic idea employed in this sort is to place successive pairs of values in a list in the proper order, as illustrated in Figure 10.2-3 for a

---

Phase I: All list elements used.

16	16	16	16
20	20	20	20
23	23	23	23
49	49	49	49
19	19	19	19

Items 1 and	Items 2 and	Items 3 and	Items 4 and
2 compared;	3 compared;	4 compared;	5 compared;
exchange	no exchange	no exchange	exchange

Phase II: List Items 1–4 used; Item 5 is largest in list.

16	16	16
22	22	22
23	23	23
19	19	19
49	49	49

Items 1 and	Items 2 and	Items 3 and
2 compared;	3 compared;	4 compared;
no exchange	no exchange	exchange

Phase III: List Items 1–3 used; Items 4 and 5 are in correct order and position in list.

16	16
22	22
19	19
23	23
49	49

Items 1 and	Items 2 and
2 compared;	3 compared;
no exchange	exchange

Phase IV: List Items 1 and 2 used; Items 3, 4, and 5 are in order and position in list.

16
19
22
23
49

Items 1 and
2 compared;
no exchange

FIG. 10.2-3. Sinking sort.

list of five positive integers. The objective is to arrange the elements in the list in ascending order. This type of sorting strategy is at least as efficient as the replacement sort, and may be considerably better if the list has some sorted sublists in it. The first pass through the list guarantees that the largest list item "sinks" to the bottom element; during each succeeding pass, the next smaller element is placed in its proper order. If no exchanges occur after element K on a given pass, then the elements K through N can be guaranteed to be in the proper order, N being the total number of elements in the list. Thus, only elements 1 through (K − 1) need be considered in the next pass. Using this approach, we obtain the sinking sort GSA SINK as illustrated in Figure 10.2-4. This GSA is capable of sorting lists of integers.

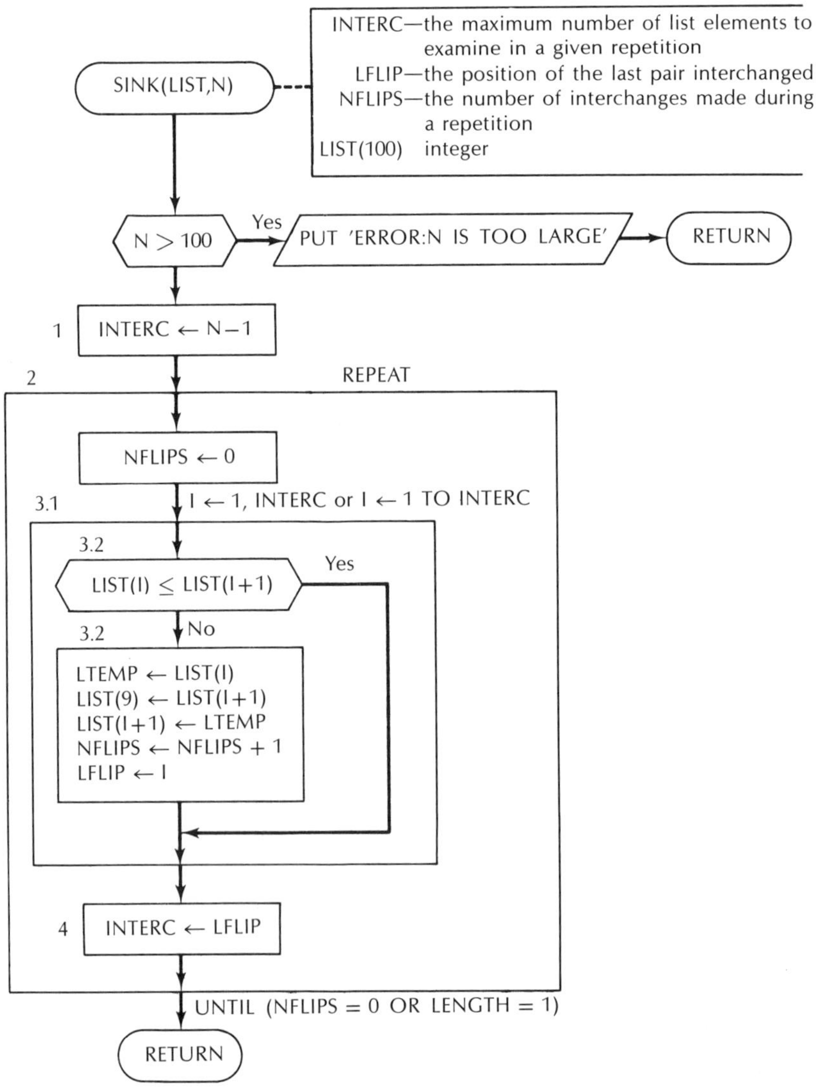

FIG. 10.2-4. Sinking sort GSA.

The related input parameters are

(a) LIST, a one-dimensional array used to communicate (to SINK), the list of integers to be sorted. Up to 100 integer numbers can be sorted into ascending order. The sorted list is returned via this parameter;

(b) N, an integer used to communicate (to SINK) the length of the list to be sorted.

The corresponding structured algorithm is as follows:

1. Test validity of the input, establish the maximum number of interchanges during the first examination of the list.
    1.1 If $N > 100$, then return.
    1.2 Establish maximum number of interchanges as $N - 1$.

2. Repeat Steps 3 and 4 until it is determined that the list is completely sorted.
    2.1 Repeat Steps 3 and 4 until no interchanges have occurred during the examination of the list, or until the unsorted sublist consists of only one element.
    2.2 Reset the counter for the number of interchanges of elements to zero.

3. Examine each pair of elements in the unsorted list and interchange if necessary.
    3.1 Repeat Step 3.2 for each pair of elements in the unsorted sublist.
    3.2 If list elements should be interchanged, then interchange them; increase the number of interchanges by 1, and record the last position of the interchange.

4. Establish the number of list elements to be examined for the next repetition from the position of the last interchange.

5. Return.

Our discussion of business data processing terminology would not be complete without elaborating, to some extent, on the notion of direct-access files. These are also known as nonsequential files. In some cases, the misnomer "random-access files" is also used. This type of file organization finds the records scattered over such storage media as magnetic disks or drums. Often it is possible to use a unique identifier of a record (for example, a person's employee number) as the address of the record in memory. The order of the file may be indicated in each record by means of a pointer (or address) that points to its successor. To illustrate, the sequential and nonsequential forms of a file of payroll information are shown in Figure 10.2-5, where the user assumes that the records illustrated in Part (c) are scattered in memory. The first element (ADAMS or ZUMB) must be identified for the programmer to access the list. The first element of the list ordered by name is ADAMS, in the lower right-hand corner of Figure 10.2-5(c). Next, the address that follows ADAMS (illustrated by the solid arrow), points to the second element, BROWN. The last element of this ordering of the list is ZUMB (in the lower left-hand corner), and is

279

10	ADAMS	2.65
11	BROWN	4.15
51	COX	6.50
42	WHITE	4.60
6	ZUMB	2.32

(a) Sequentially ordered
by name

6	ZUMB	2.32
10	ADAMS	2.65
11	BROWN	4.15
42	WHITE	4.60
51	COX	6.50

(b) Sequentially ordered
by salary

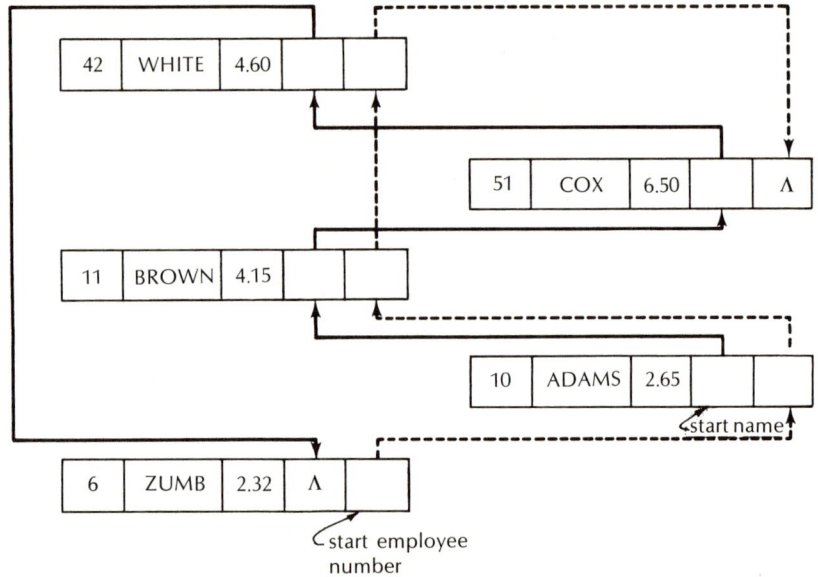

| 42 | WHITE | 4.60 |

| 51 | COX | 6.50 | | Λ |

| 11 | BROWN | 4.15 |

| 10 | ADAMS | 2.65 | | |

start name

| 6 | ZUMB | 2.32 | Λ | |

start employee
number

(c) Nonsequential organization ordered by name with one pointer
field and by employee number with a second pointer field

FIG. 10.2-5. Sequential and nonsequential file organization.

denoted by the symbol Λ in its pointer field. The second list then requires a second pointer in each record, as illustrated in Figure 10.2-5(c).

From the preceding discussion, it is apparent that the nonsequential file requires more storage than does the sequential, since the former involves a pointer (or address) in each record. However, if the file has to be organized by more than one data item, the nonsequential approach can result in substantial savings in storage. This is because the file need not be reproduced or resorted to obtain the new organization.

10.3 MANAGEMENT INFORMATION SYSTEMS

The application of computers to data processing problems can be justified in four ways, as described by Kernighan and Plauger:

Purpose	Application
1. To reduce administrative expenses	1. Administrative and accounting uses
2. To reduce cost of product sold or service rendered	2. Control systems
3. To increase revenues	3. Product or service innovation; improved customer relations
	4. Management information systems; computer simulation modeling
4. To improve staff output and management decisions	

In the following discussion, we will restrict our attention to part of the fourth application—management information systems. The term **management information system** (for which the acronym MIS is sometimes used) implies a computer system that works with an integrated data base which delivers reports, graphs, and other analytical information pertinent to decision processes.

Management information systems, in order to be operational, depend on an integrated data base of corporate information. The term *integrated* means that data are compatible, but not necessarily all in one file, available in one location. Each time a data item occurs, it must be the same type, order, and size. For example, if an employee name should appear as last, first, middle initial, separated by commas in 32 characters of information, then all employee names should appear exactly this way. The number of occurrences of an employee name should be no greater than one, except when dictated by security or efficiency considerations. In contrast, all departments would have files containing the employee name, address, social security number, etc. in a nonintegrated data base. Nonintegrated data bases are more expensive and difficult to update than are integrated ones. For example, in the former, when an employee moves, there is no single place in which to change the employee address.

Management information systems were considered to be "the thing" during the late 1960s and early 1970s. During this period, there was a tendency for many installations to merely change their current names in order to be included in the overall management information systems community. The economically trying times that followed, however, played a major role in encouraging management to view management information systems in proper perspective. As a result, the following set of criteria was formulated in connection with the use of management information systems:

☐ The system should provide timely and accurate information for control and planning needs.

☐ The system should contain information necessary to generate the vast majority of all legal and governmental operating and financial information.

☐ Ideally, the system should integrate most of the institution's information.

281

☐ The system should mechanically and functionally provide for data collection close to its source. Data items should be stored within the data base as few times as security of the information will allow.

☐ To increase compatibility with like institutions, and to ease compliance with governmental reporting needs, the data should be based upon a uniform classification scheme, whenever possible.

Here, total systems are envisioned, which perform or assist with some or all of the following operations in a timely fashion:

1. general institutional and accounting, such as payroll;
2. production and process control, such as computer control of an assembly line;
3. inventory management;
4. distribution control, including acceptance of orders from the marketing force;
5. cost accounting;
6. quality control;
7. personnel management;
8. marketing strategy;
9. financial strategy;
10. overall planning information; and
11. nonstandard report creation.

Most institutions have an information system to perform at least one, or perhaps several, of the listed operations. A desired goal, yet to be totally achieved, is to realize an economically sound management information system which is not only useful to low and middle management, but also to top management for institutional-level planning.

10.4 COMPUTER SYSTEMS

Our main objective in this section is to introduce a set of terminology that is used quite frequently in connection with computer systems. As such, some of the terminology we will consider is directly related to data processing systems.

Any device in an overall computer system that is not in direct communication with other units in the system is said to be an **offline device.** Keypunches, keystroke devices, some data collection devices, and perhaps digital incremental plotters are examples of offline devices. In contrast, online devices are capable of continual communications (electronically) with the rest of the computer system. Online computer system hardware components can be classified as fulfilling one of the four basic functions of a computer. For review purposes, these functions are shown in Figure 10.4-1.

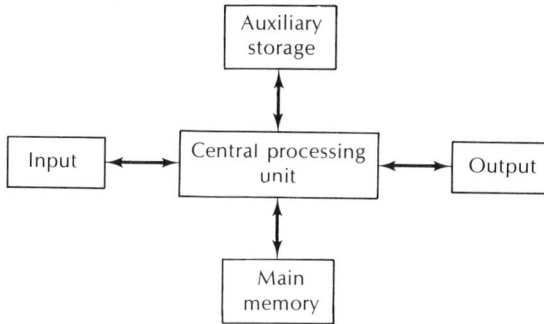

FIG. 10.4-1. Computer system functional components.

The type of computer service provided by a system can be described in terms of two modes: **batch** and **online.** In batch processing, the data to be processed are collected and grouped for processing during a computer run. This mode of operation results in a more efficient use of the computer hardware, relative to the online mode. However, the batch mode may cause personnel to wait for jobs being run, and hence lead to inefficiency in terms of overall organization. In contrast, the input data are entered directly into the computer in an online system. Such data are entered directly at their points of origin, and the results are transmitted directly to the user(s)—i.e., data are essentially processed as they are generated. Alternately, an online system may simply collect input data directly, and then act as a batch-processing system, or take on more advanced forms of online computing.

Conversational computing is a form of online computing, in which each entry from an input/output device (called a terminal) elicits a response to the terminal, and vice versa. Another form of online computing is known as **real-time computing.** It refers to the collection and processing of data arising from a physical process, so that the corresponding results can be used to influence the physical process—e.g., controlling a distillation tower in an oil refinery or a total information system.

Computer systems are designed to operate in either online, or batch, or both modes. Though the use of programs called **operating systems** that manage the computer, a single computer can provide several modes of operation to a number of users. This ability to provide concurrent use of a computer system to several users is referred to as **time-sharing.** The concurrent use of a computer system is achieved by dividing the available time into disjoint intervals, and then assigning each unit of time to a user. **Multiplexing** is one of the methods used to this end. A multiplexed operating system attempts to keep busy as much of the hardware as is possible by judiciously allocating time units to users who are ready to compute. A form of multiplexing in which several programs are placed in main memory simultaneously is called **multiprogramming.**

Multiprogramming is a common form of computing. Its main disadvantage as far as the programmer is concerned is that the amount of

283

Conversational Computing Manager
Batch user 1
Batch user 2
⋮
Student Express Service Manager
Card reader program
Line printer program
Operating system

FIG. 10.4-2. A hypothetical use of memory in a multiprogrammed computer.

memory allocated to each individual is limited, as illustrated in Figure 10.4-2. There are two schemes that are used to overcome this difficulty: **virtual storage** and **paging.** Computers with virtual storage merely have an auxiliary addressing mechanism that gives the impression that the memory is much larger than it actually is. On the other hand, a paging system causes a program to be stored in the form of fixed-size pieces called **pages,** so that, at a given time, only a few pages of the program are brought into the computer memory for execution. If a needed page of the program is not available in memory, the operating system replaces a page that is not needed by the one that is required. Virtual storage systems are commonly used with paging.

To conclude our discussion of computer systems, we turn again to the problem of job processing on a computer, noting that a user program can exist in one of three forms:

1. source form—in the language used by the programmer;
2. object or relocatable form—an intermediate form of the program produced by some language translator; or
3. machine-usable form—a form understood by the machine without resorting to translation; it is usually in binary form.

Language translation that results in the object or relocatable form is accomplished by compilers or interpreters. In the case of **compilers,** the entire program is first converted to an object program and then executed (see Figure 10.4-3). On the other hand, **interpreters** execute a program one line at a time. A line is "interpreted" and executed each time it is encountered. Consequently, compilers are more efficient when a program must be executed repeatedly, while interpreters may be more efficient in one-time executions.

Figure 10.4-4 shows the manner in which job processing is typically carried out on a medium- to large-scale computer system equipped with a multiprogrammed operating system. In this figure, there is one term that has not yet been defined; namely, *spool.* A **spool** is an intermediate storage area, often on a disk, where the input (output) from (to) a slow device is temporarily stored. It effects input/output operations only, in that it enables the CPU to deal with a relatively fast

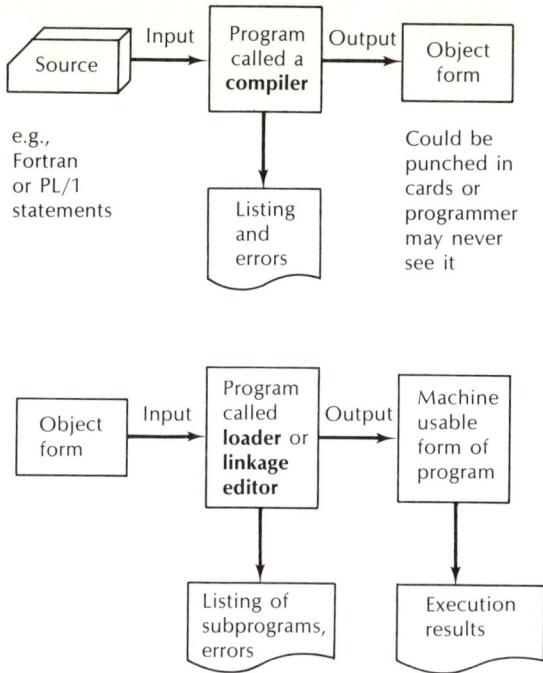

FIG. 10.4-3. A compilation process.

FIG. 10.4-4. A medium- to large-scale computer system equipped with a multiprogrammed operating system. (Courtesy of Sperry Univac Corporation.)

285

device, while **channels** deal with a slow device on an incremental basis.[1] Spooling is a means of improving the overall efficiency of a computer system.

The most common user measure of a system performance, other than cost, is the time it takes to carry out a computing request. In conversational computing, the time taken for the computer to respond line by line (of a program) is referred to as **response time.** Anywhere from 0.1 to 1 second is considered as good response time. The term **turnaround time** is used often in a batch-processing mode. It refers to the interval between the time a job is submitted to the system and the time corresponding output is ready. Good turnaround time varies with job length and user expectation.

10.5   SUMMARY

In this chapter we discussed computer and data processing systems. Data processing is concerned with the storage, processing, and retrieval of large volumes of data associated with businesses and institutions. Data records are organized into sequential or direct-access (nonsequential) files. Information stored in this manner can be retrieved through search and sort algorithms, a few of which were considered.

Our discussion of management information system concepts emphasized that such systems strive to integrate the computers, people, and procedures of an organization to provide information useful in making management decisions. We noted that since these systems are relatively new, they generally do not integrate all or even most of an institution's information on a timely basis. In the future, these systems should become increasingly more timely, more complete, and easier to use. Thus, they will probably become extremely useful to top management for formulating short-term strategy and long-range planning for growth and investment.

In the last section of the chapter, we introduced some of the terminology used in connection with computer systems.

problems for computer solution

The set of learning objectives for this set of problems is the same as that given for the problems in Chapter 9.

CS 10-1.   *Additional learning objective:*   to use a disk file.

Develop a general subalgorithm for updating the inventory file for an integrated information system. Test this subalgorithm by enclosing it in a program.

---

[1] *Channels* are special-purpose computers that are used in computer systems to provide additional processing capability, so that input/output operations can overlap with CPU operation.

The formal parameters are the quantity of the item sold and the stock number. Assume that the inventory master file is stored on magnetic disk. Each master file record contains the following:

1. stock number;
2. item description;
3. minimum quantity below which reorder is to be made;
4. quantity to reorder;
5. unit selling price;
6. unit cost;
7. quantity on order;
8. quantity on hand;
9. supplier's name and address; and
10. total cost of quantity on hand.

Each time this subalgorithm is invoked, the inventory master file should be searched until the appropriate record is found. The quantity on hand and the total cost of the quantity on hand should be reduced by the quantity and cost of the quantity sold, respectively. Then the quantity on hand plus the quantity on order should be compared with the minimum quantity below which a reorder is to be made. If the quantity on hand plus the quantity on order is less than that value, then an order should be written on a reorder file. Data Fields 1, 2, 4, 6, and 9 should be the value output to the reorder tape. Then Data Field 7 should be updated and the revised master record output to the master file. Trace a sample set of input data.

(COBOL, PL/I [business])

CS 10-2. *Additional learning objective:* to make extended use of the end-of-file condition.

A common problem in a business that manufactures, distributes, or sells items or parts is that of maintaining the inventory. The manual maintenance of stock record cards often leads to an out-of-stock condition and an extended waiting time by the customer, which is bad business practice.

The Purple Pride Sales Company has such a problem. The company has been maintaining the inventory of items on stock record cards by updating the cards manually. You have been hired as a programmer to automate the inventory process. An outline provided by the systems analyst follows:

1. Print out header line(s) of the data items to be read in from the stock records.

2. Read in the stock records and print them as they are being read in under the appropriate headings.

3. Read in sales transaction records which consist of a code (1), the part number, and the quantity sold. The program should then print a line showing all of the information in

287

the record, except the reorder information. This should include an extension of the price for multiple quantities.

4. Read in a receipt for an item into stock record, consisting of a code (2), the item number, the quantity received, and the price. If the price has changed, the new price should be posted. The program should then print a line that shows all of the information concerning that item of stock.

    Note: Steps 3 and 4 require the use of a search subroutine to find the part number in the array of structures.

5. When all transaction records have been read in, all of the inventory records should be printed.

6. Then, any item, the on-hand (O/H) quantity of which is below the reorder point, should be printed. Pertinent information about the item and the quantity of the item to order should be given. This aspect of the program should be handled by a subroutine.

All of the information that the program prints must be appropriately headed or labeled.

See the accompanying table for the data for your program. You must make up the transaction record with the following requirements:

(a) There must be at least two receipts into stock, at least one of which has a price change.

(b) There can be any number of sales transaction records, but at least three items must show up on the reorder list.

Part number	Part name	O/H	Price	Reorder Point	Reorder Quan.	Bin locations 1st	Bin locations 2nd	Bin locations 3rd
3KX6031N	Executive desk	3	199.95	1	2	A22		
3KX61009N	Bookcase	2	74.95	2	4	C13		
3KX3666	Wastebasket	15	2.89	5	12	A46	B83	C77
3KX3672	Letter tray	62	4.19	12	60	B72	B97	D93
3KX4329	Scratch pad	65	.89	50	400	D58	D96	
3KX3834L	Duplicator paper	74	3.95	100	900	A48	B04	C62
3KX4162	File cards (3×5)	86	.59	75	500	B37	C35	D95
3KX41321	Manila file folders	28	4.99	15	50	A27	A94	
3KX60889N	Four-drawer file cabinet	2	62.95	1	10	C20		
3KX61355N	Executive chair	1	149.95	2	10	D84		
3KX6188NH	Side arm chair	6	114.95	6	25	D85		
3KX5824C	Eight-digit calculator	6	49.50	4	20	A48	B52	C25

(COBOL, PL/I[business])

CS 10-3. The National Company has just hired you as a programmer. Your long-term assignment is to automate all economically justified accounting operations that are now done manually. After considerable study, you choose to initially automate the personnel files and produce the paychecks by machine.

The personnel files for this and other future applications should contain at least the following information:

(a)   employee number;

(b)   name;

(c)   department;

(d)   home address;

(e)   rate of pay;

(f)   monthly health insurance premium;

(g)   union affiliation (no more than one);

(h)   union dues;

(i)   number of dependents;

(j)   years of employment;

(k)   sex;

(l)   age;

(m)   gross pay to date; and

(n)   net pay to date.

Organize a master file record. Is anything missing? If so, include it after you have read and understood the entire problem.

Create a subalgorithm which, if called, will produce net pay as follows:

1.   (# of hours worked up to 176) × rate of pay

2.   plus (# of hours over 176) × 1.5 × rate of pay

3.   minus health insurance premium

4.   minus union dues

5.   minus federal withholding tax, FICA, which is (rate of pay × 176) × (10 − # of dependents or 10, whichever is greater) × .0135.

6.   If year to date FICA > 15% of 10 500, then FICA = 0.

Print paychecks; you may assume check forms exist.

The program should read in information for the master file containing data for 6–10 employees. This would not, of course, be a part of a program in real life; actually, the file would exist on a magnetic tape or disk.

Issue a command to update the file—i.e., add an employee, delete one, and change items in existing records. Assume the master file is not ordered. In your documentation note advantages and disadvantages of this file organization. Indicate changes that would have to be made to your program to handle a file ordered sequentially by employee number. Under what conditions would you recommend such a change?

(COBOL, PL/I [business, computer science])

CS 10-4.   In today's society credit cards and charge accounts are a part of daily life for many. In this problem you are asked to create a program that will produce a monthly statement for each account held by a credit card company. This report

289

should consist of the person's name, address, account number, old balance, payment (if any), interest charge, new balance, and payment required. Use an array of structures to store the information for up to 50 credit customers.

The new balance is equal to old balance − payment + interest charge, where interest charge = .015 × old balance (i.e., 1.5%).

The payment required is 10 percent of the new balance or $5, whichever is greater. A penalty of 2 percent is to be charged if the payment made was not at least the required amount.

The input will consist of the following:

(a) person's name—last
                       —first
                       —initial
(b) account number
(c) address—street
                    —city
                    —state
                    —zip code
(d) balance at last report
(e) payment since last report

Arrange the output in an orderly form with a separate report for each individual account. Finally, include a summary showing totals.

(PL/I [business])

CS 10-5. This problem involves updating a billing file for *Punt,* a football magazine.

Use a heterogeneous structure to store information about each subscriber. The program should be written to anticipate adding and deleting records as well as issuing bills. Use a subalgorithm for each of these functions. When the billing is issued, *all* subscribers' records should be examined, then bills should be issued to all subscribers whose date of subscription expiration is within three months of the present date.

(PL/I [business, general])

CS 10-6. *Additional learning objective:* to use Polish postfix notation.

Normally, equations and expressions are given in a notation called **infix notation.** This means that an operation to be performed on two values is always indicated by placing an operator between those values (e.g., 2 + 3, A * 6, X − Y). A problem associated with infix notation is that parenthesis must be used to group items when operations are to be performed in an order different from that defined by the precedence rules of the language. This can be remedied if the expressions can be stated in parenthesis-free fashion. One such system is postfix or Polish postfix notation. In this system, a+b is written as ab+; (a+b)/c is written as ab+c/; and a+(b/c) is abc/+. The operators are processed left to right and operate on the previous two operands. An algorithm for this can be found in many elementary compiler or translator textbooks. Develop an algorithm that will convert expressions with single-letter variable names, single-digit constants, and the operators, +, −, *, /, and that will allow parenthesis.

(PL/I [computer science])

CS 10-7. In your role as liason between the personnel office and the president's office of a small college (no more than 40 faculty members), you have been asked by the president to call a meeting of the entire faculty to discuss student complaints. Therefore, you must select a time when all the faculty can be present.

You have been given permission by the director of personnel to scan the file containing information for each faculty member. This record contains the name, social security number, a series of 10 single digits (0's and 1's only) for each of the five weekdays. This series of digits represents whether the faculty member is free (indicated by zero) or teaching (indicated by 1) for each of 10 hourly time periods during the day.

Write a program that will use a subalgorithm to select the hours during the week when all of the faculty are available. If no such times exist, find the hour(s) at which the fewest number of faculty are unavailable and print the names of those who may not attend. You are not permitted to know the social security numbers of the faculty members; therefore, your program must not read these.

Use a heterogeneous structure to store the information from the faculty members' file.

(PL/I [general])

291

11

The Development of Computer Systems— Their Impact on Society

KEYWORDS

artificial intelligence
data base
heuristic
history
impact
privacy
security

After growing wildly for years, the field of computing now appears to be approaching infancy. (U.S. President's Science Advisory Committee on Computers in Higher Education, 1967)

INTRODUCTION

The period between 1945 and the present was predicted by experts to be "the nuclear age." However, the rapid and unpredicted growth of computers, measured by the number produced, their capabilities, and the extent of their application, has caused this time period to be referred to as "the computer age." Scientists who study the impact of change on civilizations and societies generally agree that the computer age is, in essence, a revolution. The main objective of this chapter is to provide a framework that will facilitate a basic understanding of this revolution.

To understand the computer revolution, it is best to first explore the interwoven fabric of ideas, inventions, and events that have led to current computer technology. To this end, a summary of the historical milestones is presented in Section 11.1. Section 11.2 is concerned with the computer revolution in terms of current technology, and Section 11.3 with the type and breadth of computer applications. The material in these three sections provides the background that is necessary to explore the current and future impacts of the computer revolution; these are considered in Section 11.4.

Two factors must be kept in mind in order to put this introductory material into proper perspective. First, is the very brief time span, relative to similar revolutions, with respect to the conception of an idea and the changes it produces in our society. The first electronic computer was put into operation in 1945, while the first commercial electronic computer was produced in 1951. Thus, we observe that a span of just over two decades has resulted in technical, economic, political, ethical, and social impacts that have equaled or exceeded the corresponding impacts that occurred in a 100-year period following the Industrial Revolution.

The second important factor that we must not lose sight of is that the computer is the invention of human beings to enhance their capability to accomplish tasks. People must decide on the uses to which computers are put. Our designs and uses of this tool shape our lives, our institutions, and our entire society.

A HISTORICAL PERSPECTIVE

At least as early as 1000 B.C., people were using pebbles or beads on a counting board to carry out simple calculations. This tool is now called the **abacus** (see Figure 11.1-1). The abacus employs a positional number system and is still widely used as a calculating tool. The pebbles (or beads) were called *calculi* in Latin, from which we derive the words "calculate" and "calculus."

Calculating devices existed in many prehistoric civilizations. Two such devices of interest are (a) Stonehenge (1900–1600 B.C.), and (b) the quipu (1400 A.D.). **Stonehenge** is a stone monument, located in England. It was most probably erected to predict astronomical events, such as the change of seasons. The desire of people to predict natural phenomena has been a motivating source for computation in the past,

FIG. 11.1-1. An abacus. (Courtesy of International Business Machines Corporation.)

and remains so today—e.g., in weather and earthquake prediction. It is interesting to note that an electronic computer analysis of Stonehenge data assisted researchers in developing its postulated purposes.[1]

The **quipu** was a device developed by the Inca civilization for counting and recording important events. It consisted of a main cord, from which hung smaller knotted cords of various colors, each having a specific meaning. The Incan rulers used the information gathered via the quipu system in an attempt to supply the needs of all citizens. However, when the Spanish invaded this civilization, the organized information available in the quipus they captured allowed them to totally exploit the Incas.

From 1400 A.D. to 1945 A.D. many important developments took place, a sampling of which is as follows:

1617 A.D.: **Napier's bones.** John Napier, a Scottish scholar, developed a set of numbered rods in order to multiply, divide, and extract roots. These rods were the predecessor of the slide rule, invented by William Oughtred in 1630.

1642 A.D.: **Pascal's adding machine.** This device invented by 19-year-old Blaise Pascal, who was tired of adding long columns of figures in his father's tax office in Rouen, France. The device is about the size of a shoebox. It established three important principles in the development of mechanical calculators: (a) the carry operation could be done automatically, (b) subtraction could be performed by reversing the dials, and (c) multiplication could be accomplished by repeated additions. Pascal received a pa-

[1]See G. S. Hawkins, "Secret of Stonehenge," *Harper*, June 1964, pp. 96–99, and "A Scottish Stonehenge," *Science*, January 1965, pp. 127–30, also by Hawkins.

tent on this mechanical adding machine which performed only addition and subtraction. Several such devices exist today and work via rotating gears located in a box, as illustrated in Figure 11.1-2. Numbers are painted on the

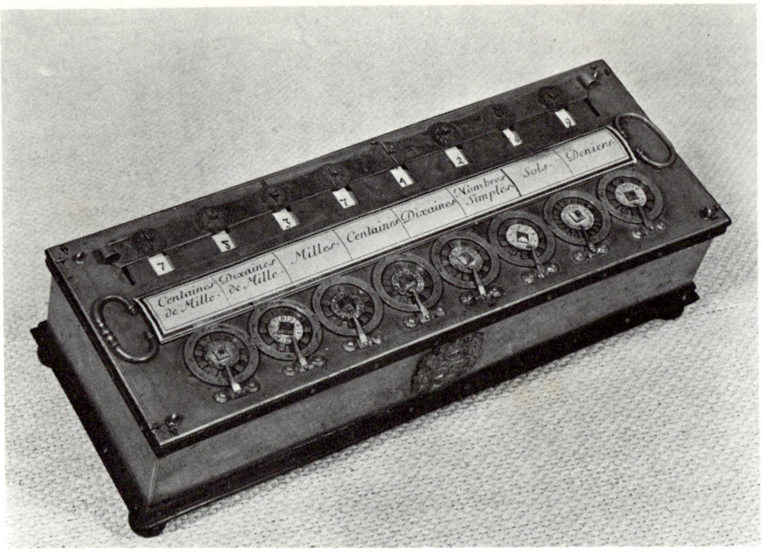

FIG. 11.1-2. Pascal's adding machine. (Courtesy of International Business Machines Corporation.)

gears and displayed through holes to indicate the sum. As the rightmost gear completes its 10-unit revolution, a pin on it moves the next gear one unit forward, etc. The basis for digital adding machines was thus established.

1671 A.D.: **Leibniz's calculator.** The first calculator that could perform multiplication and division was built by a German, Baron von Leibniz. It consisted of long gears with nine teeth, each of a different length. This design was perhaps the basis of the mechanical calculating machine called the Arithmometer, which was developed in the mid-nineteenth century. Multiplication was carried out via successive additions. A second contribution of Leibniz was that binary arithmetic was used in lieu of decimal arithmetic in this calculator.

1801 A.D.: **Jacquard's loom.** In France, Joseph Marie Jacquard invented a punched card as an accessory to the loom. This accessory helped automate the loom for the weaving of intricate patterns. His ideas evolved from a long line of automaton creators including Jacques de Vaucanson. In 1736, Vaucanson created an automaton, controlled by a revolving drum punched with holes, to simulate lip and finger action so accurately as to be able to play a flute. From these efforts came Jacquard's idea of controlling a device through the placement of holes in a card. Instructions for the loom were stored on punched pasteboard cards strung together in a long set. The holes allowed the wires, which were individually attached to the warp thread in a fabric, to pass through and cause the thread to be lifted. Thus, a pattern could be programmed as a set of punched pasteboard cards and cause the corresponding pattern to be created by the loom.

1812 A.D.:   **Babbage's engines.** In England, two calculating machines were invented by Charles Babbage. These machines were capable of computing and printing mathematical tables. They were called the Difference Engine (1812) and the Analytical Engine (1833). In essence, they led to the definition of the four basic parts of a computer system—input/output, control, arithmetic unit, and storage. Punched cards were used for input/output and control. The Analytical Engine had a memory large enough to hold a thousand 50-digit numbers. It was capable of adding or subtracting two 50-digit numbers in one second, and multiplying the same in one minute. Babbage successfully integrated existing ideas to form the basis for most of our current computing ideas, but in his own time he was a failure. Existing metallurgical and mechanical engineering

FIG. 11.1-3.   Difference Engine. (Courtesy of International Business Machines Corporation.)

could not provide the strong and precise gear systems required. The Bank of England, Babbage's backer, never regained the large financial losses incurred, even though two Difference Engines were sold after Babbage's death—one to an American observatory, the other to the British government (see Figure 11.1-3). A discussion of Babbage would not be complete without mention of Lady Augusta, Countess of Lovelace. She was a mathematician who documented and clarified Babbage's work.

1848 A.D.: **George Boole,** a British mathematician, introduced the notion of an algebra, now referred to as Boolean algebra. It is a useful tool for the analysis and design of binary computing devices.

1884 A.D.: **William Burroughs,** an American, invented the first commercially available adding machine. It was able to list the input to it and the corresponding results.

1890 A.D.: **Herman Hollerith,** an American engineer, either expanded upon Jacquard's idea of using punched cards, or may have formulated his ideas independently. He used the cards to store data, with the holes representing data values. While working for the Census Bureau, he developed a set of electromechanical tabulating machines for such punched cards (see Figure 11.1-4). All this effort was in response to a contest designed to develop a way to tabulate the 1890

Fig. 11.1-4. Hollerith tabulating machine. (Courtesy of International Business Machines Corporation.)

census before the data became obsolete. The result was that the 1890 census was processed in one-third the time of the 1880 census. In 1896 Hollerith started his own company, which later merged with a clock and butcher-scale company to become the International Business Machines (IBM) Corporation.

The use of tabulating machines grew rapidly in the early part of this century. In the late 1930s and early 1940s, the cooperation between universities and governmental and research organizations led to the solution of difficult problems faced by a growing nation with increasing computational needs. One of the significant developments that occurred in this period (1944) was the construction of MARK I, a computer built at Harvard University under the sponsorship of IBM (see Figure 11.1-5). This computer made extensive use of electronic tubes and electrical relays. It was designed to produce ballistics tables which were to be used in connection with World War II.

FIG. 11.1-5. Harvard-IBM MARK I—the Automatic Sequence Controlled Calculator. (Courtesy of International Business Machines Corporation.)

While the United States was working on MARK I, a British mathematician by the name of Alan Turing was working on wartime cryptography problems. In this connection, he created a mathematical model for computing—the **Universal Turing Machine**—still used to this day. German war messages were decoded via an implementation of this machine. One of the messages that was decoded during the earlier part of the war was that the town of Coventry would be bombed. Winston Churchill, then prime minister of England, had two choices: to either repel the attack and evacuate the town at the risk of

299

letting the Germans know that their messages were being decoded, or to let Coventry be bombed, and hence assure a continued flow of messages that could be decoded in the future. At this point, it is instructive to note that the use of computers led to the decision, which was linked to social and ethical issues. Churchill chose to let Coventry be bombed—a decision that tormented Turing, who later committed suicide in 1954.

The preceding discussion can conveniently be summarized in terms of four categories, as follows:

1. counting devices (abacus and adding machines);
2. logical ideas (boolean algebra and the Turing Machine);
3. measuring devices (Napier's bones); and
4. storage via punched cards (loom and tabulating machine).

11.2 THE COMPUTER AGE

By 1890, essentially all the theoretical ideas necessary for large-scale computing were available. However, the related technology was not sophisticated enough to produce the devices to implement large-scale computing schemes. The introduction of the vacuum tube, thousands of times faster than the earlier electromechanical devices, provided the necessary speed. The age of **electronic computers** thus began. The first electronic computer was built in 1945 by J. Presper Eckert and John Mauchly at the University of Pennsylvania. It was called the ENIAC, was capable of multiplying two 10-digit decimal numbers in 0.003 second, and filled a room 20 feet by 40 feet. It contained 18 000 vacuum tubes. Eckert and Mauchly formed their own company and built a computer called UNIVersal Automatic Computer (UNIVAC) in 1951 (see Figure 11.2-1). It was delivered to the Census Bureau, and hence became the first commercially available electronic computer.

Following the ENIAC, John von Neumann of Princeton demonstrated that instructions could be encoded, read, and stored in the memory of a computing machine. Further, such instructions could also be easily changed when desired. This important development occurred in 1946, leading to the fundamental concepts of a stored program and sequential execution of instructions—concepts that are still widely used.

The development of the transitor led to the **"second generation" of computers.** Thus, in 1957, a computer was made that consisted predominantly of transistors, rather than of vacuum tubes. Another breakthrough occurred in 1964 with solid state technology, leading to the use of integrated circuits in computers. Such computers are referred to as **"third-generation" computers.** Several other characteristics are associated with computer generations; these are summarized in Table 11.2-1.

FIG. 11.2-1. UNIVAC I. (Courtesy of Sperry Univac Corporation.)

Next, the late 1960s initiated the development of **minicomputer systems.** The use of minicomputers is growing rapidly; hundreds of thousands of them will be sold by 1980. In addition, a class of even smaller sized computers, called microcomputers, is now having a major impact in many application areas.

TABLE 11.2-1. Characteristics of computer generations

Characteristics	Generation		
	First **1945–1954**	**Second** **1954–1964**	**Third** **1964–present**
Electronic component	Vacuum tubes	Transistors	Integrated circuits
Time/operation	0.1–1.0 msec	1–10 μsec	0.1–1.0 μsec
Main memory	Electrostatic tubes	Magnetic core	Semiconductors
Programming languages	Binary and symbolic code	Procedure and problem oriented	Extensible and concurrent languages*
Operating system	None	Batch	Interactive conversational computing
User participation in running the program	User operates computer	No participation	Multiprogramming Time-sharing Central file systems Virtual memory Text editing

*Extensible languages allow the user to extend the capability of the language, e.g., add new operators. Concurrent languages allow the user to express the actions that may take place simultaneously.

A summary of some of the more significant computing events is given in Table 11.2-2. The interested reader is encouraged to refer to other publications for more detailed presentations. The book by Goldstine and that edited by Randall, published in 1972 and 1975, respectively, are highly recommended (see Bibliography).

TABLE 11.2-2. Some significant events (dates may be approximate)

1600 B.C.	Stonehenge
1000 B.C.	Abacus
1400 A.D.	Quipu
1617	Napier's bones (logarithms)
1642	Pascal—adding machine
1671	Leibniz—calculator
1801	Jacquard—loom
1822	Babbage—engines
1848	Boole—formal logic
1890	Hollerith—punched cards and tabulating machines
1937–44	Mark I—electromechanical computer

First Generation

1945	ENIAC—first electronic computer went into operation
1946	Von Neumann—stored program idea defined
1951	UNIVAC I—first commercial computer
1958	IBM 701—defense computer
	UNIVAC 1103—powerful scientific computer (could add a column of typed 10-digit numbers as high as the Statue of Liberty in one second)

Second Generation

1957	UNIVAC II—core memory
1959	IBM 1620, IBM 7090 (popular scientific machine could multiply 2 ten-digit numbers in 1/100 000 of a second)
1959	RCA 501—business-oriented computer featuring COBOL

Third Generation

1964	CDC 6600—fastest computer for many years
	GE 600—microcircuitry
1965	IBM 360 series—compatible series of computers, some capable of 2 500 000 additions per second
1966	GE 645 and MULTICS operating system
	PDP-9 features graphic capability
1967	IBM 360 model 91 (can add 16 000 000 numbers/second)
1968	E. W. Dijkstra—structured programming
	H. Mills—programming teams
1970	Minicomputers
	International computer networks
	Microcomputers—small special- or general-purpose processors to perform small tasks
	Hand calculators

The limitation on the ultimate speed of computers is the speed of light. We are close to a realization of that speed in some new computer systems. The internal speed of these computers is measured in nanoseconds. A nanosecond is one billionth of a second—the time it takes a pulse to travel 11.9 inches. This distance is the physical limitation in computer systems. Thus, most future improvement in computer systems will not come from faster CPUs and memory. It will come from better machine organization, refinement of algorithms, and quality of software available.

Machine organizations that are new (and those being proposed) include organizations that are capable of performing many tasks simultaneously. Such systems employ schemes called **parallelism** or **concurrency.** A machine may have multiple CPUs; there is one, for example, with 64 CPUs capable of operating simultaneously. Another scheme involves forming a network of several computers, attempting to maximize production over all of them. Networks are predicted to become increasingly popular at the national and international levels. In some instances they will enable communication between various geographic locations, with computing power and information transfer capabilities at each location. One such network, called SWIFT, is an international banking network. It provides electronic banking between almost all major European banks and some of the banks in the United States (see Section 11.3, page 305). When fully developed, it could link most of the banks in the world, providing information and fund transfer capabilities at electronic speeds.

The cost of computer hardware has decreased at a rate comparable to reduction in execution time. To illustrate, in 1955 it cost about $10 to accomplish a million additions on a computer (considering CPU and storage costs). In 1975 this cost had dropped to $0.00005 per million additions. By 1980, it is predicted that it will cost under $0.00001 per million additions.

The size of computers has also been drastically reduced. A computer that occupied 1000 square feet in 1955 would now occupy about 0.1 square foot due to technological advances.

The cost and size reductions, accompanied with increases in speed, have made computers more feasible for an increasing number of tasks. This trend is continuing, and it is probable that each household will have some form of computer, or computer-related service in the next few years. The notion of computer utilities (not unlike electric or gas utilities) is receiving attention in the United States.

We mentioned earlier that in addition to better machine organizations, improvements in software can lead to greater computer speed and usefulness. Programming was considered an art until the early 1970s. Since then, several efforts have been initiated to develop software in a systematic manner. Some languages such as COBOL and FORTRAN, have been standardized. Thus, programs written in such languages can be shared. The development of languages will continue. Languages which are more powerful and easier to implement will be sought. Efforts related to techniques that enable system-

303

atic software development will continue to receive attention. Early efforts, including structured programming (1969–1970) and management of programming teams (1973) have improved programs and programming management. Current research efforts are also concerned with ways to prove the correctness of programs.

The rate of growth of computers in the United States has been predicted to be about 15.5 percent. This means that the worth of installed computers will rise from 11.9 billion dollars in 1975 to 18.2 billion dollars in 1980. Furthermore, there is a predicted trend toward the development of hardware and software systems so simple to use that the ultimate consumer of computer services will be responsible for much, if not all, of the necessary programming. It follows that these consumers will have to be familiar with computing. As a consequence, the systems analyst will be called upon to comprehend larger and relatively more interrelated problems.

In concluding our discussion related to the computer age, we make the following observations:

1. The use of computers will continue to grow.
2. Ultimate consumers will require some technical knowledge of computing (perhaps gained in secondary schools).
3. Computer and data processing personnel will need to achieve significantly higher skill levels.

11.3 SOME ILLUSTRATIVE EXAMPLES OF APPLICATIONS

In this section we will describe four current applications of computers. These descriptions will provide a basis for understanding the impact of computers, to be discussed in Section 11.4. Here we will merely illustrate a few of the actual applications of computers to diverse problems.

☐ **Data bases.** A data base may be defined as a collection of interrelated data stored together without harmful or unnecessary redundancy, to serve one or more applications in an optimal fashion. The data are stored independent of the programs. A common and controlled access is used for retrieving the data. The data base is becoming a vital resource within corporations and institutions, and must be organized so as to maximize economic value. In 1974, approximately 20 percent of the United States Gross National Product (GNP) was devoted to the collection, processing, and dissemination of information. Most of the information is not yet computerized, but it appears to be only a matter of time before it will be cost-effective to store information in the form of data bases, relative to other means. Such bases are versatile in that they can store a variety of data (e.g., pictorial, speech), which can easily be put into digital form.

Data bases are commonly manipulated by programs. The combination is termed an **integrated management information (IMI) system,** or, perhaps, a **management information system** (see Chapter 10). A common misunderstanding regarding such systems is that an institution that uses one keeps all of its data in a large archive, accessible to many, to be used in any imaginable way. Although it is possible to think in terms of such a system being developed in the future, none is currently available. Most data bases now are limited to one machine, and a few well-defined uses. These systems are intended to provide standard operating information and summary reports for management.

If these systems are designed correctly, they can provide management with information required to remain competitive in a business environment. In addition, they can assist in avoiding crises within an organization.

☐ **Electronic funds transfer (EFT) systems.** These are systems involving the replacement of paper financial documents by electronic records to the greatest extent possible. For example, the United States federal reserve banks are now linked via a network of computers, to allow electronic transfer of financial transactions between them. The Bank of America has a network of minicomputers and a data base distributed among them. This network serves in excess of 11 million residents in California, and enables instantaneous information transfer of banking information. In certain cities, home banking service is also provided via systems that enable users to pay their bills, transfer money between accounts, and so on. Such tasks are accomplished by convenient means (e.g., a touchtone telephone), thereby reducing the use of paper checks.

The typical American is involved in over a thousand cash transactions annually. In addition, Americans write over 19 billion checks per year. Business and government add another 8 billion yearly checks that do not involve individuals. Payment systems have therefore received a great deal of attention. The various EFT systems may be placed into one of the following three categories: (a) preauthorization, (b) automated banking, and (c) point-of-sale devices.

Preauthorization systems involve the automatic deposit of a payment, such as a paycheck, a social security payment, or perhaps a welfare payment. These systems can also be made to handle automatic payments of an individual's debts by electronic means (payment of utility bills, for example), thereby circumventing the labor and paper work involved.

Automated banking extends banking hours and provides a greater number of bank locations for customers. This is achieved via the use of computerized tellers that are located on the exterior of a bank, or in other locations such as grocery stores. The home banking service is useful in any transaction for which a user does not need cash.

Point-of-sale device replacement of cash registers can be thought of as replacing cash, checks, and credit card purchases by obtaining all relevant user information at the point of sale by electronic means. Electronic records are created for billing, inventory, and so forth.

In the next few years, EFT systems could replace at least 17 billion of our annual financial transactions. Computer networks could be interconnected in the future to obtain a system configuration as shown in Figure 11.3-1. In fact, this system already exists in certain parts of the United States. The advantages realized by such an interconnection are best illustrated by an example. Suppose you are traveling in another part of the country, and you purchase an item from a retail outlet connected via online

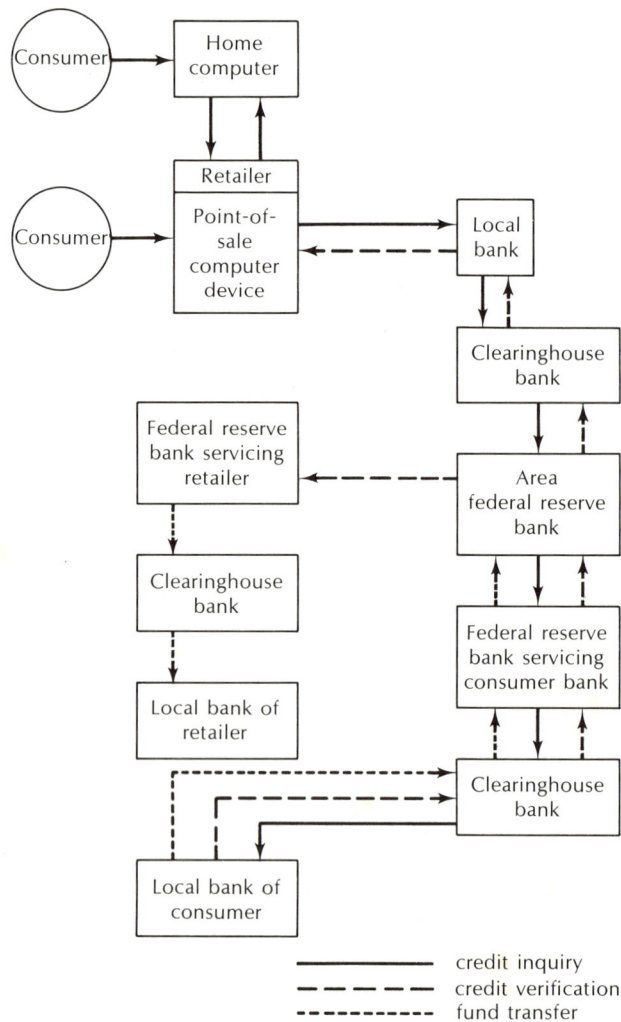

FIG. 11.3-1. A partially implemented EFT system for the United States.

point-of-sale computers to a local bank. This bank is connected online to a bank clearinghouse, which, in turn, is connected to a federal reserve bank. Through these interconnections, your transaction is transferred from the federal reserve bank to a bank clearinghouse, and then on to your home bank. Since such transfers occur almost instantaneously, it is possible for the retail store to receive a transfer of funds from your account to theirs—even before you leave the cash register. This system will bring about the "debit" card system, replacing the paper check, and causing direct withdrawal from a bank account.

☐ **Medical assistance.** The shortage of trained medical personnel, along with the rapid processing of massive amounts of information, are two motivating factors for the introduction of computers into the field of medicine. Computers are used by hospitals and groups of physicians to accomplish financial record keeping, inventory, payroll, and the storage and retrieval of patient records (see Figure 11.3-2). In some hospitals, nurses are reminded of patient medication needs by a computer at nurse stations. Hospital dietary programs are often automated to accommodate patient meal preferences and dieting needs.

Patient monitoring via real-time computer systems is currently used in many hospitals for a variety of health-care activities. In such cases, the computer monitors human physiological processes through measurements of temperature, heart rate, blood pressure, etc. It analyzes the data, records the information, notifies personnel of emergencies, and also may take direct action on a patient.

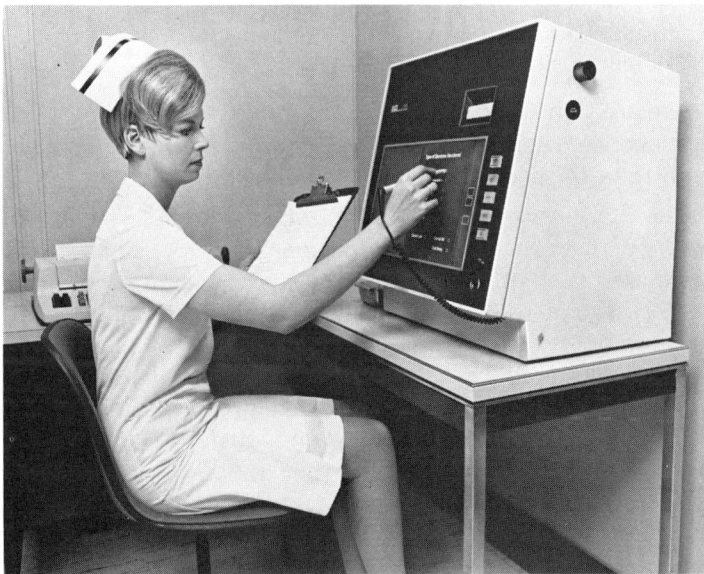

FIG. 11.3-2 A computer used for record keeping at a nurses station in a hospital. (Courtesy of International Business Machines Corporation.)

One interesting development is the CAT (computer-aided tomography) scanner, which is a unit that can scan the head of a patient in small increments. With the collected data, three-dimensional images are constructed using image-reconstruction algorithms. Such images assist a surgeon in determining the presence, size, shape, and location of a tumor, for example, thereby reducing the need for exploratory surgery.

☐ **Artificial intelligence.** The ability of computers to learn, reason, and produce behaviors normally associated with human intelligence is an intriguing notion. A. C. Clarke, who enjoys the reputation of making accurate scientific predictions, has stated that man's role as the dominant species is near an end, that in fact we could become obsolete, and that computers will come to exhibit more intelligent behavior than their creators.[2] A logical question that follows is Can computers think? This question is indeed an emotional semantic morass for most people. The word "think" connotes the notion of human intelligence; hence, to consider thinking within a nonhuman context raises a basic contradiction in our minds. To help clarify this issue, we quote Marvin Minsky, professor of Electrical Engineering and Computer Science at M.I.T.:

> As the machine improves . . . we shall begin to see all the phenomena associated with the terms "consciousness," "intuition" and "intelligence" itself. It is hard to say how close we are to this threshold, but once it is crossed, the world will not be the same. . . . It is unreasonable, however, to think the machine could become *nearly* as intelligent as we are and then stop, or to suppose that we will always be able to compete with them in wit and wisdom. Whether or not we could retain some sort of control of the machines, assuming we want to, the nature of our activities and aspirations would be changed utterly by the presence on earth of intellectually superior beings.[3]

There are some who believe that computers may surpass human intelligence. Their pertinent research efforts have been concerned with the notion of developing "intelligent computers."

There are several instances in which computers can modify future performance based upon experience; for example, in playing games such as checkers or chess. The main purpose of such activity is to investigate the intellectual processes and the further development of machine capabilities. Game playing is approached in three ways: (a) algorithmically, (b) heuristically, and (c) via learning. The entire game strategy can be embodied in an algorithm if the game is simple enough to yield a guaranteed winning strategy. This is not the case, for example, with chess or checkers; the total number of possible chess games is approximately 10^{120}! In such cases, the use of heuristics helps to reduce the number of games to be considered by selecting only those games having a high probability of success. Using a learning

---

[2] A. C. Clarke, *Profiles of the Future* (New York: Harper & Row, Publishers, 1973).
[3] Marvin Minsky, "Artificial Intelligence," *Scientific American*, September 1966, p. 260.

approach, a program alters its own execution pattern by keeping a record of successful and unsuccessful moves.

Computers can be programmed to achieve a limited understanding of human conversation, as discussed, for example, by B. Raphael in 1964, J. Weizenbaum in 1967, and I. Winograd in 1975. Mathematical proofs can be developed by computers in a limited manner by providing the program with all axioms and previously proved theorems. By randomly selecting words and phrases that meet the construction rule requirements of poetry, poems have been created. Computers have also been used to create some artwork; one example is shown in Figure 11.3-3.

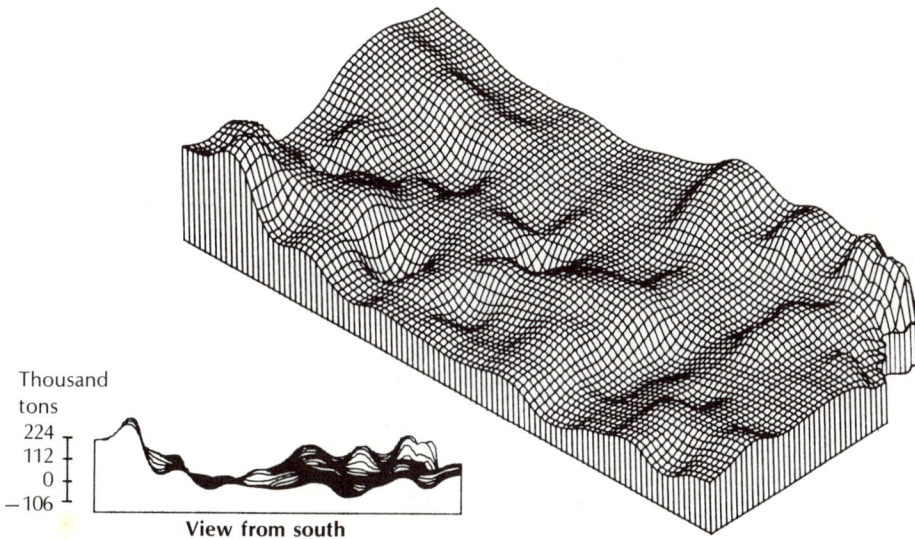

Feed grain balances in feed unit tons, Kansas, five-year average, 1970–71 to 1974–75 crop years.

FIG. 11.3-3. Computer artwork illustration. (Courtesy of Dr. John McCoy and Kim Rochat, Agricultural Experiment Station, Kansas State University.)

The use of "intelligent computers" to perform humanlike tasks introduces us to the field of robots. One robot developed at Stanford Research Institute, called Shakey, has a TV camera and touch and distance sensors for input [see Part (a) of Figure 11.3-4]. He is able to roll about on wheels in a laboratory of several interconnected rooms populated by large wooden blocks. Given the command "PUSH THE BOX OFF THE PLATFORM," Shakey has evolved to the point where he is able to realize that, since he has no appendage to reach the box, he must gain access to the platform. To this end, he finds a ramp and pushes it against the platform [see Part (b) of the figure]. Next, he rolls up the ramp and then shoves the box onto the floor [see Parts (c) and (d) of the figure].

309

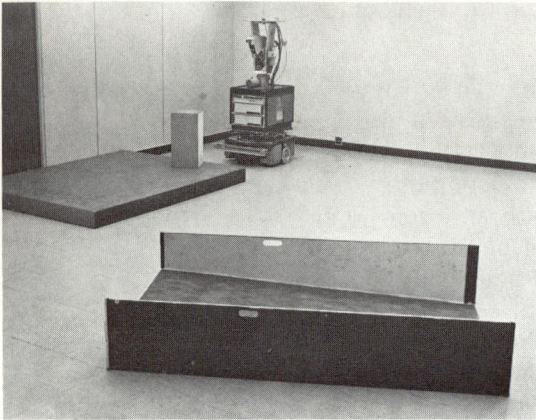

(a)

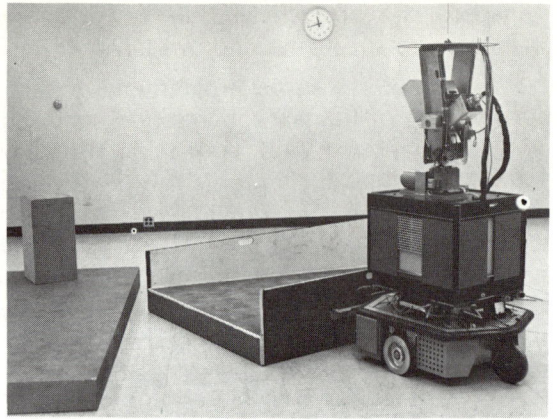

(b)

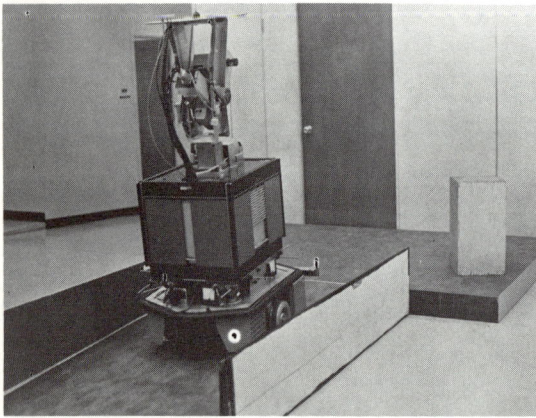

(c)

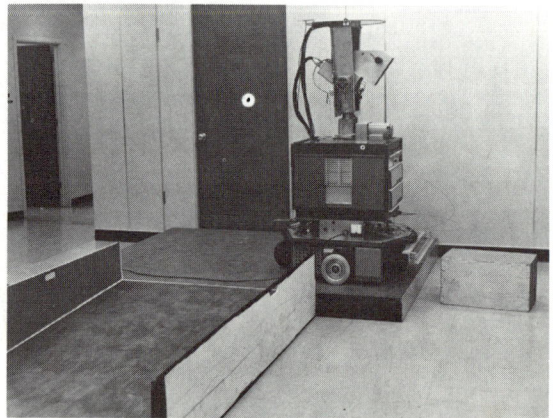

(d)

FIG. 11.3-4. Stanford Research Institute's Shakey, a robot with visual and tactile input. (Courtesy of Stanford Research Institute.)

We can see that machines do exhibit capabilities associated with human intelligence. However, whether these present capabilities necessarily lead to the conclusion that computers are capable of "thinking" is indeed an open question.

11.4 IMPACTS ON SOCIETY

Study of the computer revolution only from a computer system development point of view is a very narrow study. The use of computers has caused significant changes in our society. In this section we will consider some of the positive and negative aspects of the impact computers have had. By no means will the discussion be complete. It is mainly intended to provide a background that will enable the interested reader to pursue further study of the area.

Economic reasons generally serve to justify increasing the use of computer technology. Such economic reasons have the most obvious impact. By introducing computers judiciously, the productivity per worker can be increased dramatically. If the cost of the workers displaced due to automation is less than the cost of the automation, then lower consumer prices result, which, in turn, creates a demand for more goods and services. The overall result is an increase in the standard of living.

The displacement of workers resulting from increased productivity per worker contributes to other problems. Unemployment caused by automation via computers is a complex issue. Most people believe that, on an overall basis, the use of computers actually results in an increase of available jobs. However, the skills required to secure such jobs are, in general, significantly higher. There have been limited opportunities for some of the displaced workers to increase their job skills. This is clearly a problem that concerns the field of education. Computer utilities that provide information service, computation capability, and a full complement of communication facilities are a possibility. Such utilities could provide (a) news; (b) instant stock quotations; (c) schemes for shopping without having to visit a specific local store; (d) education (for example, computer-aided instruction—see Figure 11.4-1); (e) the capability to register opinions on local, state, regional, or national issues; and function in a variety of other ways. Just the economic impacts of such services on society would be overwhelming.

FIG. 11.4-1. An illustration of computer-aided instruction. (Courtesy of International Business Machines Corporation.)

We now turn our attention to some aspects of the social and personal impact of the computer revolution. One aspect that has received a great deal of scrutiny in recent years is individual privacy. Ramsey Clark, former attorney general of the United States, cautioned that once a right or protection is lost it may never be regained. He was referring to the fact that rather detailed records on many people in the United States are currently available in various computer data bases. As a consequence, several basic questions are raised in connection with invasion of privacy of an individual. These questions were discussed in an international conference held in late 1977. Some of them are as follows:

- ☐ What sort of information should be stored?
- ☐ Who should have access to such information, and under what circumstances?
- ☐ How secure are the data bases; i.e., to what extent are they protected?
- ☐ Who is liable for information leaks?
- ☐ What happens to those individuals who choose not to participate in the system?

New legislation concerning the privacy issue is forthcoming. An example of a recent United States law is the Privacy Act of 1974 (Public Law 93–479). This law makes it unlawful for any federal, state, or local government agency to withhold services from anyone who refuses to disclose his/her social security number. Thus, for example, students can refuse to reveal their social security numbers to schools, or individuals can refuse to reveal them for acquiring driver's licenses. In other words, only certain agencies authorized by law can ask for social security information—the Internal Revenue Service, for example. The Privacy Act of 1974 is also a deterrent to the collection of national dossiers on individuals. However, it sidesteps the real issue—that is, under what conditions can particular items of information related to individuals be collected and stored?

The impacts of EFT systems are already being felt in three broad areas of concern, with further impacts projected. Let us now briefly consider these three areas.

The first impact deals with white collar jobs lost as a consequence of automation. Fewer retail organizations will have their own credit systems as new clearinghouse organizations develop. These organizations will service small banks and retail outlets providing network EFT service, accounting, inventory control, and other services predicated upon transaction data files. Savings to the consumer should be realized through cost reduction in the payments systems. Such systems will eliminate jobs of lower skill level white collar workers, such as bookkeepers. On the other hand, they will create jobs requiring greater technical skill and training in the areas of data bases, network communication, and legal affairs associated with business institutions.

A second impact arises from the potential of the mass of information collected in machine-readable form. Computers provide people with the ability to organize and use this information. Conceivably, a record of every item purchased, donation made, or dollar earned could be stored and retrieved. This could clearly involve the privacy of individuals and organizations. On the other hand, it is possible that several benefits could result; for example, improved economic and product planning along with marketing projections, law enforcement, and income tax calculation and verification. As a consequence, organizations that are able to utilize such masses of information effectively will have an advantage over their competitors.

The third impact includes a variety of service-oriented agencies. For example, it is estimated that there will be a drop of 6 billion pieces of mail prior to 1981. For the most part, this drop is attributed to EFT systems.[4] It is clear that the side effects of this type of impact will be felt by a variety of people—stationery makers, ink makers, lumber workers who feed lumber to paper pulp factories, and so on. Again, fewer tellers and bookkeepers in banks will be required, and the need for richly furnished bank lobbies will diminish. A little thought on the reader's part will lead to the conclusion that several other impacts can be added to our list, resulting from just one partially implemented application—EFT systems.

We have already mentioned that computer technology has created many jobs, requiring considerably higher skill levels than those of persons displaced as a result of automation. Whether there is a net increase or decrease in the number of jobs is a subject of debate. But clearly, the retraining programs currently available for displaced workers are not adequate.

Changes have been predicted that will dwarf those already observed, when current technology is fully implemented. For example, most production plants could be completely automated with computers; information systems based upon large computer data banks conceivably could replace newspapers, mail, magazines, and books as communication means; and a virtually cashless society based upon the EFT systems is a not-too-distant possibility. We would point out but one drastic impact of such change—the impact on employment. In 1974 it took only 38 percent of the world's population to produce all goods and services; it has been predicted that eventually only 10 percent of the population, with support of automated systems, could produce essentially all required goods and services.

The trend towards larger and fewer institutions will continue, and the use of well-developed management information systems will lead to fewer people being required for the management of more resources.

SUMMARY 11.5

It is apparent from the foregoing discussion that computers have freed people from the drudgery of computations and the toil associated with service-oriented tasks. However, we still should keep in mind the fears

---

[4] *Time Magazine,* 15 March 1976, p. 71.

voiced by some prominent scholars. Norbert Wiener, widely known for his contributions in systems and cybernetics, has stated, "The world of the future will be an ever more demanding struggle against the limitations of our intelligence, not a comfortable hammock in which we can lie down to be waited upon by our robot slaves."[5] To the contrary, there are other experts who predict a life of leisure as a result of automation.

Regardless of who proves correct, we must strive to maintain a realistic perspective of the future and remain in control of the forces that change our society, ideas, and lives.

problems

11-1. Write a term paper, 10–15 pages long typed and double spaced, which is a concise and indepth discussion of some aspect of the use of computers or the impacts of the use of computers on our society. Suggested topics are

☐ The use of computers in _____ (for example, accounting, inventory control, weather forecasting, manufacturing processes).

☐ The impacts of computers on _____ (for example, leisure, the press, education, politics, legislation).

☐ The positive and negative aspects of computers (for example, robots, computer crime).

☐ Managing computer systems.

☐ The reintroduction of the town meeting—Everyone votes by computers.

☐ The computer as an adjunct doctor.

☐ Computer management report: How to reduce volume and increase effectiveness.

☐ Computer revolution forces intellectual stimulation through games.

☐ Understanding the human mind through computers.

☐ Dictatorship—Computers—Watergate.

You must have three references, and at least one of them must come from an accepted computer science source; for example, (a) *Surveys of the ACM* (Association for Computing Machinery); (b) *Communications of the ACM;* (c) newspapers—*Computer World;* (d) magazines—*Nations Business, Scientific American, Business Week, Datamation;* (e) books written by computer scientists.

The use of newspaper articles, except from *Computer World,* must be accompanied by the clipping.

[5]Norbert Wiener, *God & Golem, Inc.* (Cambridge, Mass.: M.I.T. Press, 1964), p. 69.

Appendices

A.1

Structured Flowcharts

In this book we have used flowcharts to present structured algorithms in graphical form. There is another graphical form, known as a **structured flowchart.**[1] Although not as widely used as are conventional flowcharts, structured flowcharts have the potential to gain popularity as time goes on. Therefore, the main objective of this appendix is to introduce the reader to the notion of structured flowcharts. Such an introduction is best achieved by means of examples.

Symbols used in structured flowcharts are based on the familiar rectangular symbol that is used in conventional flowcharts to represent processing statements. In structured flowcharts, however, a rectangular symbol may represent a single step, or a sequence of steps, in the related algorithm. These symbols are summarized in Table A.1-1.

In what follows, we present structured flowcharts corresponding to the algorithms we have developed in various chapters of the book. For convenience, the corresponding conventional flowcharts (developed earlier) are also included.

Example A.1-1   Concerns the structured algorithm in Example 3.6-6, and presents the corresponding structured flowchart.

☐   The structured and conventional flowcharts are shown in Figures A.1-1(a) and A.1-1(b), respectively (page 319).

---

[1] Refer to I. Nassi and B. Schneiderman, "Iteration Graphs," *ACM SIGPLAN* notices, 8, no. 8 (August 1973): 12–56.

TABLE A.1-1. Summary of structured flowchart symbols

Symbol	Name
IF (condition) THEN action$_1$ | ELSE action$_2$	Binary decision symbol
WHILE or UNTIL (condition) loop body	Indeterminate loop symbol
DO index ← initial to final by increment loop body	Determine loop symbol
name (parameters) subalgorithm body	Subalgorithm symbol

Example A.1-2 Concerns the structured algorithm in Example 7.5-1, and presents the corresponding structured flowchart.

☐ The structured and conventional flowcharts are shown in Figures A.1-2(a) and A.1-2(b), respectively (page 320).

Example A.1-3 Concerns the structured algorithm in Example 7.5-3, and presents the corresponding structured flowchart.

☐ The structured and conventional flowcharts are shown in Figures A.1-3(a) and A.1-3(b), respectively (pages 321–22).

Example A.1-4 Concerns the structured algorithm in Figure 9.2-2, and presents the corresponding structured flowchart.

☐ The structured and conventional flowcharts are shown in Figures A.1-4(a) and A.1-4(b), respectively (pages 323–24).

Example A.1-5 Concerns the structured algorithm in Figure 10.2-1, and presents the corresponding structured flowchart.

☐ The structured and conventional flowcharts are shown in Figures A.1-5(a) and A.1-5(b), respectively (page 325).

INPUT ORDER,NOSOLD	
IF NOSOLD < 100	
THEN ORDER ← 1.1*ORDER	ELSE ORDER ← ORDER/2.
COST ← ORDER*.50	
OUTPUT 'COST =', COST, '# OF ROSES ORDERED', ORDER	

FIG. A.1-1(a). Structured flowchart (see Example A.1-1).

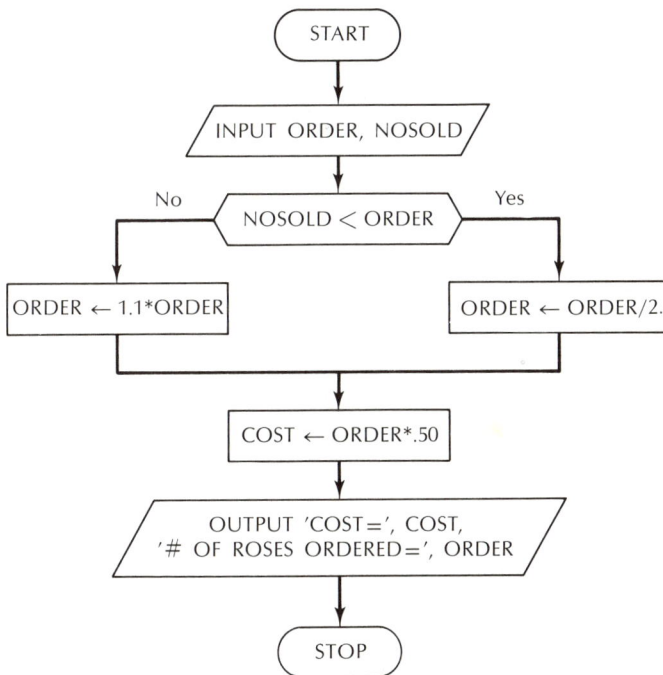

FIG. A.1-1(b). Conventional flowchart (see Example A.1-1).

SUM ← 0.
DO I ← 1, 10 or DO I ← 1 TO 10
INPUT SCORES(I) SUM ← SUM+SCORES(I)
AVERAGE ← SUM/10.
DO I ← 1,10 or DO I ← 1 TO 10
DIFFERENCE ← SCORES(I) − AVERAGE OUTPUT SCORES(I), DIFFERENCE

FIG. A.1-2(a).   Structured flowchart (see Example A.1-2).

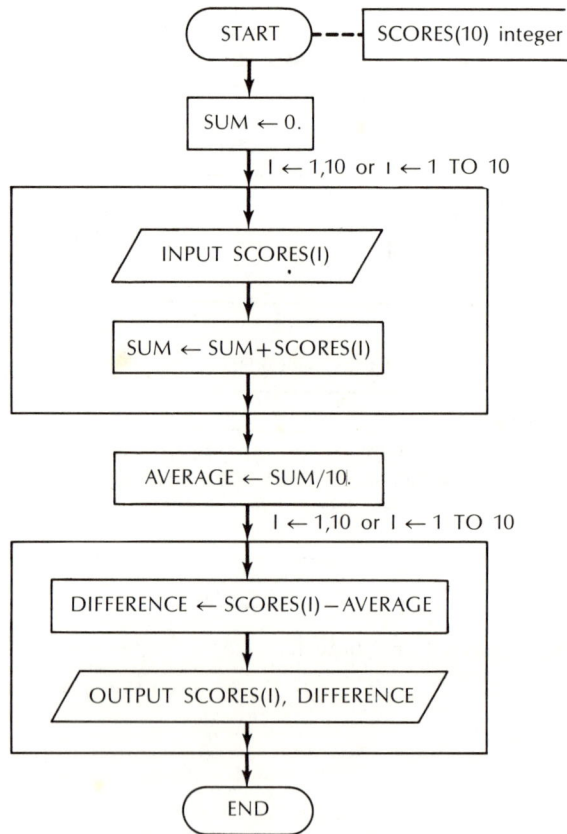

FIG. A.1-2(b).   Conventional flowchart (see Example A.1-2).

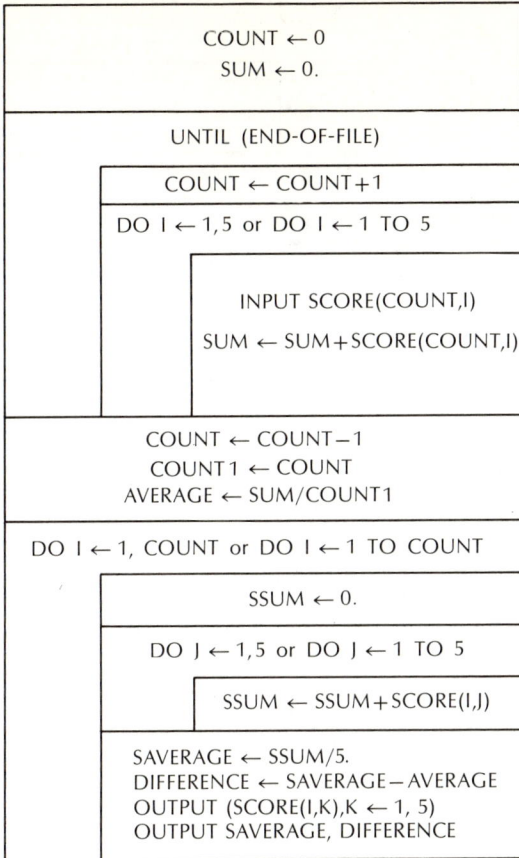

```
┌─────────────────────────────────────────────────────┐
│                    COUNT ← 0                          │
│                    SUM ← 0.                           │
├─────────────────────────────────────────────────────┤
│              UNTIL (END-OF-FILE)                      │
│   ┌───────────────────────────────────────────────┐  │
│   │             COUNT ← COUNT+1                     │  │
│   ├───────────────────────────────────────────────┤  │
│   │  DO I ← 1,5 or DO I ← 1 TO 5                    │  │
│   │     ┌──────────────────────────────────────┐   │  │
│   │     │                                       │   │  │
│   │     │       INPUT SCORE(COUNT,I)            │   │  │
│   │     │                                       │   │  │
│   │     │    SUM ← SUM+SCORE(COUNT,I)           │   │  │
│   │     │                                       │   │  │
│   └─────┴──────────────────────────────────────┘   │  │
│                                                     │  │
│              COUNT ← COUNT−1                        │  │
│              COUNT1 ← COUNT                         │  │
│            AVERAGE ← SUM/COUNT1                     │  │
├─────────────────────────────────────────────────────┤
│  DO I ← 1, COUNT or DO I ← 1 TO COUNT               │
│   ┌───────────────────────────────────────────────┐  │
│   │               SSUM ← 0.                         │  │
│   ├───────────────────────────────────────────────┤  │
│   │  DO J ← 1,5 or DO J ← 1 TO 5                    │  │
│   │     ┌──────────────────────────────────────┐   │  │
│   │     │   SSUM ← SSUM+SCORE(I,J)              │   │  │
│   │     └──────────────────────────────────────┘   │  │
│   ├───────────────────────────────────────────────┤  │
│   │  SAVERAGE ← SSUM/5.                             │  │
│   │  DIFFERENCE ← SAVERAGE−AVERAGE                  │  │
│   │  OUTPUT (SCORE(I,K),K ← 1, 5)                   │  │
│   │  OUTPUT SAVERAGE, DIFFERENCE                    │  │
└───┴───────────────────────────────────────────────┴──┘
```

FIG. A.1-3(a).   Structured flowchart (see Example A.1-3).

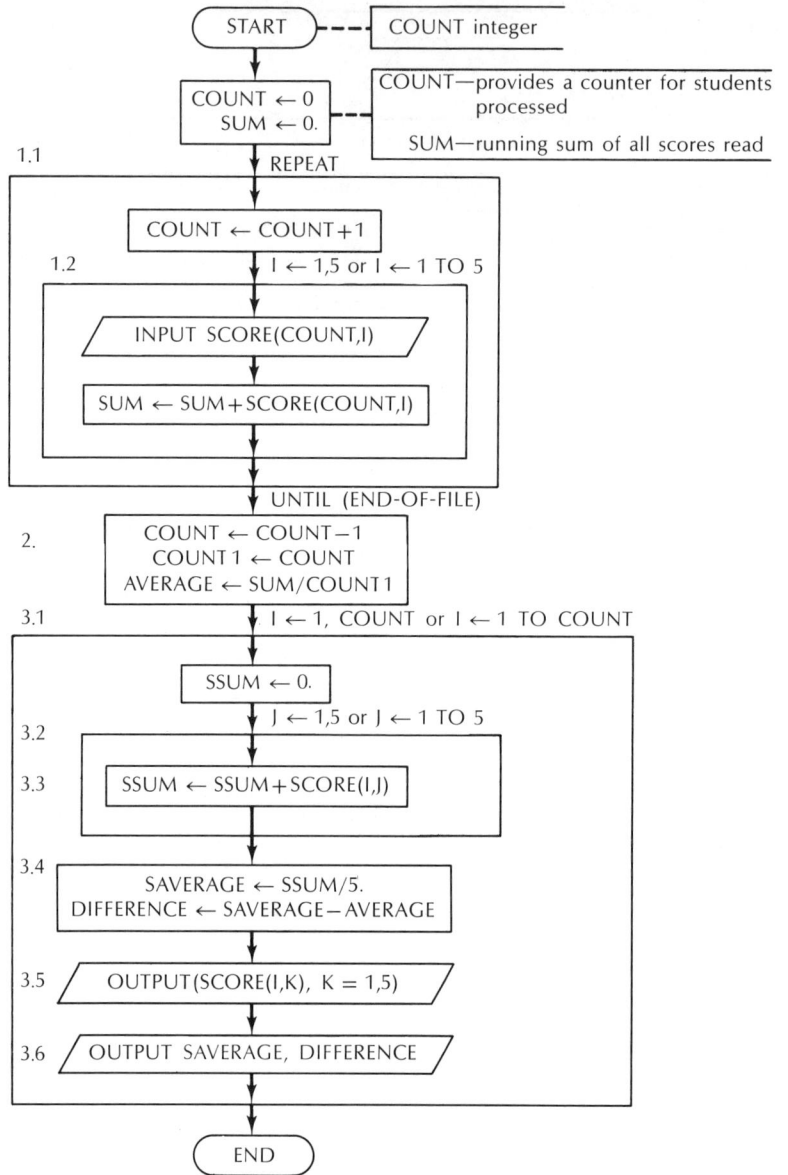

FIG. A.1-3(b). Conventional flowchart (see Example A.1-3).

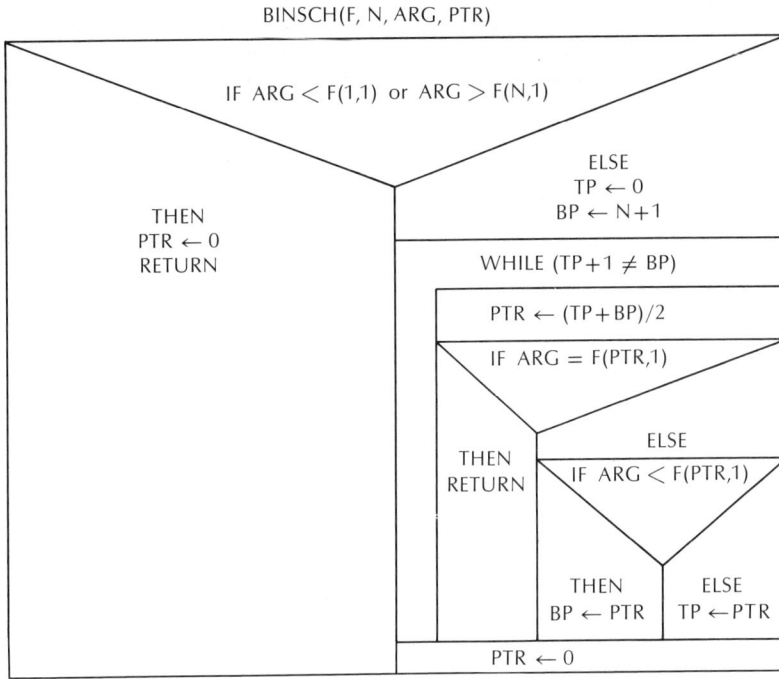

FIG. A.1-4(a). Structured flowchart (see Example A.1-4).

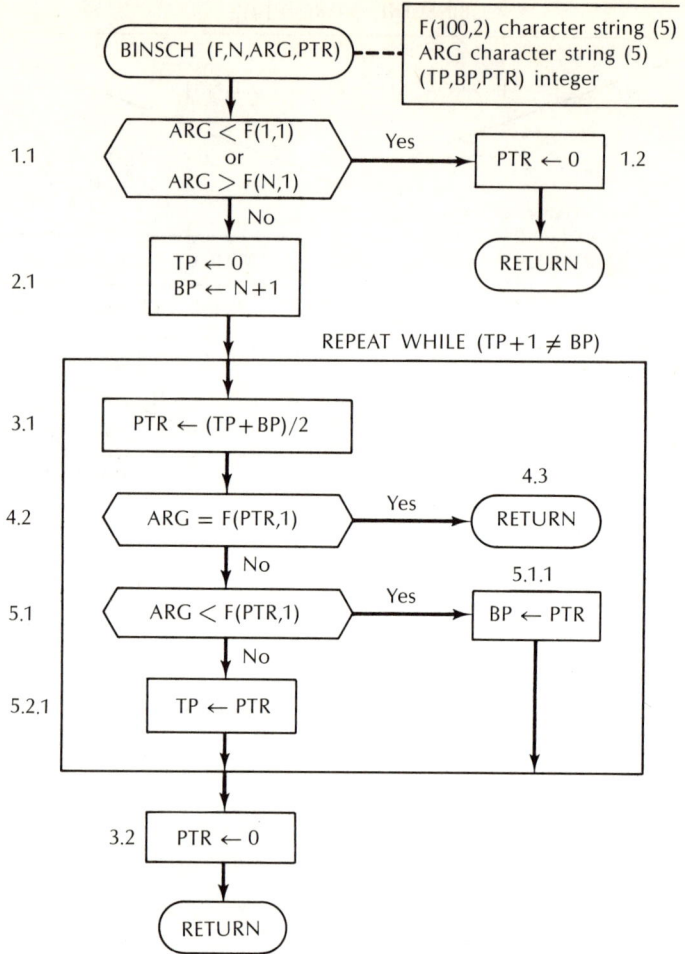

FIG. A.1-4(b). Conventional flowchart (see Example A.1-4).

TWOREP (LIST, N, LSORTED)

DO J ← 1, N or J ← 1 TO N

MIN ← 100000

DO L ← 1, N or L ← 1 TO N

IF LIST(L) > 0 or
LIST(L) < MIN

THEN
K ← L
MIN ← LIST(L)   ELSE

LSORTED(J) ← LIST(K)

LIST(K) ← 1

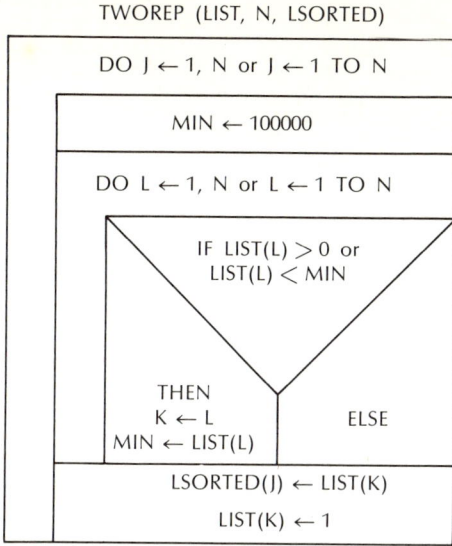

FIG. A.1-5(a).   Structured flowchart (see Example A.1-5).

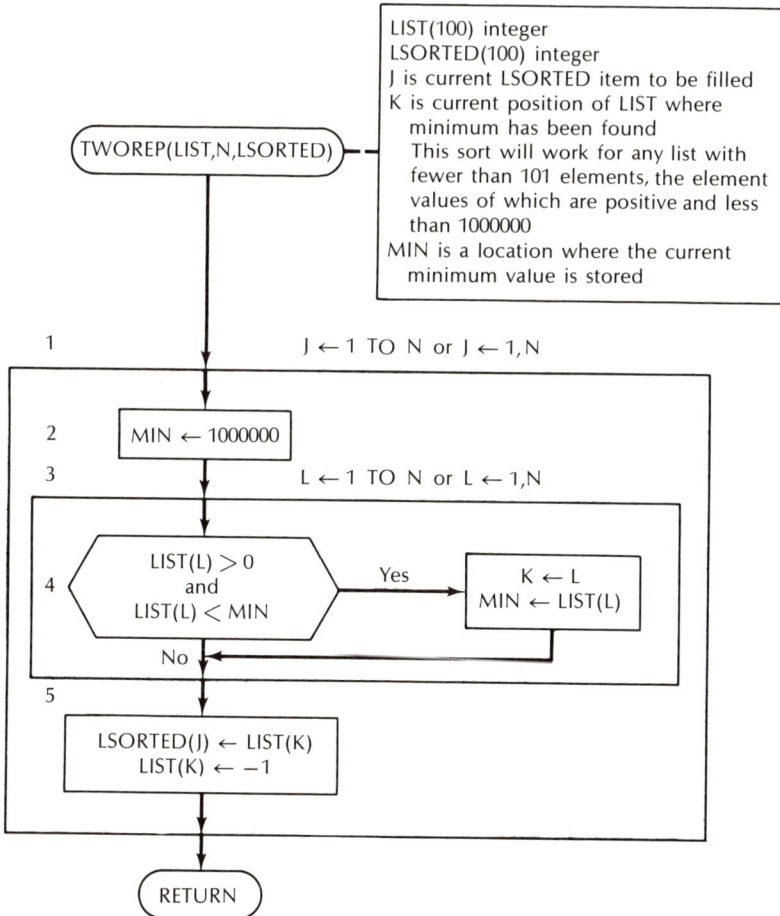

LIST(100) integer
LSORTED(100) integer
J is current LSORTED item to be filled
K is current position of LIST where
   minimum has been found
This sort will work for any list with
   fewer than 101 elements, the element
   values of which are positive and less
   than 1000000
MIN is a location where the current
   minimum value is stored

TWOREP(LIST,N,LSORTED)

1   J ← 1 TO N or J ← 1,N

2   MIN ← 1000000

3   L ← 1 TO N or L ← 1,N

4   LIST(L) > 0
and
LIST(L) < MIN   Yes   K ← L
MIN ← LIST(L)

No

5   LSORTED(J) ← LIST(K)
LIST(K) ← −1

RETURN

FIG. A.1-5(b).   Conventional flowchart (see Example A.1-5).

325

A.2

Computer Representation of Data

The purpose of this appendix is to provide a brief introduction to some of the basic concepts of data representation within a computer. Our focus will be the storage of numeric data and character string data.

Computers which store data in binary form are called **binary computers.** In such computers, data is stored in terms of binary digits called **bits,** where the term bit (binary digit) represents either of the digits, 0 or 1. Again, a group of adjacent bits is called a **word,** and the number of bits in a word denotes its **word length.** The word length used generally depends on the size of the computer. For example, minicomputers may use 16 bits per word, while large general-purpose computers may employ word lengths of 32 to 60 bits. Such word lengths are also referred to in terms of **bytes**—a byte denoting a group of bits. For instance, a byte consists of 8 bits in an IBM computer.

The term **decimal machine** (computer) implies that bits are grouped together into fixed units, each of which is used to represent a decimal digit. Although most computers are basically binary machines, both binary and decimal representation schemes are available.

327

To introduce the notion of number systems, let us consider the **decimal number system,** which is defined with respect to a **base** or **radix** of 10. This means that any decimal number can be expressed in terms of integer powers of 10, as in the following examples, where 123_{10}, 0.15_{10}, and 123.15_{10} denote the decimal numbers 123, 0.15, and 123.15:

(a) $\quad 123_{10} = 1 \times 10^2 + 2 \times 10^1 + 3 \times 10^0$

(b) $\quad 0.15 = 1 \times 10^{-1} + 5 \times 10^{-2}$

(c) $\quad 123.15 = 1 \times 10^2 + 2 \times 10^1 + 3 \times 10^0 + 1 \times 10^{-1}$
$\quad\quad\quad + 5 \times 10^{-2}$

From the preceding examples, it follows that, in general, a decimal number can be expressed in the form

$$N_{10} = \pm(d_0 \times 10^0 + d_1 \times 10^1 + d_2 \times 10^2 + \cdots + d_n \times 10^n$$
$$+ \cdots + d_{-1}10^{-1} + d_{-2}10^{-2} + \cdots + d_{-m}10^{-m} + \cdots) \quad \text{(A.2-1)}$$

where each d_i is one of the digits, 0 through 9.

Next, we consider the binary number system. This number system uses the base (or radix) 2, as illustrated in the following examples:

(a) $\quad 1001_2 = 1 \times 2^3 + 0 \times 2^2 + 0 \times 2^1 + 1 \times 2^0$
$\quad\quad\quad = 8_{10} + 0_{10} + 0_{10} + 1_{10}$
$\quad\quad\quad = 9_{10}$

(b) $\quad 1111011_2 = 1 \times 2^6 + 1 \times 2^5 + 1 \times 2^4 + 1 \times 2^3$
$\quad\quad\quad\quad + 0 \times 2^2 + 1 \times 2^1 + 1 \times 2^0$
$\quad\quad\quad = 64_{10} + 32_{10} + 16_{10} + 8_{10} + 0_{10} + 2_{10} + 1_{10}$
$\quad\quad\quad = 123_{10}$

(c) $\quad .01_2 = 0 \times 2^{-1} + 1 \times 2^{-2}$
$\quad\quad\quad = 0_{10} + .25_{10}$
$\quad\quad\quad = .25_{10}$

(d) $\quad -11.01 = -(1 \times 2^1 + 1 \times 2^0 + 0 \times 2^{-1} + 1 \times 2^{-2})$
$\quad\quad\quad = -(2_{10} + 1_{10} + 0_{10} + .25_{10})$
$\quad\quad\quad = -3.25_{10}$

Thus, corresponding to (A.2-1) we have

$$N_2 = \pm(b_0 \times 2^0 + b_1 \times 2^1 + b_2 \times 2^2 + \cdots + b_n \times 2^n + \cdots$$
$$+ b_{-1} \times 2^{-1} + b_{-2} \times 2^{-2} + \cdots + b_{-m} \times 2^{-m} + \cdots) \quad \text{(A.2-2)}$$

where each b_i denotes a 0 or 1.

In the preceding examples of the binary number system, we observe that for a given number there is a corresponding decimal equivalent. As this is indeed the case, in general, a systematic method (algorithm) is available to convert a given decimal number into binary form. The basic idea of this conversion process should be apparent to the reader from the illustrative examples given in Figure A.2-1. We add that *exact*

Example A.2-1 $123_{10} = (?)_2$

$$
\begin{array}{rl}
2)\overline{123} & \textit{Remainder} \\
2)\overline{61} & 1 \leftarrow \text{least significant bit} \\
2)\overline{30} & 1 \\
2)\overline{15} & 0 \\
2)\overline{7} & 1 \\
2)\overline{3} & 1 \\
2)\overline{1} & 1 \leftarrow \text{most significant bit} \\
\,0 &
\end{array}
$$

Therefore, $123_{10} = 1111011_2$.

Example A.2-2 $0.3125_{10} = (?)_2$

$$
\begin{array}{c|c}
 & .3125 \\
 & \times 2 \\
0 & .6250 \\
 & \times 2 \\
1 & .2500 \\
 & \times 2 \\
0 & .5000 \\
 & \times 2 \\
1 & .0000
\end{array}
$$

Therefore, $0.3125_{10} = .0101_2$.

Example A.2-3 $3.25_{10} = (?)_2$

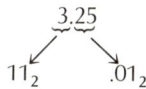

$$3.25$$
$$11_2 \qquad .01_2$$

Therefore, $3.25_{10} = 11.01_2$.

FIG. A.2-1. Decimal to binary conversion examples.

binary representation is not always possible. For example, it is straightforward to verify that

$$.4_{10} = .011\underline{0011}\,\underline{0011}\,\underline{0011}\cdots$$

which does not terminate, thereby implying that the decimal number 0.4 does not have an exact terminating binary representation.

There are two other number systems that are also employed—the **octal** and **hexadecimal number systems.** They are associated with the radices 8 and 16, respectively. The relationship between these two number systems and the binary and decimal systems is summarized in Table A.2-1. It follows from this table that while the octal system has only

329

TABLE A.2-1. Four number systems

Decimal number	Binary representation	Octal representation	Hexadecimal representation
0	0000	0	0
1	0001	1	1
2	0010	2	2
3	0011	3	3
4	0100	4	4
5	0101	5	5
6	0110	6	6
7	0111	7	7
8	1000	10	8
9	1001	11	9
10	1010	12	A
11	1011	13	B
12	1100	14	C
13	1101	15	D
14	1110	16	E
15	1111	17	F

eight distinct symbols (i.e., the digits 0, 1, 2, . . . , 7), the hexadecimal system consists of the symbols 0, 1, 2, . . . , 9, A, B, C, D, E, F. Thus, given an octal or hexadecimal number, we can find the corresponding decimal number via expansions of the form given in Equation (A.2-1) or Equation (A.2-2). This is illustrated by the following examples:

(a) $201_8 = 2 \times 8^2 + 0 \times 8^1 + 1 \times 8^0$
$= 128_{10} + 0_{10} + 1_{10}$
$= 129_{10}$

(b) $12.1_8 = 1 \times 8^1 + 2 \times 8^0 + 1 \times 8^{-1}$
$= 8_{10} + 2_{10} + .125_{10}$
$= 10.125_{10}$

(c) $3CF_{16} = 3 \times 16^2 + 12 \times 16^1 + 15 \times 16^0$
$= 768_{10} + 192_{10} + 15_{10}$
$= 975_{10}$

(d) $CF.8 = 12 \times 16^1 + 15 \times 16^0 + 8 \times 16^{-1}$
$= 192_{10} + 15_{10} + .5_{10}$
$= 207.5_{10}$

In closing our brief discussion of the number systems, we add that it is a simple matter to convert a given binary number to the corresponding octal or hexadecimal equivalents. It essentially involves grouping the 0's and 1's in the given binary number into group sizes of 3 and 4 with respect to the octal and hexadecimal conversions, respectively, as illustrated in Figure A.2-2. Each of these groups of 0's and 1's is then expressed as the corresponding octal or hexadecimal symbol, using the information in Table A.2-1.

Example A.2-4 $101100_2 = (?)_8$

$$\underbrace{101}_{\displaystyle 5} \quad \underbrace{100}_{\displaystyle 4}$$

Therefore, $101100_2 = 54_8$.

Example A.2-5 $1100_2 = (?)_8$

leading
zeros
inserted

$$\underbrace{001}_{\displaystyle 1} \quad \underbrace{100}_{\displaystyle 4}$$

Therefore, $1100_2 = 14_8$. *Note:* Leading zeros are inserted if necessary, as is the case in this example.

Example A.2-6 $101001110010_2 = (?)_{16}$

$$\underbrace{1010}_{\displaystyle A} \quad \underbrace{0111}_{\displaystyle 7} \quad \underbrace{0010}_{\displaystyle 2}$$

Therefore, $101001110010_2 = A72_{16}$.

Example A.2-7 $11110.11110100_2 = (?)_{16}$

$$\underbrace{0001}_{\displaystyle 1} \quad \underbrace{1110}_{\displaystyle E} . \underbrace{1111}_{\displaystyle F} \quad \underbrace{0100}_{\displaystyle 4}$$

Therefore, $11110.11110100_2 = 1E.F4$.

FIG. A.2-2.   Binary to octal and hexadecimal conversion examples.

NUMERIC DATA CONSIDERATIONS

The storage of most numeric data within computers is accomplished either by **fixed-point** or **floating-point representation.** The former is used for integer numbers; the latter, for real numbers.

Binary fixed-point schemes are fixed-length representations, the lengths of which are some multiple of the word length. One common scheme is the sign-magnitude form, in which the leftmost bit (the most significant) is generally called the **sign bit,** since its value indicates the sign of the number being represented. Usually, if the sign bit is 0, then the number is considered to be positive. Conversely, a negative number is indicated by a 1 in the sign bit position. The remaining bits are used to represent the magnitude portion of the number. For instance, the 16-bit fixed-point representations of the integers $+123_{10}$ and -123_{10}, respectively, are as follows:

331

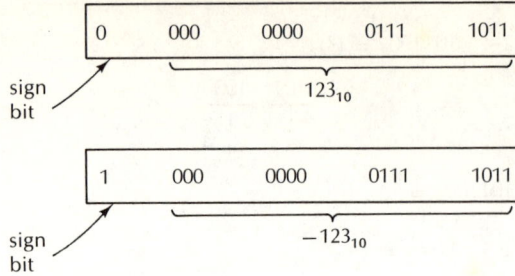

Thus we note that the 32-bit fixed-point representations of the largest (positive) and smallest (negative) integers are

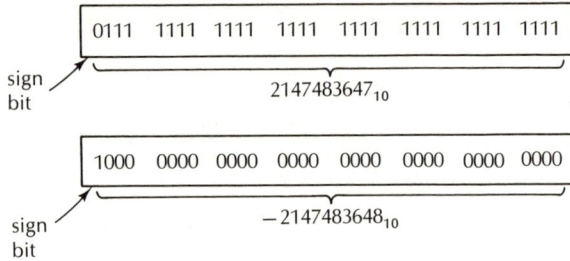

From the preceding illustrations it is apparent that the range of integers that can be represented via the 32-bit representation is given by

$$-2147483648 \leq N_{10} \leq 2147483647$$

or

$$10^{-8.332} \leq N_{10} \leq 10^{8.332} \qquad \textbf{(A.2-3)}$$

A little thought on the reader's part will lead to the conclusion that the range indicated in (A.2-3) is rather restricted, since there are numerous applications involving integers outside this range. Any attempt to store an integer consisting of more digits than can be accommodated in a word results in a condition known as **integer overflow.** Some computers generate error messages when such integer overflows occur; other do not.

Next, we consider some aspects of representing numeric data using a floating-point scheme. The general form of this representation is as follows:

$$\boxed{\text{Sign} \mid \text{Fraction} \mid} \times \text{Radix}^{\text{sign, exponent}} \qquad \textbf{(A.2-4)}$$

For discussion purposes, let us consider the case in which the radix in (A.2-4) equals 10. Then, for example, the floating-point representation of the decimal numbers .0012, -16305.732, and 101.721 are as follows:

(a) $+.12 \times 10^{-2} = .0012$

(b) $-.16305732 \times 10^{+5} = 16305.732$

(c) $+.101721 \times 10^{+3} = 101.721$

332

Similarly, if the radix in (A.2-4) equals 16, real numbers can be represented in floating-point form via the hexadecimal system, as illustrated in the following example:

$$-.215 \times 16^{+3} = -880.64_{10}$$

The floating-point method is extremely useful in that it enables us to represent a wide range of numbers—a range much greater than that possible via the fixed-point approach. The radix employed depends on the particular computer being considered.

A form of floating-point representation is called **binary floating-point representation.** For example, consider the 32-bit representation shown in Figure A.2-3.

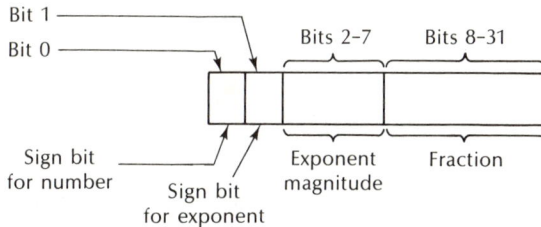

FIG. A.2-3. A binary floating-point representation.

From this figure it is clear that the exponent is stored in 7 bits. One of these bits is allocated to be the sign bit, which means that the magnitude of the exponent is encoded using 6 bits. With this information, it can be shown that numbers that can be stored via the 32-bit floating-point representation in Figure A.2-3 lie in the range

$$2^{-64} \leq N_{10} \leq 2^{64}$$

or

$$10^{-19} \leq N_{10} \leq 10^{19} \qquad \textbf{(A.2-5)}$$

Comparing (A.2-5) with (A.2-3), we see that the floating-point range is far greater than the fixed-point range. However, it is important to note that in the case of floating-point representations, the accuracy (or **precision**) with which a number can be stored is limited by the number of bits allocated to the fractional part of the number. For example, consider the representation of the decimal number 0.1, which is as follows:

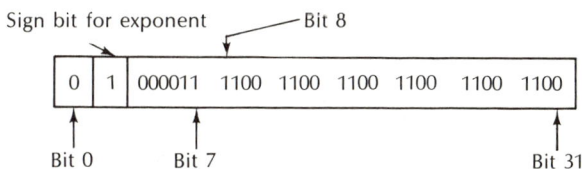

Converting this representation back to decimal form results in 0.099999999302..., which is not exactly 0.1.

Overflow can occur in floating-point representation, just as it can in fixed-point. This can be seen by examining (A.2-5). Any attempt to store a number N_{10}, the representation of which requires an exponent larger than $+19$, results in a condition known as a **floating-point overflow.** On the other hand, if N_{10} is such that its representation requires an exponent that is less than -19, then a **floating-point underflow** is said to occur. In most computers, messages are generated in the event of either of these occurrences.

It should be added that hexadecimal floating-point representations are also used. For example, the IBM System/370 employs 32-bit hexadecimal representation. It is capable of handling decimal numbers N_{10} in the following range:

$$10^{-79} \leq N_{10} \leq 10^{76}$$

CHARACTER STRING CONSIDERATIONS

To close our discussion related to computer representation of data, we will briefly discuss the manner in which character string information is stored.

TABLE A.2-5. Eight-bit EBCDIC code (see page 335)

Space	0100 0000					0	1111 0000
¢	0100 1010	a	1000 0001	A	1100 0001	1	1111 0001
.	0100 1011	b	1000 0010	B	1100 0010	2	1111 0010
<	0100 1100	c	1000 0011	C	1100 0011	3	1111 0011
(	0100 1101	d	1000 0100	D	1100 0100	4	1111 0100
+	0100 1110	c	1000 0101	E	1100 0101	5	1111 0101
|	0100 1111	f	1000 0110	F	1100 0110	6	1111 0110
&	0101 0000	g	1000 0111	G	1100 0111	7	1111 0111
!	0101 1010	h	1000 1000	H	1100 1000	8	1111 1000
$	0101 1011	i	1000 1001	I	1100 1001	9	1111 1001
*	0101 1100	j	1001 0001	J	1101 0001		
)	0101 1101	k	1001 0010	K	1101 0010		
;	0101 1110	l	1001 0011	L	1101 0011		
¬	0101 1111	m	1001 0100	M	1101 0100		
^	0110 0000	n	1001 0101	N	1101 0101		
/	0110 0001	o	1001 0110	O	1101 0110		
,	0110 1011	p	1001 0111	P	1101 0111		
%	0110 1100	q	1001 1000	Q	1101 1000		
—	0110 1101	r	1001 1001	R	1101 1001		
>	0110 1110	s	1010 0010	S	1110 0010		
?	0110 1111	t	1010 0011	T	1110 0011		
:	0111 1010	u	1010 0100	U	1110 0100		
#	0111 1011	v	1010 0101	V	1110 0101		
@	0111 1100	w	1010 0110	W	1110 0110		
'	0111 1101	x	1010 0111	X	1110 0111		
=	0111 1110	y	1010 1000	Y	1110 1000		
"	0111 1111	z	1010 1001	Z	1110 1001		

In a computer, character strings are stored on a character-by-character basis, using a fixed number of bits per character. There are two coding schemes that are commonly used:

1. ASCII (American Standard Code for Information Interchange)

2. EBCDIC (Extended Binary Coded Decimal Interchange Code)

For purposes of illustration, an 8-bit EBCDIC code is given in Table A.2-5. Using the information in this table, it follows that the character string AB12 is represented in the form of the string of 0's and 1's (i.e., binary string) as follows:

$$\underbrace{1100\ 0001}_{A}\quad \underbrace{1100\ 0010}_{B}\quad \underbrace{1111\ 0001}_{1}\quad \underbrace{1111\ 0010}_{2}$$

For a given computer, communication of data with I/O devices takes place using one of the character coding schemes, such as EBCDIC. We remark that all numeric data received from an input device must be converted to fixed/floating-point, binary or decimal form, prior to processing in a computer. The results are then converted back into EBCDIC or ASCII form and then forwarded to an output device (see Figure A.2-4).

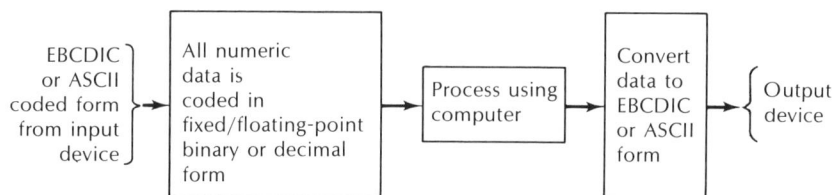

FIG. A.2-4. On handling numeric data.

335

A.3

A Glossary of Computer Science Terms

A

Actual parameter   A positional entity in the invocation of a routine that either contains input to be communicated to a subalgorithm or is to be used to communicate the results of the subalgorithm back to the invoking routine.

Address   A means of identifying a register, main memory location, auxiliary memory location, or computer-related device via a group of symbols.

Algorithm   A set of well-defined rules or processes for the solution of a problem in a finite number of steps.

Alphabet   An ordered set of symbols used in a language.

Alphameric   A generic term describing the alphabetic letters, numeric digits, and special characters that are available in the alphabet of a given language; also called alphanumeric.

Analog — A term relating to the representation of data by means of continuously variable physical quantities.

Analog computer — A computer that represents data in analog form, and operates on such data.

Annotation — A descriptive comment or explanatory note.

Array — A group of elements constituting an arrangement in one or more dimensions.

Artificial intelligence — The capability of a device to perform functions that are normally associated with human intelligence, such as reasoning, learning, and self-improvement.

Ascending sequence — The order of a set of items in which the largest item is last and the smallest is first.

ASCII — Acronym for American Standard Code for Information Interchange—the standard code used to represent characters for storage in a computer memory; also used for interchange between computer systems and their associated devices.

Assignment — An operator in programming languages that stores the value on its right in the memory location(s) named on its left; e.g., X ← 2.

Auxiliary storage — Computer storage other than the main memory of the machine; e.g., magnetic tapes and magnetic disk.

B

Batch processing — A technique for executing a set of computer programs such that each is completed before the next program in the set is begun.

Baud — A signaling unit used in data transmission that equals 1 bit per second.

Binary search — A search process in which the number of items in a set is divided into two equal parts at each step in the process. Appropriate adjustments are usually made for dividing an odd number of items.

Bit — A contraction of the two words binary digit; either of the numbers 0 or 1.

Block — A sequence of data items, symbols, records, program statements, or other objects that are treated as a unit.

Blocking of records — The storage of n logical records in one physical record (n may be less than 1 but is usually greater). The blocking factor is also called n.

Branch An instruction that can be used to change the normal sequential flow of a program. There are two types: (a) an unconditional branch, which is always executed, and after which execution continues at another point in the algorithm, and (b) a conditional branch, which allows for a decision to select one path for continued execution from a set of such paths.

Bug An error in an algorithm or program.

Buffer A memory area or device that is used to compensate for any differences in rates of data flow.

Byte A grouping of adjacent bits operated on by the computer as a unit. The most common size of byte contains eight bits.

Channel A special-purpose computer used to execute I/O instructions and relieve the CPU of such tasks.

Character A letter, digit, punctuation mark, or other symbol used in the representation of information. Each character is represented by a series of bits.

COBOL An acronym for COmmon Business Oriented Language, a procedure-oriented programming language that has diverse business applications.

Coder One who is mainly involved in the writing of programs, but who is not responsible for their design.

Compile To generate an object program (in a machine-oriented language) from a source program (in a procedure-oriented language).

Compiler A program that compiles.

Computer A data processor capable of automatically performing computations that include numerous arithmetic and logical operations.

Computer generation A chronological classification of computer systems based upon hardware, facilities, and software characteristics. There are three such generations.

Computer programming language An artificial language designed to convey instructions to a computer.

Concatenate To set two things beside one another (juxtapose) and treat the combination as a unit.

339

Constant A literal value that does not change.

Control unit Those parts of a computer that effect the retrieval of instructions in proper sequence and pass signals to other parts of the computer in accordance with these instructions.

Conversational computing A form of computing in which the user has a dialogue with the machine. Each user entry elicits a response from the machine. This provides a facility for users to interact with their programs from such remote equipment as terminals.

CPU (Central Processing Unit) The portion of a computer that is used to interpret and execute instructions, and also to perform arithmetic and logical operations.

Data A representation of facts, concepts, or instructions in a manner that is suitable for communication, interpretation, or processing.

Database A comprehensive collection of data; also called a data bank.

Data processing The area of computing that deals primarily with problems characterized by a predominance of input/output, as contrasted with scientific computation.

Data verification The determination of whether the transcription of data to machine-readable form has been done correctly.

Debug To remove errors from a program.

Decision table A table containing all contingencies and related decision rules that cause various actions to be taken for each of these contingencies. Such tables can be used in place of flowcharts for problem description and documentation.

Declare statements Computer programming language statements that provide information to the compiler, especially about variables. Such statements are not executed.

Default conventions The assumptions made by a language if no conditions are specified by the programmer. In our flowchart language, for example, variable names starting with any of the letters I, J, K, L, M, or N are treated as integer variables, unless declared otherwise.

Definite loop A form of loop in which the number of repetitions is guided by a loop counter; e.g., $I \leftarrow 1$ TO 12 BY 2.

Deque A list to which items may be added or deleted at either end.

Descending sequence An ordered sequence of items in which the largest item value occurs first and the smallest item value occurs last.

Digital computer A computing device in which discrete representation of data is used.

Direct-access storage Pertaining to data storage in which the position from which information is to be obtained does not depend on the location of previous information obtained. Magnetic drums, magnetic disks, magnetic core memory are examples.

Documentation A description of a program or computer system made up of one or more of the following: user guides, flowcharts, decision tables, source program listings, program comments, and the like.

EBCDIC An acronym for Extended Binary Coded Decimal Interchange Code, an eight-bit code used for representing data.

Echo check A method for checking the accuracy of data transmission in which the received and original data are compared.

ENIAC An acronym for Electronic Numerical Integrator And Calculator, the first operational electronic computer (1946).

Execution time (a) Time required by a CPU to execute an instruction once it has been moved to the CPU, or (b) the total time a CPU spends in executing a program.

Field A single position or group of consecutive positions on a record, allocated for a specific category of data.

File A collection of related records, treated as a unit.

File maintenance The process of keeping files up to date by adding, changing, or deleting data.

File organization A general term that refers to the logical organization of a file on a given storage device. Sequential and direct-access are two file organization schemes.

File security Pertaining to the techniques employed to physically protect a file from destruction and unauthorized use.

341

Firmware An integral combination of computer hardware and software.

Fixed-decimal representation Pertaining to numbers represented as signed integers, with no integers to the right of the decimal point.

Fixed-point representation Pertaining to numbers represented as signed digits, with no digits to the right of the radix point.

Floating-point representation Pertaining to numbers represented as a value multiplied by a fixed base (radix) raised to some power. In memory, only the value and the exponent of a floating-point number are stored.

Flowchart A graphical representation of the definition, analysis, or solution of problems, in which symbols are used to represent operations, data flow, declaration, etc.

Formal parameter A positional entity in the definition of a subroutine which must be a simple variable name. It is used to symbolically indicate where and how the actual parameter will be used when the subalgorithm is invoked.

Format A descriptor of the arrangement of data on input/output media.

FORTRAN An acronym for FORmula TRANslation, a procedure-oriented language that is mainly used for solving scientific problems.

Function subalgorithm A form of subalgorithm that returns exactly one value to be inserted in an expression in place of its name.

General-purpose computer A computer that is designed to handle a wide variety of problems.

General subalgorithm A form of subalgorithm that returns any number of values (or none) through its parameter list. It is invoked by a separate statement.

Generation backup A method of providing backup for a file by keeping old copies that are not up to date.

Global variable A variable that can be used to communicate between an invoked and invoking routine without using the parameter list.

Hardware The physical components of a computer.

Heuristic	A term applied to exploratory methods of problem solving in which solutions are discovered by evaluating the progress made toward a desired end result.
High-level algorithm	An English language form of an algorithm, best expressed as a structured algorithm.
Hollerith punched card	A punched card utilizing 12 rows and usually 80 columns. Hollerith was an American engineer who devised punched card data storage, and developed several punched card computing devices.
Hybrid computer	A computer for data processing using both continuous (analog) and discrete (digital) representation of data.

Implicit or implied loop	A form of loop that does not require the programmer to express completely the syntax usually necessary for a loop.
Indefinite loop	A form of the loop in which some condition dependent on processing in the loop is used for termination.
Index	(a) Any symbol (usually an integer) that is used to uniquely identify an element in an array, or (b) an ordered list that contains references to the contents of a file or document, together with pointers to the location of those contents.
Information	The meaning assigned to data via some system of representation.
Information retrieval	The methods and procedures for recovering specific data.
Input	Information to be processed that is received from outside a computer system.
Input/Output	A term applied to input of data to the main memory, or output of data from main memory, or both.
Instruction	A sequence of symbols defining an operation to be performed, and the data or unit of equipment to be used.
Integrated management information (IMI) system	A computer-based management system that is designed by considering all the resources and needs pertaining to an organization.
Invoke	To cause an algorithm or subalgorithm to execute.
Invoke by reference	To communicate between an invoking and invoked subalgorithm by referencing the same location in memory for the corresponding formal and actual parameters; also referred to as call by location.

343

Invoke by value   To communicate between an invoking and invoked subalgorithm by transferring a value to the subalgorithm.

Item   A group of characters, treated as a unit of data.

Iteration   A single cycle of a series of algorithm steps that are executed repeatedly.

Iterative calculation   A form of calculation in which a series of steps is executed repeatedly.

J

Job   A set of logically related tasks that constitutes a unit of work for a computer.

L

Label   One or more characters used to identify an item of data or a statement in a computer program.

Language   An alphabet, a syntax, describing correct constructions in the language, and a set of semantic rules to interpret the meaning of the allowed constructions.

List   An ordered collection of items.

Local variable   A variable that can be used only within a subalgorithm.

Logical record   A record that is defined in terms of the information it contains, rather than by its physical attributes.

Loop   A sequence of computer instructions that is executed repeatedly until a terminal condition prevails.

Lower bound   The first value for a subscript of an array in a declaration; by default this value is 1 in flowchart language.

M

Machine language   A language that is used directly, without translation, by the computer.

Magnetic core   A type of memory medium capable of storing one binary digit (i.e., a 0 or 1).

Magnetic disk	A memory medium on which data are stored in concentric circles.
Magnetic drum	A cylindrical device on which data are stored magnetically in the form of bands.
Magnetic tape drive	A device used to read/write data that are recorded on magnetic tape.
Main memory	That part of computer memory allowing for direct access by the CPU.
Management Information System (MIS)	A computer system integrating hardware, software, and procedures to provide analysis important to the management function.
Master file	A file that contains relatively permanent data.
Memory	A set of computer components used to store data and instructions.
Memory or storage	A device into which data can be entered and held, and from which they can be retrieved at a later time.
Microsecond	A unit of time equal to one millionth of a second (μs).
Millisecond	A unit of time equal to one thousandth of a second (ms).
Minicomputer	A relatively small, inexpensive general-purpose computer, compared to large computers, such as the CDC 6600 or IBM 370/148.
Modem	Acronym for Modulater-Demodulator—a device capable of converting digital data into analog form suitable for transmission over a communication line, such as a telephone line, and then converting the data back to digital form at the receiving end.
Multiplexer	A device that combines signals received from low-speed lines, and interleaves them for transmission over a high-speed line.
Multiprocessing	Simultaneous execution of more than one program or sequences of instructions by a computer system. Multiprocessing requires more than one CPU.
Multiprogramming	Pertaining to the concurrent execution of two or more programs; usually this is implemented by having more than one program or parts of more than one program in the main memory.

Nanosecond	A unit of time equal to one billionth of a second, or 10^{-9} second (ns).
Natural language	A language, the rules of which are determined by current usage as opposed to being predefined on a logical basis.

Nested loops   Loops imbedded within loops, so that each inner loop is executed to completion for each iteration of the outer loop.

Numeric   Pertaining to numerals or representation by means of numerals.

Object program   The output of a language translator (i.e., the translated version of a source program).

Offline   Pertaining to equipment or devices not under control of the CPU.

Online   (a) Pertaining to equipment or devices under control of the CPU, or (b) a system in teleprocessing in which input enters the computer directly from the point of origin or in which output is transmitted directly to where it is used.

Operand   A quantity (object) that is operated upon during the execution of a computer instruction.

Operating system   A collection of computer programs enabling algorithms to be processed. These programs also schedule the resources of a computer system among its users.

Output   Information a computer program transmits as the result of processing.

Overflow   A condition in which the result of some operation is too large to be stored in the space allocated to the value.

Page   A group of adjacent memory locations, the contents of which can be moved in and out of main memory as a unit. Its length, which is usually fixed, depends on the machine.

Paging   The process of transferring pages (parts of programs) into and out of main memory from an auxiliary storage device.

Parameter   A variable, constant, or expression used to communicate between an invoking and invoked algorithm See *Actual parameter* and *Formal parameter.*

Physical record   A record from the standpoint of the manner in which it is stored and retrieved. It may contain all or part of one or more logical records.

Picosecond A unit of time equal to 10^{-12} second, or one thousandth of a nano-second (ps).

PL/I Acronym for Programming Language/I—a procedure-oriented language.

Pointer A data item that indicates the memory location of an item or record.

Pop To remove an item from a stack.

Precedence rules The set of rules defining the order in which operators are to be executed in an expression.

Privacy Pertaining to the issue of what type of data should be collected and made available.

Problem-oriented language A programming language designed for convenient handling of a class of problems; e.g., RPG, a language designed to produce reports.

Procedure A finite sequence of instructions which can be systematically carried out. A procedure that always stops is an algorithm.

Procedure-oriented language A programming language designed for handling procedures used in the solution of a wide class of problems.

Program A set of coded instructions to direct the computer in the solution of a predefined problem.

Protection A means to prevent unauthorized or accidental access to a device, program, or data.

Punched card A card in which data are represented by patterns of holes.

Punched paper tape A roll of paper tape in which data are represented by patterns of holes.

Push To add an item to a stack.

Queue A form of list in which items can be added only at one end and removed only at the other end.

Radix The base of a number system; e.g., 2 and 10 are the radices of the binary and decimal number systems, respectively.

Random access	See *Direct access*.
Real-time system	A system in which outputs are available in time to affect the process that causes future actions.
Reasonableness check	A data verification technique wherein tests are made on the data reaching a system, or being output, to ensure that the values lie in a reasonable, predetermined range.
Record	A collection of related items of data, treated as a unit.
Response time	The time it takes a system to react to an input; usually applied to conversational systems in which response time is the time elapsed between pressing the last key of the input and receiving the first character of the reply from the computer.
Return	In most programming languages, a statement that causes flow of execution to return to the invoking routine.
RPG	An acronym for Report Program Generator—a problem-oriented language.

S

Scan	To examine a file or record sequentially, item-by-item.
Scientific computation	The area of computing that deals primarily with problems characterized by a predominance of arithmetic calculations.
Search	The process of examining a set of items for one or more having a desired property.
Security	Pertaining to the prevention of access to or use of data or programs without authorization.
Semantics	The set of rules which give the meaning of programming language statements.
Sequential file	A file, the records of which are organized on the basis of their successive physical positions; e.g., a magnetic tape file.
Serial access	The sequential transmission or access of data. The time required for access is dependent upon the location of the last information obtained.
Serial search	A form of search that consecutively compares the search argument with the list arguments, starting with the first list argument.

348

Side effect An action which occurs without a programmer specifying it, and which can result in both desirable and undesirable effects.

Software Consists of computer programs and documentation which describe the use and operation of these programs.

Source program A program written in a programming language that must be translated into machine language prior to execution by a computer.

Spooling The process of reading and writing of input and output streams on auxiliary storage devices during job execution.

Stack A form of list in which elements can be added and deleted only at one end.

String A linear sequence of entities, such as characters.

Stored-program computer A computer designed to store a program in the same way it stores data. The program in stored form is then executed by the machine.

Structured programming A problem-solving methodology employing well-organized, top-down, stepwise refinement procedures.

Subalgorithm A named collection of programming language statements which can be executed by referencing the name (i.e., by invoking). There are two types—function and general subalgorithms.

Subscript variable The variable(s) used to indicate which element of an array is to be used.

Subscripted variable The name of an array in the flowchart language; e.g., A(I) indicates A is a subscripted variable.

Syntax The rules governing the construction of legal statements in a language.

Systems analyst An individual concerned with the development of suitable algorithms to solve problems. Seldom would a systems analyst write programming language statements.

```
┌─────────┐
│ ┌─────┐ │
│ │  T  │ │
│ └─────┘ │
└─────────┘
```

Table lookup A procedure for obtaining a function value (from a table) corresponding to a given argument.

Tag A field of one or more characters attached to a record for identification purposes.

349

Terminal A device by which data are entered into the system and by which results are communicated by the system. Terminals include typewriter-like devices and cathode ray tubes.

Time-sharing The division of time of a computer system among several users in such a way that the computer appears to be dedicated to each one.

Total system An integrated management system that operates in real time.

Trace A debugging method involving manual execution of the steps of an algorithm with sample data.

Transactions file A file containing data generated during some time period. It is processed in conjunction with a master file for producing reports, statements, etc., and also for updating the master file.

Tree A data structure that provides a topology for data, such that each item with the exception of the root has only one predecessor.

Turing machine A mathematical model for computers developed by the British scientist, Alan Turing.

Turnaround time The elapsed time between the submission of a job to a computer and the availability of the output.

Underflow The condition that occurs when a floating-point result is smaller in magnitude than the smallest value that a specific computer can represent.

Upper bound The ending value for a subscript variable of an array.

Variable An entity in a program that has a name, a value, and a descriptor of legal contents; i.e., a type.

Virtual memory Addressable memory that appears to the programmer as real (or actual) main memory, but in fact may be several times larger than the main memory.

350

Word   A group of adjacent bits, operated upon as a unit.

Word length   The storage capacity of a word. It is expressed in terms of the number of bits, bytes, or characters that a word may contain.

Bibliography

Adams, J. M., and Haden, D. H. *Social Effects of Computer Use and Misuse*. New York: Wiley, 1976.

Banks, P. M., and Doupnik, J. R. *Introduction to Computer Science*. New York, Wiley, 1976.

Bartee, T. *Introduction to Computer Science*. New York: McGraw-Hill, 1975.

Bohl, M., and Walter, A. *Introduction to PL/I Programming and PL/C*. Chicago: Science Research Associates, 1973.

Booth, T. L., and Chien, Y. T. *Computing Fundamentals and Applications*. Santa Barbara, Calif.: Hamilton, 1974.

Caruth, D. L., and Rachel, F. M. *Business Systems*. San Francisco: Canfield Press, 1972.

Cashman, T. J., and Keys, W. J. *Essentials of Information Processing*. San Francisco: Canfield Press, 1974.

Consequences of Electronic Funds Transfer, The (NSF/RA/X-75-015). Washington, D.C.: National Science Foundation, 1975.

Conway, R., and Gries, D. *An Introduction to Programming*. Cambridge, Mass.: Winthrop Publishers, 1973.

Cotterman, W. W. *Computers in Perspective*. Belmont, Calif.: Wadsworth, 1974.

Cress, P.; Dirksen, P.; and Graham, J. W. *Fortran IV with Watfor or Watfiv*. Englewood Cliffs, N.J.: Prentice-Hall, 1970.

Dahl, O. J.; Dijkstra, E. W.; and Hoare, C. A. R. *Structured Programming*. New York: Academic Press, 1972.

Data Processing Glossary. IBM manual GC20–1699, 5th ed., 1973.

Dijkstra, E. W. *A Discipline of Programming*. Englewood Cliffs, N.J.: Prentice-Hall, 1976.

Dorf, R. C. *Computers and Man*. San Francisco: Boyd & Fraser, 1977.

Forsythe, A. I.; Keenan, T. A.; Organick, E. I.; and Stenberg, W. *Computer Science: A First Course*. New York: Wiley, 1975.

Gear, C. W. *Introduction to Computer Science: Short Edition*. Chicago: Science Research Associates, 1973.

Goldstine, H. H. *The Computer from Pascal to von Neumann*. Princeton, N.J.: Princeton University Press, 1972.

Gotlieb, C. C., and Borodin, A. *Social Issues in Computing*. New York: Academic Press, 1973.

353

BIBLIOGRAPHY

Hetzel, C. *Program Test Methods*. Englewood Cliffs, N.J.: Prentice-Hall, 1973.

Kennedy, M., and Solomon, M. B. *Ten Statement Fortran Plus Fortran IV,* 2nd ed. Englewood Cliffs, N.J.: Prentice-Hall, 1975.

Kernighan, B. W., and Plauger, P. J. *Software Tools*. Reading, Mass.: Addison-Wesley, 1976.

Kraus, L. I. *Administrating and Controlling the Company Data Processing Function*. Englewood Cliffs, N.J.: Prentice-Hall, 1969.

Martin, J. *Security, Accuracy and Privacy in Computer Systems*. Englewood Cliffs, N.J.: Prentice-Hall, 1973.

Martin, J., and Norman, A. R. D. *The Computerized Society*. Englewood Cliffs, N.J.: Prentice-Hall, 1970.

McCameron, F. A., *Cobol Logic and Programming,* 3rd ed., Homewood, Ill.: Ireven Publishing, 1976.

McDaniel, H., ed. *Applications of Decision Tables*. Princeton, N.J.: Brandon/Systems Press, 1970.

Murrill, P. W., and Smith, C. L. *Introduction to Computer Science*. New York: Intext Educational Publishers, 1968.

Nassi I., and Schneiderman, B. "Iteration Graphs." *ACM SIGPLAN* notices 8(1973):12–56.

Orlecky, J. *The Successful Computer System*. New York: McGraw-Hill, 1969.

Page, R., and Didday, R. *Watfiv for Humans*. St. Paul, Minn.: West Publishing, 1976.

Pollack, S. V., and Sterling, T. D. *A Guide to PL/I*. New York: Holt, Reinhart and Winston, 1969.

Pylyshyn, Z. W. *Perspectives on the Computer Revolution*. Englewood Cliffs, N.J.: Prentice-Hall, 1970.

Randall, B., ed. *The Origins of Digital Computers*, 2nd ed. New York: Springer-Verlag, 1975.

Raphael, B. *The Thinking Computer*. San Francisco: W. H. Freeman, 1976.

Rothman, S., and Mosman, C. *Computer and Society,* 2nd ed. Chicago: Science Research Associates, 1976.

Sackman, H., and Borko, H. *Computers and the Problems of Society*. Montrale, N.J.: AFIPS Press, 1972.

Scheid, F. *Introduction to Computer Science*. New York: Schaum's Outline Series, 1970.

Scott, R. C., and Sondak, N. E. *PL/I for Programmers*. Reading, Mass.: Addison-Wesley, 1970.

354

Silver, G. A., and Silver, J. B. *Simplified ANSI Fortran IV Programming,* 2nd ed. New York: Harcourt Brace Jovanovich, 1976.

Spencer, D. D. *Introduction to Information Processing.* Columbus, Ohio: Charles E. Merrill, 1974.

Squire, E. *The Computer an Everyday Machine.* Reading, Mass.: Addison-Wesley (Canada) Ltd., 1972.

Stark, P. A. *Digital Computer Programming.* New York: Macmillan, 1967.

Struble, G. *Assembler Language Programming the IBM System/360.* Reading, Mass.: Addison-Wesley, 1969.

Taviss, I., ed. *The Computer Impact.* Englewood Cliffs, N.J.: Prentice-Hall, 1970.

Teaque, R. *Computing Problems for Fortran Solution.* San Francisco: Canfield Press, 1972.

Van Tassel, D. L. The Complete Computer. Chicago: Science Research Associates, 1976.

Unger, E. A., and Ahmed, N. "An Instructionally Acceptable Cost-Effective Approach to a General Introductory Course." *ACM SIGCSE Bulletin* 8(1976):28-31.

Unger, E. A., and Ahmed, N. *Classroom Instructional Notes—Fundamentals of Computer Programming.* Dubuque, Iowa: Kendall/Hunt, 1976.

Vazsonyi, A. *Problem Solving by Digital Computers with PL/I Programming.* Englewood Cliffs, N.J.: Prentice-Hall, 1970.

Walker, T. M. *Introduction to Computer Science: An Interdisciplinary Approach.* Boston: Allyn & Bacon, 1972.

Walker, T. M., and Cotterman, W. W. *An Introduction to Computer Science and Algorithmic Processes.* Boston: Allyn & Bacon, 1970.

Wilde, D. V. *An Introduction to Computing.* Englewood Cliffs, N.J.: Prentice-Hall, 1973.

Wirth, N. *Systematic Programming.* Englewood Cliffs, N.J.: Prentice-Hall, 1973.

Wu, M. S. *Introduction to Computer Data Processing.* New York: Harcourt, Brace, Jovanovich, 1975.

Answers to Selected Exercises

chapter 1

1-1. Some pocket calculators may vary slightly from the following general description:
- ☐ Data representation: digital
- ☐ Hardware features: electronic
- ☐ Algorithm implementation: manual (some have stored programs or hardware implemented algorithms)
- ☐ Scope of application: special purpose
- ☐ Internal mode of operation: sequential
- ☐ Performance: millisecond
- ☐ Size: calculator (very small)

1-3. Applications are primarily justified by the reasons given in Section 1.4.
- (a) Point-of-sale computers are justified by the first reason in Section 1.4—repetitive task.
- (b) Determination of the path and contour of a highway is justified by the second reason—the solution of a problem must be repeated until an acceptable answer is obtained.
- (c) Weather prediction is justified by the third reason—the solution must be obtained rapidly to be of any value.
- (d) Automation of retrieval of Library of Congress information is justified by the fourth reason—large volumes of information must be stored.
- (e) Chess-playing computer systems are justified by the fifth reason—a clearer understanding of the thinking process is desired.

Comment: In this chapter, there are many acceptable solutions for each problem. The authors have presented but one correct solution to each—and it is not necessarily the optimal one, in any sense.

2-1. Problem: Changing a flat tire.

Step 1. Define the problem: Once a flat tire has been recognized by the driver, the car must be parked, the flat replaced with a spare, the car made ready to continue the trip. Assume that the car has a good spare in the trunk and that the driver is capable of changing the tire.

357

Step 2.   A good spare tire is one worthy of continuing the trip. A driver capable of changing the tire has the knowledge and appropriate tools to accomplish this task.

Step 3.   The inputs to the problem are the spare tire, tools, and car with a flat tire.

Step 4.   The output consists of the car ready to resume the trip.

Step 5.   1. Halt the vehicle.
           2. Prepare to change the tire.
           3. Change the tire.
           4. Ready the car for the trip continuation.

Step 6.   1. Halt the vehicle.
               1.1 Select a place to stop.
               1.2 Proceed to place.
               1.3 Apply brakes.
               1.4 Stop the engine.
           2. Prepare to change the tire.
               2.1 Obtain tools.
               2.2 Obtain spare tire.
           3. Change the tire.
               3.1 Loosen the lug-nuts.
               3.2 Remove the flat tire.
               3.3 Put spare tire on the car.
           4. Ready the car for trip continuation.
               4.1 Replace tools.
               4.2 Stow the flat tire.
               4.3 Return to driver's seat.

Step 7.   More detail can be added to each of these steps. A reasonably intelligent person could proceed from the algorithm given; however, for clarity, let's refine Steps 2 and 3 to another level.
           2. Prepare to change the tire.
               2.1 Obtain tools.
                    2.1.1 Proceed to trunk and open it.
                    2.1.2 Remove car jack, lug-nut wrench, and hubcap pry.
                    2.1.3 Take these tools to location of flat tire.
               2.2 Obtain the spare tire.
                    2.2.1 Proceed to location of spare tire.
                    2.2.2 Remove spare from car.
                    2.2.3 Roll spare tire to location of flat tire.
           3. Change the tire.
               3.1 Loosen the lug-nuts.
                    3.1.1 Pry off the hubcap.
                    3.1.2 Repeat Steps 3.2.3–3.1.4 for each lug-nut.
                    3.1.3 Place wrench on a lug-nut.

3.1.4 Turn nut until it comes off.
3.2 Remove flat tire.
   3.2.1 Jack up car.
   3.2.2 Pull wheel off.
3.3 Put spare tire on the car.
   3.3.1 Position wheel on bolts.
   3.3.2 Repeat Steps 3.3.3–3.3.4 for each lug-nut.
   3.3.3 Place lug-nut on bolt.
   3.3.4 Tighten nut by hand.
   3.3.5 Lower car.
   3.3.6 Repeat Step 3.3.7 for each lug-nut until each is sufficiently tightened.
   3.3.7 Place wrench on nut and tighten slightly; proceed to next nut.
   3.3.8 Replace hubcap.

2-2.

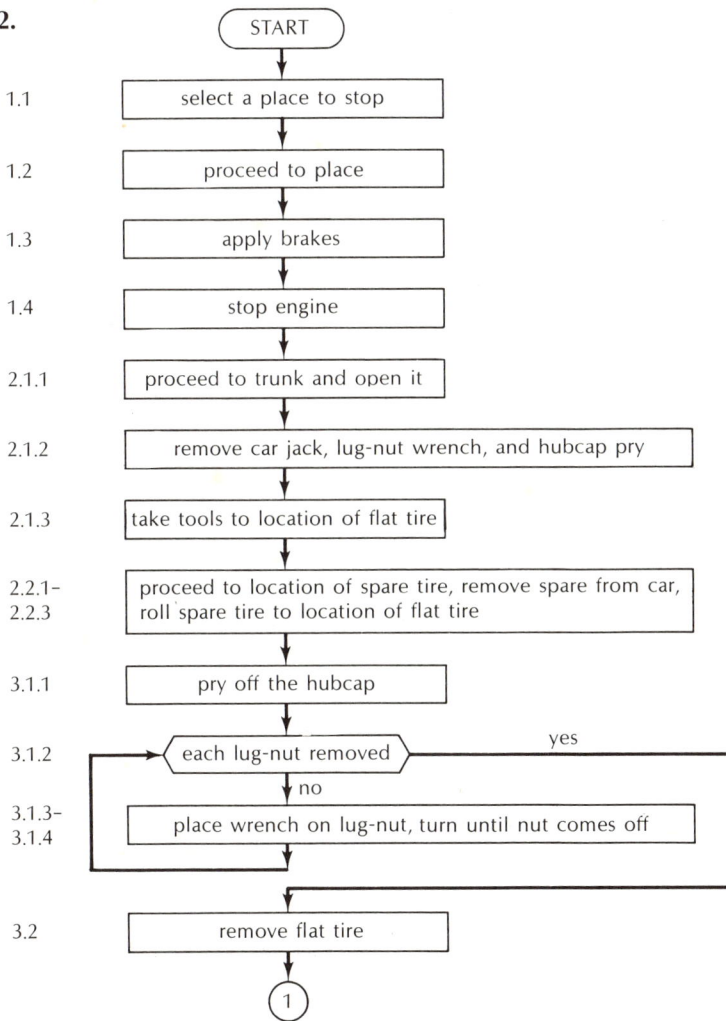

	START
1.1	select a place to stop
1.2	proceed to place
1.3	apply brakes
1.4	stop engine
2.1.1	proceed to trunk and open it
2.1.2	remove car jack, lug-nut wrench, and hubcap pry
2.1.3	take tools to location of flat tire
2.2.1–2.2.3	proceed to location of spare tire, remove spare from car, roll spare tire to location of flat tire
3.1.1	pry off the hubcap
3.1.2	each lug-nut removed — yes / no
3.1.3–3.1.4	place wrench on lug-nut, turn until nut comes off
3.2	remove flat tire

1

①

3.3.1	position wheel on bolts

repeat for each lug-nut

3.3.2	
3.3.3– 3.3.4	place lug-nut on bolt tighten by hand

repeat until each lug-nut is tight

3.3.5	
3.3.6	place wrench on lug-nut, tighten slightly
3.3.7	replace hubcap
4	ready car for trip continuation

STOP

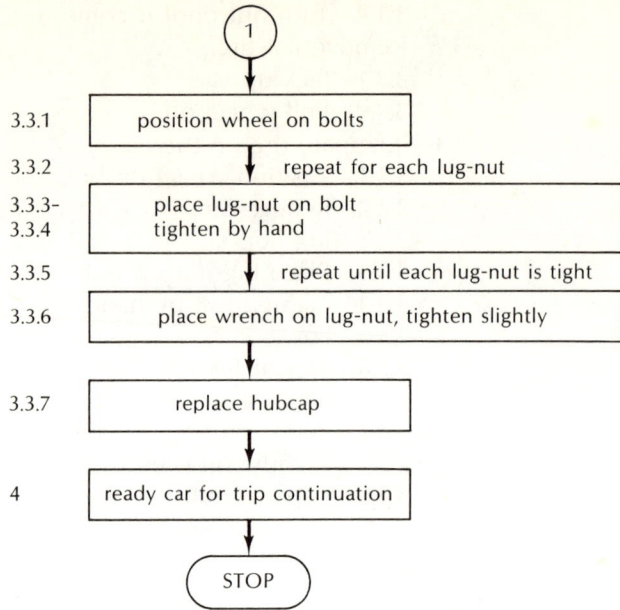

2-6. 1 & 2.   Accept the problem as defined but clarify that a count of each vowel is desired. Consider the sentence to consist of a series of characters, the last of which is a period. Assume the vowels are a, e, i, o, u and the capitals of these.

3.   Input is the sentence.

4.   Output is the counts for the five vowels.

5.   1.   Input the sentence.
    2.   Examine each character; add one to a vowel counter if appropriate.
    3.   Output the counts.

6.   1.   Input the sentence.
    2.   Examine each character; add one to a vowel counter if appropriate.
      2.1   Set counters to zero.
      2.2   Repeat Step 2.3 until character is a period.
      2.3   Examine each character; if vowel, add one to counter.
    3.   Output the counts.

7.   1.   Input the sentence.
      1.1   Input the sentence as a set of characters. (In later chapters the methods of accomplishing this are discussed in detail.)
    2.   Examine each character; add one to a vowel counter if appropriate.
      2.1   Set counters to zero.
        2.1.1 $Ca \leftarrow 0$ $CO \leftarrow 0$ $CU \leftarrow 0$ $CE \leftarrow 0$
                $CI \leftarrow 0.$
        2.1.2 $i \leftarrow 1.$

2.2   Repeat Step 2.3 until character is a period.

2.3   Examine each character; if vowel, add 1 to counter.

  2.3.1   If $character_i = \text{'a'}$ or $character_i = \text{'A'}$, then add 1 to CA.

  2.3.2   If $character_i = \text{'e'}$ or $character_i = \text{'E'}$, then add 1 to CE.

  2.3.3   If $character_i = \text{'i'}$ or $character_i = \text{'I'}$, then add 1 to CI.

  2.3.4   If $character_i = \text{'o'}$ or $character_i = \text{'0'}$, then add 1 to CO.

  2.3.5   If $character_i = \text{'u'}$ or $character_i = \text{'U'}$, then add 1 to CU.

  2.3.6 $i \leftarrow i + 1$.

3. Output the counts.

  3.1   OUTPUT 'THE NUMBER OF A''s IS', CA.

  3.2   OUTPUT 'THE NUMBER OF E''s IS', CE

$$\vdots$$

  3.5   OUTPUT 'THE NUMBER OF U''s IS', CU.

  3.6   OUTPUT 'THE LENGTH OF SENTENCE IS', i.

2-7.

```
            ( START )
                |
                v
   / INPUT CHARACTERS / - - - - |  Inputs a sentence
                |
                v
   +-------------------------------+
   | CA ← 0        CU ← 0          |
   | CO ← 0        i ← 1           |
   | CE ← 0        CI ← 0          |
   +-------------------------------+
                |
                v
   < CHARACTER_i = '.' >  ---- Yes ---->
                | No
                v
   < CHARACTER_i = 'A' or >  --- Yes --->  [ CA ← CA+1 ]
   < CHARACTER_i = 'a'     >
                | No
                v
   < CHARACTER_i = 'E' or >  --- Yes --->  [ CE ← CE+1 ]
   < CHARACTER_i = 'e'     >
                | No
                v
   < CHARACTER_i = 'I' or >  --- Yes --->  [ CI ← CI+1 ]
   < CHARACTER_i = 'i'     >
                | No
                v
   < CHARACTER_i = 'O' or >  --- Yes --->  [ CO ← CO+1 ]
   < CHARACTER_i = 'o'     >
                | No
                v
   < CHARACTER_i = 'U' or >  --- Yes --->  [ CU ← CU+1 ]
   < CHARACTER_i = 'u'     >
                | No
                v
          [ i ← i+1 ]
                |
                v
  / OUTPUT 'THE NUMBER OF A''S IS' CA /
                :
                v
  / OUTPUT 'THE LENGTH OF SENTENCE IS', i /
                |
                v
            ( END )
```

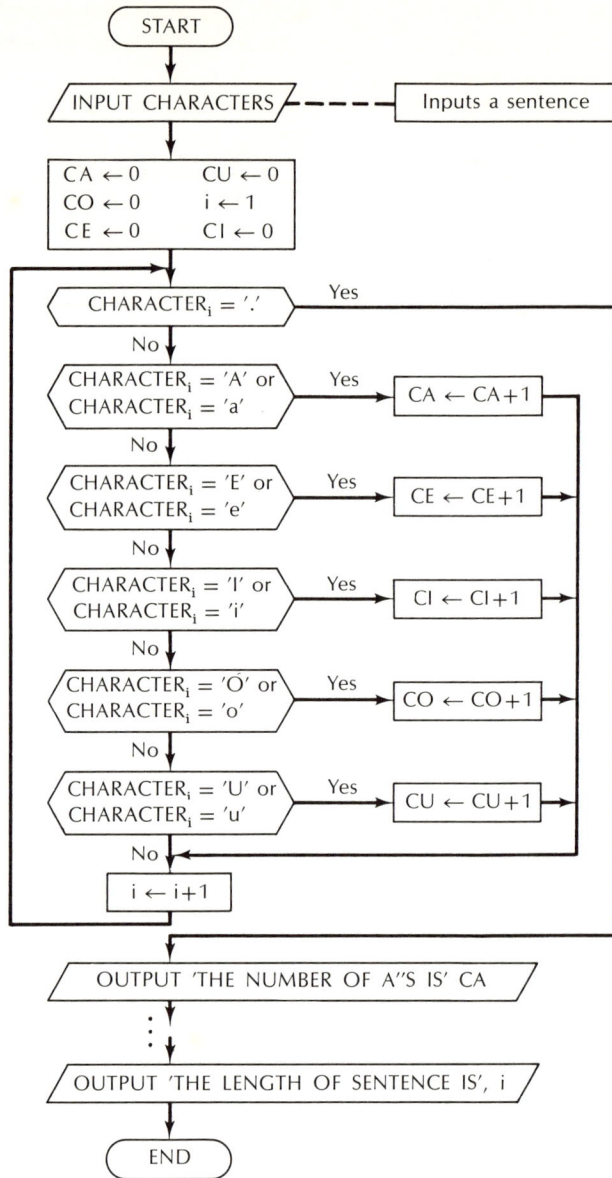

2-12.
1. Input the two numbers.
   1.1 Input the two numbers, each as four single digits.
2. Add the two numbers.
   2.1 Add the rightmost digits.
   2.2 If sum of rightmost digits is greater than 9, then reduce by 10 and set carry to one; else set carry to zero.
   2.3 Add second digits.
   2.4 If sum of second digits is greater than 9, then reduce by 10 and set carry to one; else set carry to zero.
   2.5 Add third digits.

2.6 If sum of third digits is greater than 9, then perform carry process.

2.7 Add fourth digits.

3. Output result.

```
                          ┌─────────────┐
                          │    START    │
                          └──────┬──────┘
                                 ▼
          ╱─────────────────────────────────╲
         ╱ INPUT N1D1, N1D2, N1D3, N1D4       ╲ ─ ─ ─ [ (RD1, RD2, RD3, RD4, Carry) integer ]
         ╲       N2D1, N2D2, N2S3, N2D4       ╱
          ╲─────────────────┬─────────────────╱
                            ▼
                 ┌─────────────────────┐
                 │  RD4 ← N1D4+N2D4     │
                 └──────────┬──────────┘
                            ▼                      Yes    ┌──────────────────┐
                      ◇ RD4 > 9 ◇ ───────────────────────│ RD4 ← RD4−10     │
                            │ No                          │ Carry ← 1        │
                            ▼                              └────────┬─────────┘
                   ┌──────────────┐                                │
                   │  Carry ← 0   │                                │
                   └──────┬───────┘                                │
                          ▼◄───────────────────────────────────────┘
                 ┌─────────────────────────┐
                 │ RD3 ← N1D3+N2D3+Carry    │
                 └───────────┬─────────────┘
                             ▼                     Yes    ┌──────────────────┐
                       ◇ RD3 < 9 ◇ ──────────────────────│ RD3 ← RD3−10     │
                             │ No                         │ Carry ← 1        │
                             ▼                             └────────┬─────────┘
                    ┌──────────────┐                               │
                    │  Carry ← 0   │                               │
                    └──────┬───────┘                               │
                           ▼◄──────────────────────────────────────┘
                 ┌─────────────────────────┐
                 │ RD2 ← N1D2+N2D2+Carry    │
                 └───────────┬─────────────┘
                             ▼                     Yes    ┌──────────────────┐
                       ◇ RD2 > 9 ◇ ──────────────────────│ RD2 ← RD2−10     │
                             │ No                         │ Carry ← 1        │
                             ▼                             └────────┬─────────┘
                    ┌──────────────┐                               │
                    │  Carry ← 0   │                               │
                    └──────┬───────┘                               │
                           ▼◄──────────────────────────────────────┘
                 ┌─────────────────────────┐
                 │ RD1 ← N1D1+N2D1+Carry    │
                 └───────────┬─────────────┘
                             ▼
          ╱─────────────────────────────────╲
         ╱ OUTPUT ← RD1, RD2, RD3, RD4        ╱
          ╲─────────────────┬─────────────────╱
                            ▼
                     ┌─────────────┐
                     │     END     │
                     └─────────────┘
```

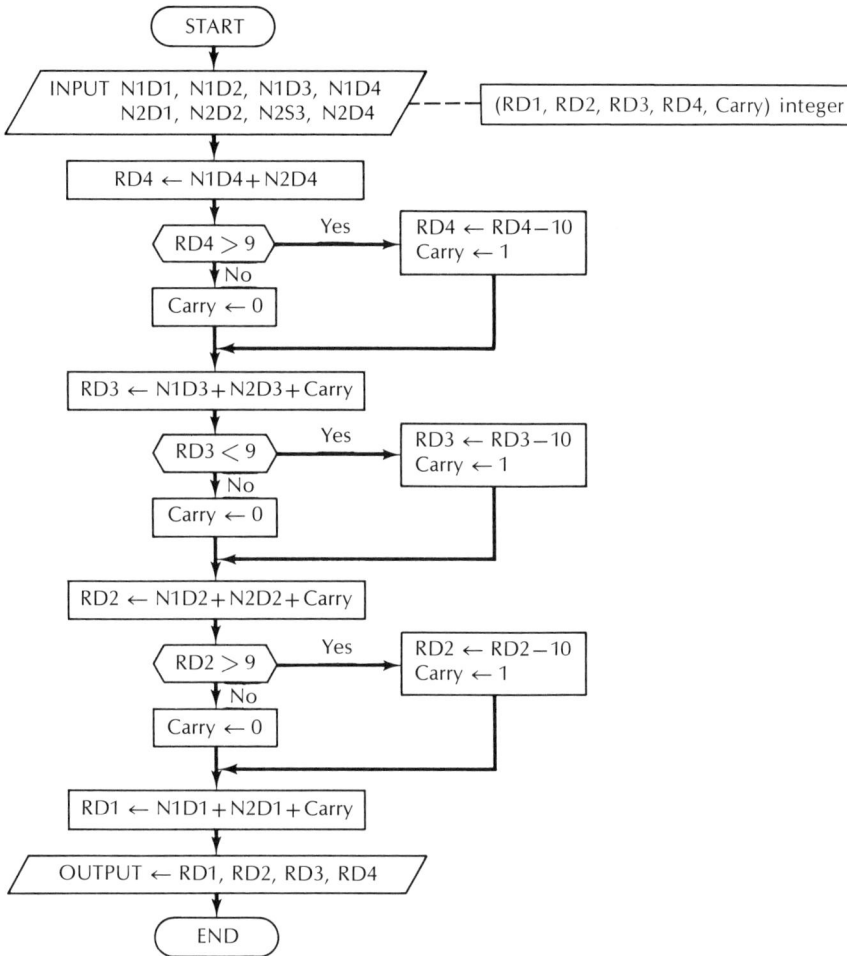

2-16. Input: Purchase amount, Payment
Output: Change amount

1. Input Purchase amount, Payment.
   1.1 Input values in PUR, PAY.

2. Calculate change and check for underpayment.
   2.1 Subtract PUR from PAY to get change.
   2.2 If Change < 0, indicate error and stop.

3. Output Change.

363

```
                    ( START )
                        |
                        v
        /  INPUT PUR, PAY  /------  PUR—holds purchase value
                        |           PAY—holds payment value
                        v
        / Change ← PAY − PUR /
                        |
                        v                 Yes
        <   Change < 0   >---------->  / OUTPUT 'PAYMENT /
                        |                 /    TOO LOW    /
                  No    |                        |
                        v                        v
        / OUTPUT Change /                   ( STOP )
                        |
                        v
                    ( STOP )
```

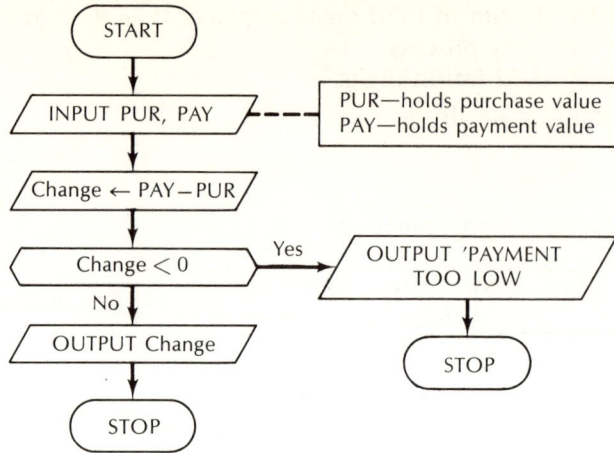

2-17. Input: Purchase amount, payment
  Output: Change expressed in minimum number of bills and coins

1. Input purchase amount and payment.

2. Calculate change, check for underpayment.
    2.1 Change is payment less purchase.
    2.2 If change is less than zero, then indicate error and stop.

3. Output number of dollars.

4. Calculate minimum coins.
    4.0 Set numbers of coins to zero.
    4.1 Determine hundredths of dollar; call it hchange.
    4.2 If hchange > 0.50, then no50 is set to 1 and hchange is reduced by 0.50.
    4.3 If hchange > 0.25, then no25 is set to 1 and hchange is reduced by 0.25.
    4.4 If hchange > 0.10, then no10 is set to 1 and hchange is reduced by 0.10.
    4.5 If hchange > 0.10, then no10 is set to 2 and hchange is reduced by 0.10.
    4.6 If hchange > 0.05, then no05 is set to 1 and hchange is reduced by 0.05.

5. Output minimum number of coins.
    5.1 Output values of no50, no25, no10, no05, hchange times 100.

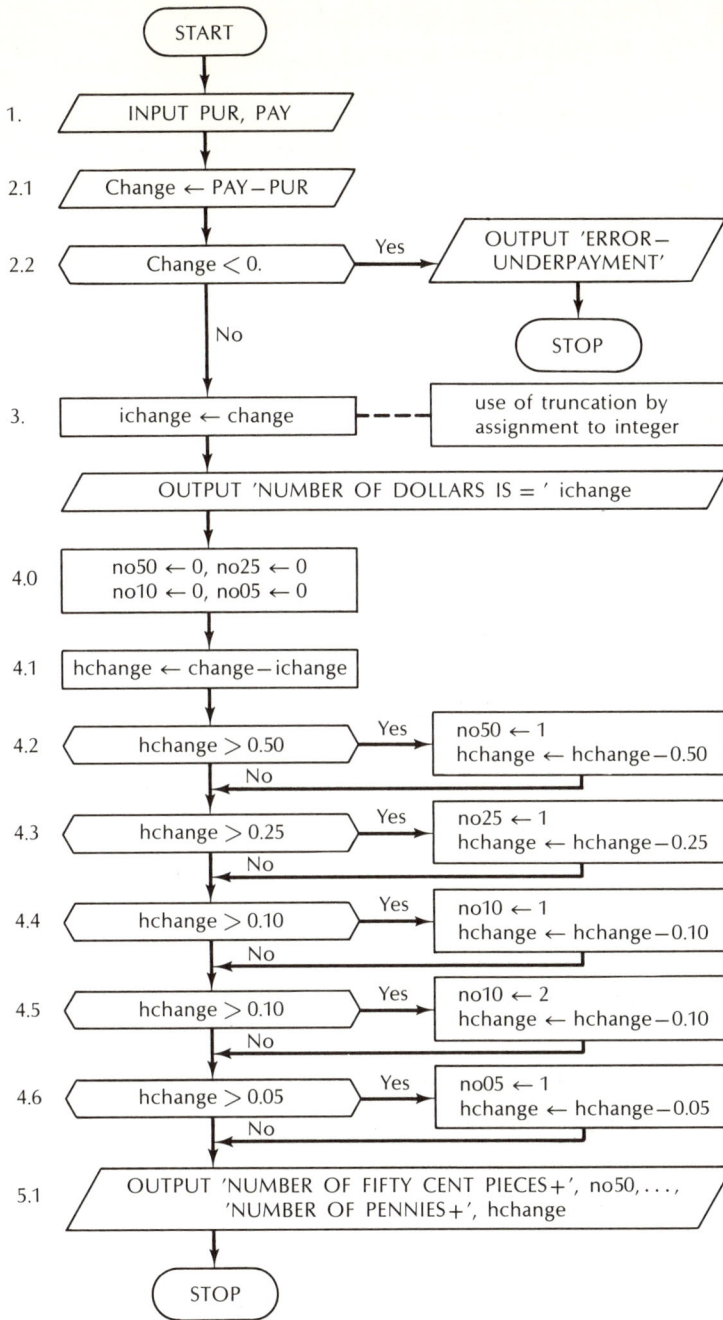

2-20.   1.   Input the three values; call them A, B, C.

2.   Find the largest value of A, B, C.
    2.1   Assume A is largest.
    2.2   If B is larger than largest, then largest takes value of B.
    2.3   If C is larger than largest, then value of C becomes largest.

3.   Output the value of largest.

```
                    ┌──────────────┐
                    │    START     │
                    └──────┬───────┘
                           ↓
                   ╱──────────────╱
                  ╱ INPUT  A, B, C╱
                 ╱───────┬────────╱
                         ↓
                  ┌──────────────┐
                  │  LARGEST ← A │
                  └──────┬───────┘
                         ↓
                  ⟨ B > LARGEST ⟩──Yes──→  ┌──────────────────┐
                         │                  │  LARGEST ← B     │
                        No                  └──────────────────┘
                         ↓
                  ⟨ C > LARGEST ⟩──Yes──→  ┌──────────────────┐
                         │                  │  LARGEST ← C     │
                        No                  └──────────────────┘
                         ↓
          ╱─────────────────────────────────────────╱
         ╱ OUTPUT 'LARGEST VALUE IS', LARGEST       ╱
        ╱──────────────────┬────────────────────────╱
                           ↓
                    ┌──────────────┐
                    │     STOP     │
                    └──────────────┘
```

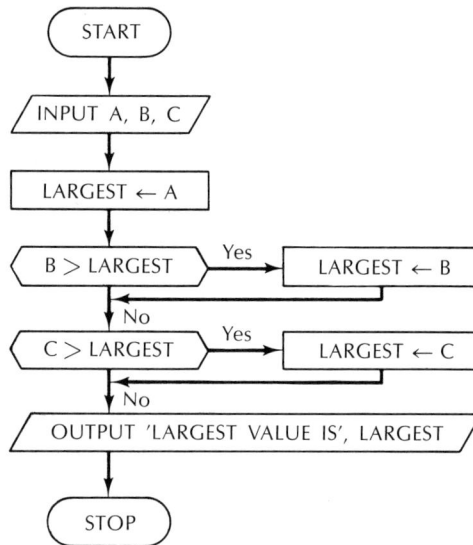

2-21.   Assume your university has a four-point system.

Input: Semester grades
Output: Grade-point average

1.   Input semester grades as A, B, C, D, F, or I.

2.   Count the number of A's, B's, C's, D's, F's.

3.   Calculate grade-point average.
    3.1   Multiply number of A's times 4.
    3.2   Multiply number of B's times 3.
    3.3   Multiply number of C's times 2.
    3.4   Add the previous three values to the number of D's; divide by total number of grades

4.   Output grade-point average.

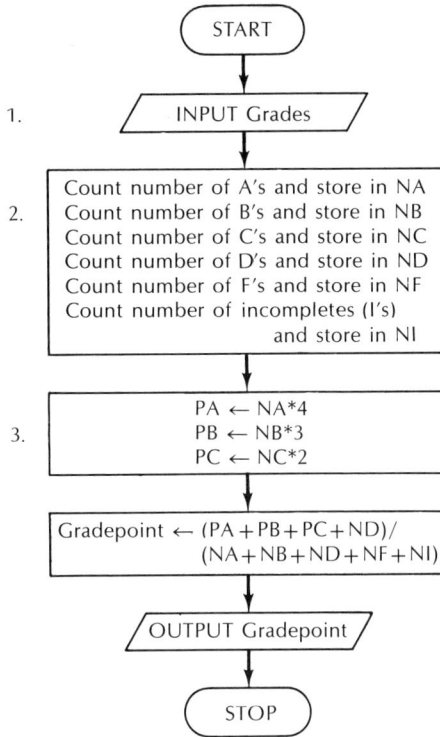

1.

```
                    ( START )
                        |
                        v
          /        INPUT Grades        /
                        |
                        v
```

2.
```
| Count number of A's and store in NA    |
| Count number of B's and store in NB    |
| Count number of C's and store in NC    |
| Count number of D's and store in ND    |
| Count number of F's and store in NF    |
| Count number of incompletes (I's)      |
|                      and store in NI   |
```

3.
```
|            PA ← NA*4            |
|            PB ← NB*3            |
|            PC ← NC*2            |
```

```
| Gradepoint ← (PA+PB+PC+ND)/     |
|              (NA+NB+ND+NF+NI)   |
```

```
          /   OUTPUT Gradepoint   /
                     |
                     v
                  ( STOP )
```

3-1. 1. Input the two numbers and echo.

    1.1 Input NUM1LD, NUM1RD, NUM2LD, NUM2RD.

    1.2 Echo.

2. Add the two right digits.

    2.1 RD ← NUM1RD + NUM2RD.

3. Check for carry.

    3.1 If RD > 10, then RD is reduced by 10 and carry ← 1;
        else carry ← 0.

4. Add the two left digits.

    4.1 LD ← NUM1LD + NUM2LD + carry.

5. Output result.

    5.1 Output LD, RD.

chapter 3

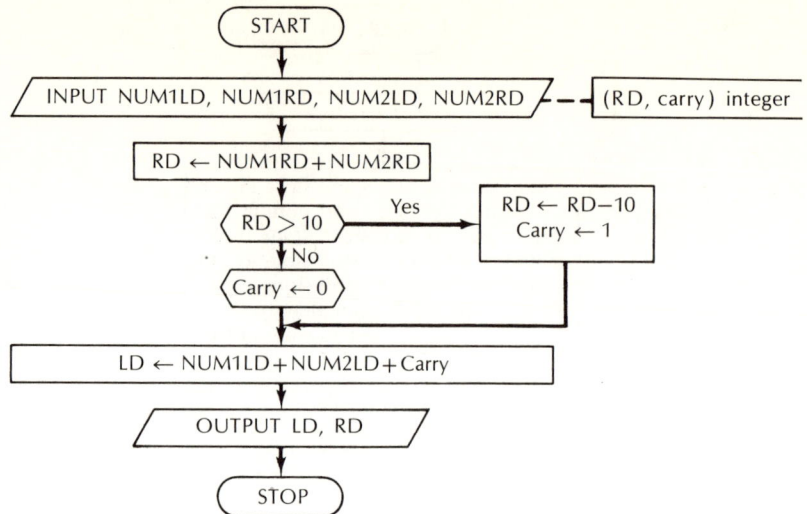

START

INPUT NUM1LD, NUM1RD, NUM2LD, NUM2RD — — (RD, carry) integer

RD ← NUM1RD+NUM2RD

RD > 10 — Yes → RD ← RD−10, Carry ← 1

No

Carry ← 0

LD ← NUM1LD+NUM2LD+Carry

OUTPUT LD, RD

STOP

3-3.
(a) invalid, no enclosing quotes
(b) numeric—real
(c) numeric—integer
(d) invalid, space between digits of a number
(e) character string
(f) numeric—integer
(g) invalid, single internal quote in string
(h) numeric—real
(i) numeric—real
(j) invalid, special character within a number
(k) invalid, space between an internal pair of quote marks
(l) character string

3-5.
(a) label constants: $1, 2, 4, 5$
(b) label variables: J
(c) string constants: I =
(d) string variables: none
(e) numeric variables: I, L
(f) 3
(g) 4
(h) $1, 3, 5, 20$

3-7. TEMPERATURE77.OF$_\wedge$=$_\wedge$25.OC$_\wedge$IT'S$_\wedge$A$_\wedge$NICE$_\wedge$DAY

3-9. $X = 22.73$
$Y = 6$
CAT = LITTLE$_\wedge$TIME
STRING = NO$_\wedge$DEAL

3-11. (a) label constants—$4, 5, 1$
label variables—T, L
character string constants—A =
character string variables—B
numeric variables—A
numeric constants—$1, 6, 3$

368

(b) 7

(c)

Symbol	Times executed
1	1
2	3
3	2
4	1

3-13.

Symbol #	X	Y	Z	SUM
1	7	10	Arb.	
2				
3				
6	17			27.
4				*

*Error—Z has no value provided through this program.

3-15. MOVE
MOVE
TURN
MOVE
MOVE
MOVE
MOVE
MOVE

3-17. MOVE
TURN
MOVE
TURN
MOVE
TURN
MOVE

3-19. IF WALL THEN STOP
         ELSE INPUT A
TURN
IF A = 'TURN∧LEFT' THEN TURN
IF A = 'TURN∧LEFT' THEN TURN
MOVE

Note: The preceding sequence must be repeated two additional times

4.1 (a) 6   (b) 1   (c) 256
     (d) 8   (e) −1   (f) 0
     (g) 0   (h) ½   (i) 8

4-3. (a) 4   (b) 3½   (c) 2½
     (d) 5   (e) 3   (f) 1½

chapter 4

4-5. (a) 3   (b) 'OP'   (c) 'RAPPD$_\wedge$IT'
(d) 0   (e) 'OPPY'   (f) 'POPPY'

chapter 5

5-1. (1)

5-3. (3)

5-5. (1)

5-7. A = 129   B = 3   C = −10.21

5-9. X = 'AI'   Y = '37'

5-11. ANSWER = $_\wedge$123.932

5-13. A = 9
B = 'BC'
C = 1.26
D = 91.91
E = 'NUTTS$_\wedge$'
F = 678

chapter 6

6-1. (a) 10   (b) 6   (c) 11   (d) 121
(e) 10   (f) 6   (g) 11   (h) 363

6-3. -- | STUDENT (4,600)

6-5. -- | WAGES (130, 6:9) FIXED DECIMAL (8.2)

chapter 7

7-1. (a) 5
(b)   2   3
4   5
6   7
8   9
10   11

(c)

K	ITEM
1	1
2	3
3	3
4	5
5	5
6	7
7	7
8	9
9	9
10	11
11	11
12	12

7-3. FEB
JUN
OCT

7-5.

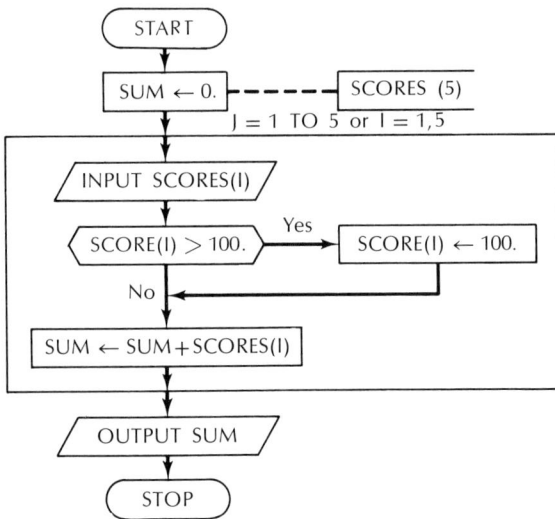

7-7.

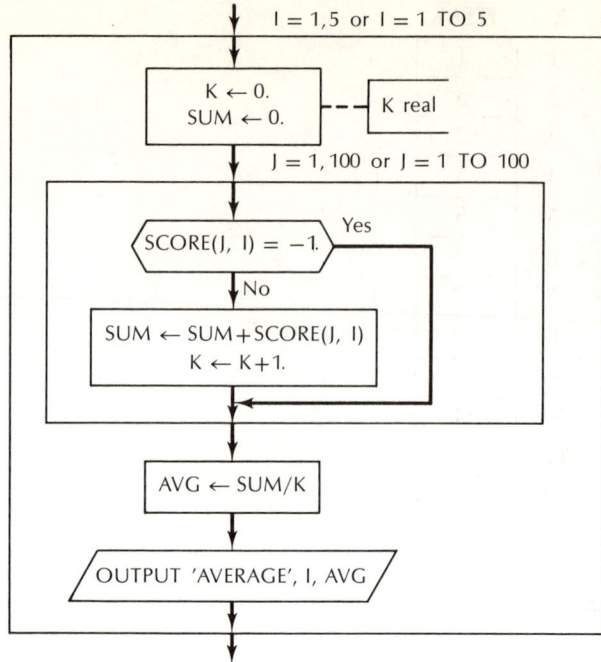

7-9. Change $\langle$ BIG $<$ LIST(I) $\rangle$

to $\langle$ BIG $>$ LIST(I) $\rangle$

and $/$ OUTPUT 'MAXIMUM VALUE =', BIG $/$

to $/$ OUTPUT 'MINIMUM VALUE =', BIG $/$

7-11. DO 6 TIMES BEGIN
        IF WALL THEN STOP
        INPUT COLOR
        IF COLOR = 'RED' THEN WAIT
        MOVE
        END

7-13. DO 48 TIMES BEGIN
        IF exit THEN STOP
        IF wall THEN BEGIN
                TURN
                TURN
                MOVE
                TURN
                MOVE
                TURN
                END
        MOVE
        END

7-15. DO 10 TIMES BEGIN
            IF exit THEN STOP
            IF wall THEN BEGIN
                        TURN
                        TURN
                        END
         MOVE
         END

7-17. DO 16 TIMES BEGIN
            IF exit THEN STOP
            IF wall THEN BEGIN
                        TURN
                        TURN
                        END
            INPUT COLOR
            IF COLOR = 'RED' THEN WAIT
         MOVE
         END

7-19.   DO 16 TIMES BEGIN
             IF exit THEN STOP
             IF wall THEN BEGIN
                         TURN
                         TURN
                         END
             INPUT COLOR
             IF COLOR = 'RED' THEN DO 10 TIMES
                         BEGIN
                         WAIT
                         INPUT COLOR
                         IF COLOR =
                           'GREEN' THEN MOVE

                         END
          END

8-1. (a) A and B
         (b) D and E
         (c) FSA (returns value through its name, MARY)
         (d) 12

8-3. (a) COLE has a local parameter R.
              The formal parameters are U and X in COLE.
              The global variable is T.
              The actual parameters are B and C.

chapter 8

373

(b) B ⟶ | 6̷ 7 | ⟶ U

C ⟶ | 1̷4̷ −4 | ⟶ X

T ⟶ | 4̷ −4 |

R ⟶ | 3 |

} Known to COLE
when it is executing

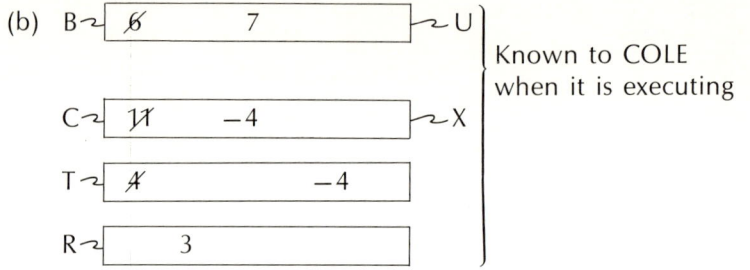

8-5. Input: R%, L dollars, N months
Output: A dollars, the total amount owed

1. Input R, L, and N.
    1.1   Receive R, L, N as formal input parameters.
2. Calculate total amount owed.
    2.1   Amount is L.
    2.2   Repeat Steps 2.3–2.4 N times.
    2.3   Calculate simple interest as Amount times R times 1
    (month).
    2.4   Amount is Amount previously calculated plus simple
    interest.

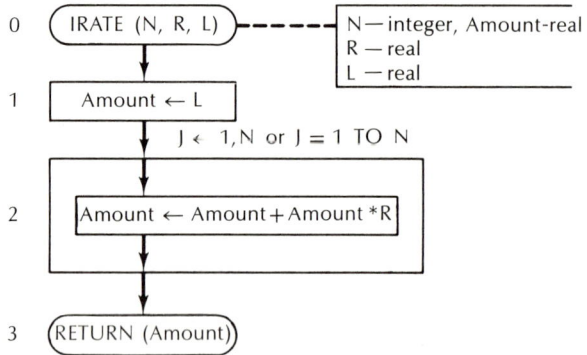

0   (IRATE (N, R, L))--------- N—integer, Amount-real
                                  R — real
                                  L — real

1 | Amount ← L |
                        J ← 1,N or J = 1 TO N

2 | Amount ← Amount + Amount *R |

3   (RETURN (Amount))

A loan of 100 dollars at 1 percent interest for 12 months
generates the following trace table:

Symbol #	N	R	L	Amount	J
0	12	0.01	100.00	Arbitrary	Arbitrary
1				100.00	
2				101.00	1
2				102.01	2
2				103.03	3
2				104.07	4
2				105.11	5
2				106.16	6
2				107.22	7
2				108.29	8
2				109.37	9
2				110.46	10
2				111.56	11
2				112.67	12
3					13

8-7. 1. Input X, Y, Z.
    1.1 Accept the values through formal parameters X, Y, Z.
  2. Shuffle values.
    2.1 Place smallest value in Z.
      2.1.1 If X < Y, then interchange values of X and Y.
      2.1.2 If Y < X, then interchange values of Y and Z.
    2.2 Place middle value in Y.
      2.2.1 If X < Y, then interchange values of X and Y.
  3. Output X, Y, Z.
    3.1 Output result of shuffle through formal parameters X, Y, Z.

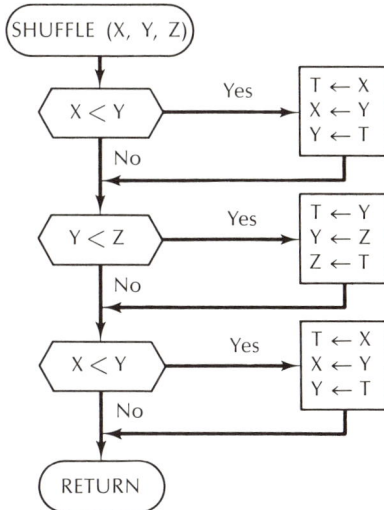

8-9. (a)

```
        START
          │
  1   / INPUT Z / ------- (Z, W) REAL
          │
  2   | W ← SORT(Z) |
          │
  3   / OUTPUT W /
          │
  4      STOP
```

(b)

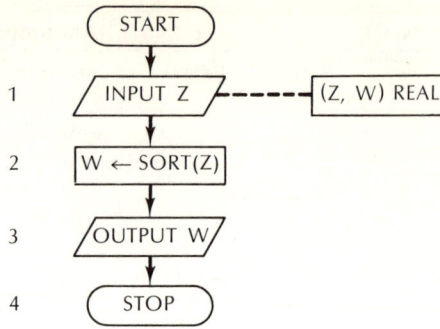

	Z	W	X	P	Q
1	18.	Arb.	Arb.	Arb.	Arb.
2			18.		
				1.0	18
				9.500	1.895
				5.698	3.263
				4.480	4.018
				4.249	4.236
				4.242	4.2433
				4.2426	4.2429
		4.2426			
3					

8-11. Input: FLAG a character string array of 100 elements, where each element has one of the values 'RED', 'WHITE', or 'BLUE'

Output: FLAG with the elements arranged such that all 'RED' come first, all 'WHITE' next, and finally all 'BLUE'

Solution 1:

1. Input FLAG.
   1.1 Input received through FLAG, a formal parameter.
2. Migrate all 'BLUE' to bottom and 'WHITE' below 'RED'.
   2.1 Establish an indicator of an interchange.
   2.2 Repeat 2.3–2.5 until no more position changes occur.
   2.3 Repeat 2.4 for all elements not in proper position.
   2.4 If $FLAG_i$ = 'BLUE' and $FLAG_{i+1}$ ≠ 'BLUE' OR $FLAG_i$ = 'WHITE' and $FLAG_{i+1}$ = 'RED', then interchange $FLAG_i$ and $FLAG_{i+1}$ and note a position change.
3. Return FLAG.

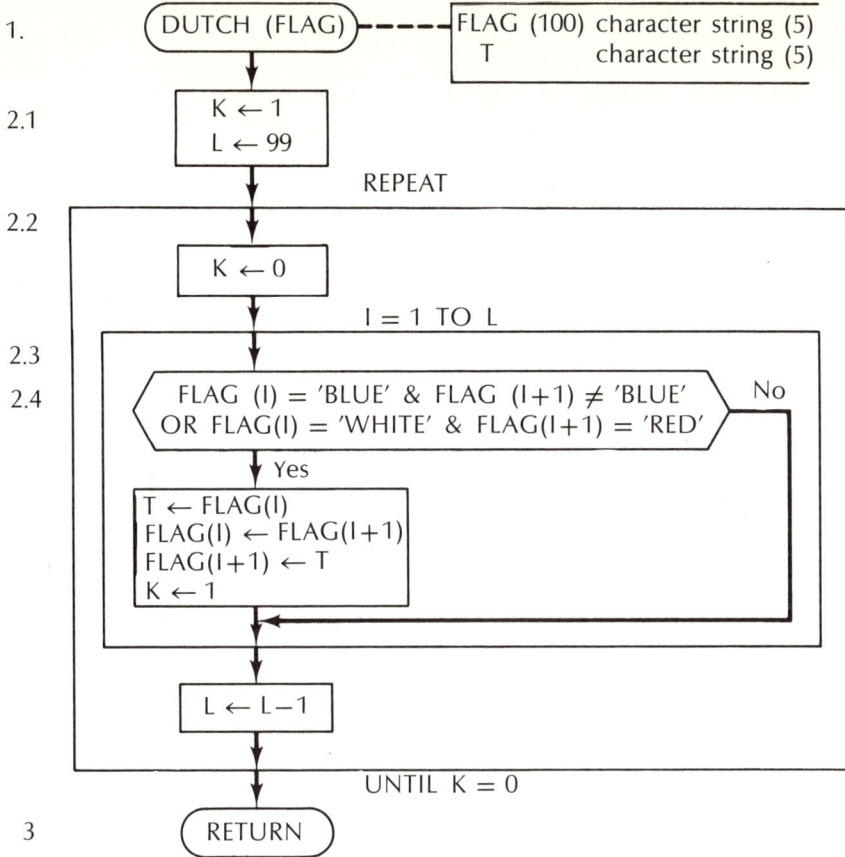

1.   DUTCH (FLAG) ------ FLAG (100) character string (5)
                          T   character string (5)

2.1   K ← 1
      L ← 99

                    REPEAT

2.2   K ← 0

                    I = 1 TO L

2.3
2.4   FLAG (I) = 'BLUE' & FLAG (I+1) ≠ 'BLUE'   No
      OR FLAG(I) = 'WHITE' & FLAG(I+1) = 'RED'

                    Yes

      T ← FLAG(I)
      FLAG(I) ← FLAG(I+1)
      FLAG(I+1) ← T
      K ← 1

      L ← L−1

                    UNTIL K = 0

3   RETURN

Solution 2:

1. Input FLAG.

2. Count number of RED, WHITE, and BLUE.

3. Place correct colors into FLAG.

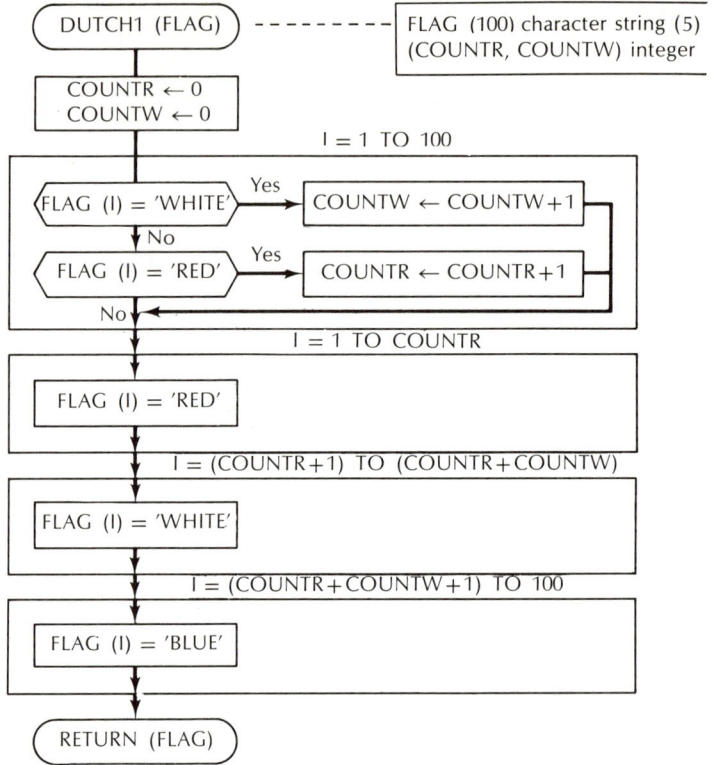

9-1. Input: The list to be sorted, A, and the length of list, I
Output: The sorted list, A

1. Input A and I as formal parameters.

2. Sort A using a replacement-type sort.
   2.1 For each element of the list, search the list to find the largest element not yet sorted.
      2.1.1 Repeat 2.1.2–2.1.3 for each element in list.
      2.1.2 Find largest element in list not yet sorted.
      2.1.3 Put element found in next position in list to be sorted.

3. Return results in formal parameter.

```
        ┌──────────────────┐      ┌─────────────────────────┐
        │  SORTER (A, I)    │------│ A(I) integer            │
        └──────────────────┘      │ SORTELEMENT integer     │
                 │                 └─────────────────────────┘
                 │          K = 1 TO (I−1)
        ┌────────▼───────────────────────────────────────────┐
        │  ┌──────────────────────────┐                       │
        │  │ SORTELEMENT ← K           │                       │
        │  │ LARGELEMENT ← A(K)        │                       │
        │  │ LARGEINDEX ← K            │                       │
        │  └──────────────────────────┘                       │
        │              │         J = SORTELEMENT TO I          │
        │  ┌───────────▼─────────────────────────────────┐    │
        │  │                            Yes ┌──────────────┐ │
        │  │ ⟨ A(J) > LARGELEMENT ⟩─────────│LARGELEMENT ← A(J)│ │
        │  │                                │LARGEINDEX ← J    │ │
        │  │           No │ ◄────────────────────────────┘   │
        │  └──────────────┼──────────────────────────────┘    │
        │  ┌──────────────▼─────────────────────────┐         │
        │  │ T ← A(SORTELEMENT)                       │         │
        │  │ A(SORTELEMENT) ← A(LARGEMENT)            │         │
        │  │ A(LARGEMENT) ← A(SORTELEMENT)            │         │
        │  └──────────────────────────────────────────┘       │
        └────────────────┼───────────────────────────────────┘
                  ┌───────▼──────┐
                  │   RETURN     │
                  └───────┬──────┘
                  ┌───────▼──────┐
                  │    STOP      │
                  └──────────────┘
```

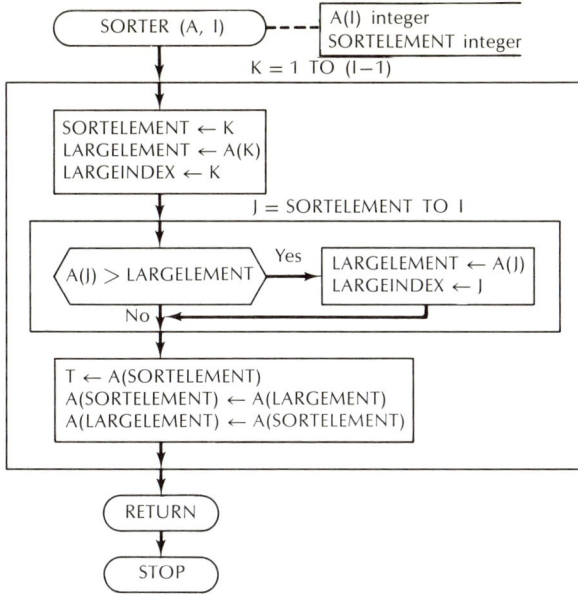

9-3. Change the input to read the 13th column. Add a step after each input to test value of 13th-column input, (J); for example,

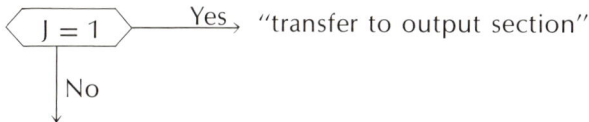

```
⟨ J = 1 ⟩────Yes──→ "transfer to output section"
     │
     │ No
     ▼
```

9-5. In order to accomplish this task, we must make Step 6 a part of one of three subalgorithms. We choose to incorporate Step 6 with the subalgorithm for Step 3, as follows (compare with Figure 9.4-1):

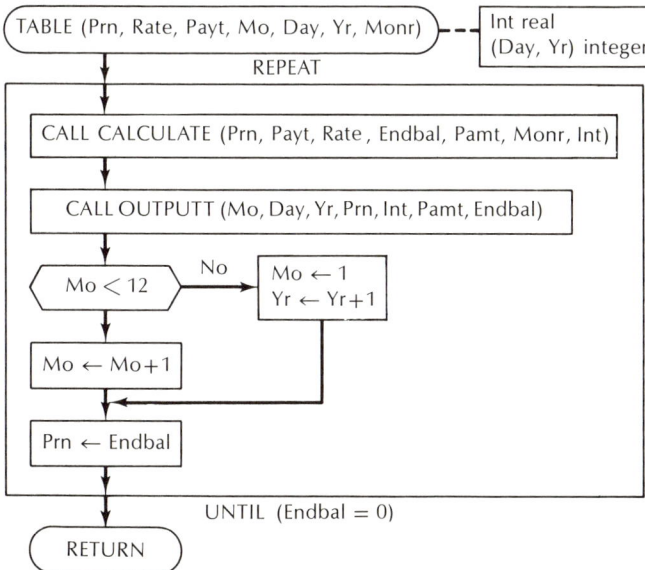

```
   ┌────────────────────────────────────────────────────┐   ┌──────────────────┐
   │ TABLE (Prn, Rate, Payt, Mo, Day, Yr, Monr)          │---│ Int real         │
   └────────────────────────────────────────────────────┘   │ (Day, Yr) integer│
              │              REPEAT                           └──────────────────┘
   ┌──────────▼───────────────────────────────────────────────────┐
   │ ┌──────────────────────────────────────────────────────────┐ │
   │ │ CALL CALCULATE (Prn, Payt, Rate , Endbal, Pamt, Monr, Int)│ │
   │ └──────────────────────────────────────────────────────────┘ │
   │              │                                                 │
   │ ┌────────────▼─────────────────────────────────┐             │
   │ │ CALL OUTPUTT (Mo, Day, Yr, Prn, Int, Pamt, Endbal)│        │
   │ └──────────────────────────────────────────────┘             │
   │              │                No  ┌──────────────┐            │
   │        ⟨ Mo < 12 ⟩────────────────│ Mo ← 1        │           │
   │              │                    │ Yr ← Yr+1     │           │
   │              │                    └──────┬───────┘            │
   │ ┌────────────▼─────┐                     │                    │
   │ │ Mo ← Mo+1        │                     │                    │
   │ └────────┬─────────┘ ◄──────────────────┘                    │
   │ ┌────────▼─────────┐                                          │
   │ │ Prn ← Endbal     │                                          │
   │ └────────┬─────────┘                                          │
   └──────────┼──────────────UNTIL (Endbal = 0)───────────────────┘
   ┌──────────▼──────┐
   │    RETURN        │
   └─────────────────┘
```

379

Subalgorithm for Step 4:

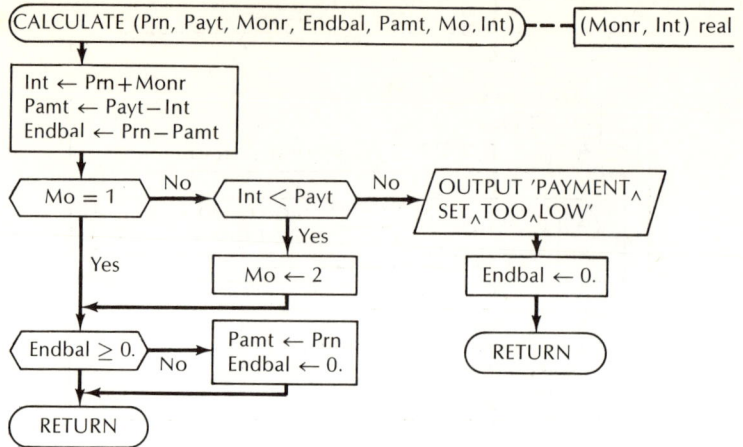

CALCULATE (Prn, Payt, Monr, Endbal, Pamt, Mo, Int) --- (Monr, Int) real

Int ← Prn + Monr
Pamt ← Payt − Int
Endbal ← Prn − Pamt

Mo = 1 → No → Int < Payt → No → OUTPUT 'PAYMENT˄ SET˄TOO˄LOW'

Yes (from Int < Payt): Mo ← 2

Endbal ← 0.

RETURN

Mo = 1 Yes → Endbal ≥ 0. → No → Pamt ← Prn / Endbal ← 0.

RETURN

Subalgorithm for Step 5:

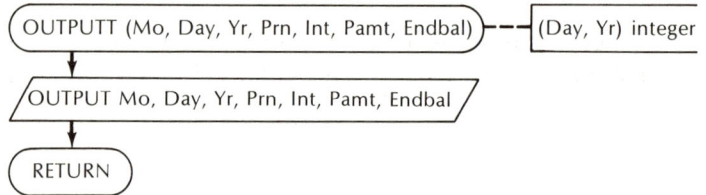

OUTPUTT (Mo, Day, Yr, Prn, Int, Pamt, Endbal) --- (Day, Yr) integer

OUTPUT Mo, Day, Yr, Prn, Int, Pamt, Endbal

RETURN

Index

Index

Page numbers for definitions are in italics

A-field, 108
Abacus, 294
Actual parameter, *183,* 187
Addition, 76
Address, 130
ALGOL, *28*
Algorithm, *10*
Algorithm design, 18, *19*
Alphabet, 36
Alphameric, 76, 109
Alphanumeric, 76, 109
ALU, 3, 7
Amortization table, 242
Analog, 10
APL, *28*
Annotation, *23*
Arbitrary, *27,* 45
Arithmetic expression, 76
Arithmetic logic unit, 7
Array, *123,* 129, 169
Artificial intelligence, 308
Ascending sort (*See* Sort, ascending)
ASCII, 335
Assembly language, *28*
Assignment statement, *46*
Auxiliary storage, 3

Babbage, Charles, 297
Bacon, Francis, 75
BASIC, *28*
Batch computing, 283

Binary computer, 327
Binary tree, 133, 224
Binomial coefficients, 228
Bisection method, *257*
Bit, 327
Block, 269
Boole, George, 298
Branch, 133
Branching, *52*
    unconditional, 52
    conditional, 53
Buffer, 102
Burroughs, William, 298
Byte, 327

Calculator, *13,* 14
Call by location, 196
Call by value, 196
Catenation (*See* String operators, concatenation)
Channel, 286
Character string, 41
Child, 133
Chip, 10, 11
Churchill, Sir Winston, 300
Clark, Ramsey, 312
Clarke, A. C., 308
Classification and merge sort (*See* Sort, classification and merge)
COBOL, *29,* 303
Coder, *14*
Coding, *28*

383